SMALL ANTENNA HANDBOOK

SMALL ANTENNA HANDBOOK

ROBERT C. HANSEN

ROBERT E. COLLIN

IEEE Antennas and Propagation Society, *Sponsor*

IEEE

WILEY

A JOHN WILEY & SONS, INC., PUBLICATION

Published by John Wiley & Sons, Inc., Hoboken, New Jersey
Published simultaneously in Canada

Library of Congress Cataloging-in-Publication Data:

Hansen, Robert C.
Small antenna handbook / R.C. Hansen and R.E. Collin.
p. cm.
Includes bibliographical references and index.
ISBN 978-0-470-89083-7
1. Antennas (Electronics) 2. Miniature electronic equipment. I. Collin, Robert E. II. Title.
TK7871.6.H35157 2011
621.382′4–dc23

2011013601

oBook ISBN: 9781118106860
ePDF ISBN: 9781118106839
ePub ISBN: 9781118106853

10 9 8 7 6 5 4 3 2 1

This book is dedicated to these intrepid
warriors against nut-house antennas

Dick Adler
Jack Belrose
Ben Dawson
Jim Hatfield
Larry Hurzon
Rick Ridgley

IN MEMORIUM

Robert Emanuel Collin
1928–2010

Bob Collin was born in Alberta, Canada. He received the B.Sc. degree in engineering physics from the University of Saskatchewan in 1951. He attended Imperial College in England for graduate work and obtained the Ph.D. degree in electrical engineering from the University of London in 1954. From 1954 to 1958, he was a Scientific Officer at the Canadian Armament Research and Development Establishment, where he worked on guided missile antenna, randome, and radar system evaluations. He joined the Electrical Engineering Department at Case Institute of Technology in 1958, now Case Western Reserve University. During his tenure there, he served as Chairman of the Department of Electrical Engineering and Applied Physics for five years and as Interim Dean of Engineering for two years. He has been an Invited Professor at Catholic University in Rio de Janeiro, at Telebras Research Center, Campinas, Brazil, and at Beijing University, People's Republic of China. He was also Distinguished Visiting Professor at the Graduate School, The Ohio State University, during the 1982–1983 academic year.

Professor Collin authored more than 150 technical papers and 5 books. Two books, *Field Theory of Guided Waves* and *Foundation of Microwave Engineering*, became classics and were republished by IEEE Press.

He was a fellow of IEEE, a member of URSI Commission B, and a member of the Geophysical Society. Other honors include the Dickman Award from Case Western

Reserve University for distinguished graduate teaching, the IEEE APS Distinguished Career Award (1992), the IEEE Schelkunoff Prize Paper Award (1992), and the IEEE Electromagnetics Award (1998). In 1990, Professor Collin was elected to the National Academy of Engineering. His insight into the mathematics of electromagnetics was unparalleled.

Books by R. C. Hansen

Microwave Scanning Antennas (3 volumes)

Significant Phased Array Papers

Geometric Theory of Diffraction

Moment Methods in Antennas and Scattering

Phased Array Antennas, second edition

Electrically Small, Superdirective, and Superconducting Antennas

Books by R. E. Collin

Antenna Theory, with F. J. Zucker (2 volumes)

Field Theory of Guided Waves, second edition

Foundations of Microwave Engineering, second edition

Antennas and Radiowave Propagation

Principles and Applications of Electromagnetic Fields, with R. Plonsey

CONTENTS

PREFACE

Harold Wheeler, in his classic 1947 papers (Wheeler, 1947a, 1947b), created the field of electrically small antennas (ESA), although, as shown in the history (Appendix A), the early antennas were all electrically small. This field has long been important for frequencies below roughly 1 GHz, where a half-wavelength is about 6 in. These electrically small antennas have characteristics in common that limit performance: low radiation resistance, high reactance, low efficiency, narrow bandwidth, and increased loss in the matching network. Most of these limitations are shared by two other classes of antennas: superdirective antennas and superconducting antennas.

The original intent was to update *Electrically Small, Superdirective, and Superconducting Antennas* with a second edition. Sufficient material became available to broaden the scope of that book; the result is this book with a new title. This book provides mathematical foundations for important topics including ENG shells, ESA with μ or ε cores, including lossy cores, and focused/subwavelength imaging.

Chapter 1 contains detailed electromagnetic derivations of Chu and Thal Q formulations, formulas for Q when μ or ε cores are used, and effects of loss on Q. Finally, work on fundamental limitations using spheroids instead of the radian sphere is covered. Chapter 2 is a treatise on bandwidth and matching, and Foster's reactance theorem. Precise data are given for Fano's matching limitations and bandwidth improvement. The effects of loss in matching circuits and cables on VSWR are shown. The relatively new field of non-Foster matching is next discussed. Finally, performance of a short monopole that is matched is compared with that of monopole with high-impedance preamp. Chapter 3 provides updated coverage on canonical antennas: dipoles, loops, self-resonant ESA, PIFA, and dielectric resonator antennas. A comparison is made of Q of various antennas with the Chu–Thal fundamental limitation. In Chapter 4 are presented six ideas that are clever but with performance numbers that are poor. Eighteen nutty antenna ideas are detailed in Chapter 5; the list keeps growing! Chapter 6 on superdirective antennas has been updated with recent data on the use of self-resonant elements. These allow modest supergain but bandwidth and tolerance limitations remain. Also included is new material on maximum directivity of arrays. The last chapter, on superconducting antennas, has been updated with new delay line data. The earlier conclusion that superconductors should be considered for the matching network, but not the antenna, still holds.

A world history of ESA is presented in Appendix A. Antenna terms important to ESA are defined in Appendix B. Appendix C contains a paper by Karawas and Collin (2008) on NIM shells enclosing an ESA dipole. The analytical errors associated with perfect lenses and subwavelength focusing are treated in detail by Collin (2010) in Appendix D. Each chapter includes extensive references. An overall subject index and an author index are provided at the end of the book.

We antenna engineers have done what is possible by rearranging the wires; future significant advances will come through the use of new low-loss magnetic materials and through the use of circuits to compensate for impedance deficiencies.

R. C. Hansen and R. E. Collin

REFERENCES

Collin, R. E. Frequency Dispersion Limits in Veselago Lens, *Prog. Electromagnetics Res. B*, Vol. 19, 2010, pp. 233–261.

Karawas, G. K. and Collin, R. E. Spherical Shell of ENG Material Surrounding a Dipole Antenna, *IEEE MILCOM*, 2008, pp. 1–15.

Wheeler, H. A. Fundamental Limitations of Small Antennas. *Proc. IRE*, Vol. 35, December 1947a, pp. 1479–1484.

Wheeler, H. A. A Helical Antenna for Circular Polarization. *Proc. IRE*, Vol. 35, December 1947b, pp. 1484–1488.

CHAPTER 1

QUALITY FACTORS OF ESA

1.1 INTRODUCTION

In a 1947 paper, Harold Wheeler defined an electrically small antenna (ESA) as an antenna that could be enclosed within a radian sphere (Wheeler, 1947). The radian sphere was a sphere of radius equal to $\lambda/2\pi$, where λ is the wavelength. The antennas used by Marconi and Fessenden in the early years of wireless telegraphy were electrically small antennas even though they were very large physical structures, often involving wires strung as an inverted fan or cone from masts several hundred feet tall. These antennas were electrically small antennas since in order to achieve long-distance transmission the wavelengths used, typically greater than 3000 m, were much longer than the antenna heights. These electrically small antennas were characterized by a very low radiation resistance and a large capacitive input reactance. The purpose of the large inverted fans and cones was to increase the antenna capacitance and thus reduce the capacitive reactance. Wheeler introduced the radiation power factor (RPF) as a figure of merit for these electrically small antennas. He considered two basic antenna types: the magnetic dipole or loop antenna consisting of a solenoid coil with N turns, length b, and radius a; and a short electric dipole antenna consisting of a thin wire of length b and with capacitive loading at each end by means of circular conducting disks of radius a. The radiation power factor was defined as the ratio of the radiation resistance to the reactance of the antenna. For the solenoid loop antenna, the radiation resistance is given by $R_m = 320N^2\pi^6(a/\lambda)^4$. An approximate expression for the inductance of the solenoid

Small Antenna Handbook, Robert C. Hansen and Robert E. Collin.

coil is $L = \mu_0 N^2(\pi a^2/b)$. For the small electric dipole, the radiation resistance is given by $R_e = 80\pi^2(b/\lambda)^2$ and the capacitance between the two circular plates is $C = \varepsilon_0 \pi a^2/b$ when the fringing effects are neglected. From these expressions, we can calculate the radiation power factors, p_m for the magnetic dipole and p_e for the electric dipole, as follows:

$$p_m = \frac{R_m}{\omega L} = \frac{4}{3}\frac{\pi^3 a^2 b}{\lambda^3} \tag{1.1a}$$

$$p_e = R_e \omega C = \frac{4}{3}\frac{\pi^3 a^2 b}{\lambda^3} \tag{1.1b}$$

Wheeler modified Equation 1.1 by magnetic and electric shape factors derived from statics. For ESA, the radiation power factors are very small. Also see Wheeler (1975).

Wheeler's radiation power factors are related to the Q parameter introduced by Kenneth S. Johnson, an authority on wire transmission at Bell Telephone Laboratories. Initially, Johnson used the symbol K to represent the ratio of the inductive reactance to the resistance of a coil, $K = \omega L/R$. In 1920, while working on wave filters, invented by G. A. Campbell, he replaced the symbol K by the symbol Q and introduced the lowercase symbol q for the analogous quantity $\omega C/G$ for a capacitor C, where G is the parallel conductance of a capacitor (called condenser in those days). Later on in 1927 he used Q for both in his U.S. Patent No. 1,628,983. His introduction of this symbol was adopted by most people working with tuned circuits in radio receivers in the early days of radio broadcasting.

It is easily shown that the response of a tuned circuit, consisting of a parallel or series connection of an inductor and a capacitor, is reduced by the factor $1/\sqrt{2}$ when the circuit is detuned by a fractional amount $\Delta\omega/\omega = 1/2Q$, provided the Q is equal to 10 or more. Thus, $1/Q$ is the 3 dB bandwidth (BW) of the tuned circuit. If a resistive load is connected across the tuned circuit such that a maximum amount of power can be obtained from the circuit, the Q of the loaded circuit is reduced by a factor of 2 and the 3 dB bandwidth is increased by a factor of 2. Tuned circuits with high values of Q were needed in order to achieve high selectivity in the tuned radio frequency radios. This led to extensive efforts to design radio frequency coils with low loss resistance. During the World War II years, the Q became widely used to describe the sharpness of the resonance curve of both electric and mechanical resonators such as microwave cavities, quartz crystal resonators, and so on. If we have a tuned circuit with a capacitor C in parallel with an inductor L, it is known that the resonant frequency of the circuit is given by $\omega_0 = 1/\sqrt{LC}$ and that at resonance the time-averaged energy stored in the capacitor is equal to that in the inductor. If the inductor has a series resistance R and the current in the inductor is I, then the average energy stored in the magnetic field around the inductor is given by $W_m = I^2L/4$ and the average power dissipated in the resistor is $P_L = I^2R/2$. When we introduce these into the definition of Q, as used by Johnson, we can express Q in the form

$$Q = \frac{\omega L}{R} = \frac{2\omega(I^2L/4)}{I^2R/2} = \frac{2\omega W_m}{P_L} = \frac{\omega(W_m + W_e)}{P_L}$$

$$Q = \frac{\omega(\text{average energy stored in resonant circuit})}{(\text{average energy dissipated per second})} \tag{1.2a}$$

This latter definition of the quality factor or Q of a resonant circuit is the most commonly used one.

For ESA, it is approximately

$$Q \simeq \frac{\omega\, dX/d\omega}{2R} \tag{1.2b}$$

Harrington (1965) and Rhodes (1966, 1974) extended the bandwidth relationship for circuits, $\text{BW} = 1/Q$, to dipole-type antennas. Dipole bandwidth and $1/Q$ were compared by Hansen (2007); the match was excellent for $ka \leq 0.3$ and good for $ka \leq 0.5$.

The Q of electrically small antennas represents a fundamental limit on the performance of these antennas, in particular their bandwidth. If the designer of electrically small antennas is cognizant of this fundamental limit, he (she) will not expend excessive time designing an antenna to achieve what cannot be achieved. Although the operational bandwidth of electrically small antennas can be increased by the use of multiple tuned circuits in the matching network or by inclusion of magnetic materials, this invariably introduces significant additional loss and a reduced efficiency of the antenna. Fano's theory of broadband matching, which will be briefly discussed in a later section, shows that a maximum increase in the 3 dB bandwidth is by a factor of 3.2. A realistic goal is to increase the bandwidth by a factor of 2. The double-tuned coupled circuits used in intermediate-frequency (IF) transformers in superheterodyne radio receivers, to increase the audio fidelity of these amplifiers, increased the bandwidth by a factor of around 2 (see Terman, 1943). Active circuits, called non-Foster networks, can in principle provide broadband matching. These are discussed in Chapter 2 but have their own limitations.

Canonical types of ESA are loaded dipoles, patch antennas with uncommon substrates, loop antennas with air or magnetic cores, dielectric resonator antennas, as well as bent conductors of unusual shapes, and antennas incorporating metamaterials. Many of the latter cannot be realistically fabricated and do not work according to the theories proposed for them. These antennas are discussed in later sections. Some clever ideas have been found to be impractical. A long list of antenna ideas that resemble science fiction is given in Chapter 5.

L. J. Chu (1948) applied the concept of Q to small antennas and used the definition given by Equation 1.2a to find a lower bound for Q of small antennas whose radiated fields could be expressed in terms of spherical waves. The results obtained by Chu provide a more accurate measure of the limitations of ESA than Wheeler's power factors do. Since the pioneering work of Chu, many other authors have contributed to the evaluation of antenna Q.

1.2 CHU ANTENNA Q

Chu considered a hypothetical antenna that was contained entirely within a sphere of radius a. The electromagnetic field outside this sphere can be described in terms of infinite series of spherical transverse electric (TE) and transverse magnetic (TM) modes. The vector wave functions in a spherical coordinate system were derived by W. W. Hansen (1935). These modes consist of two sets of transverse (divergence-free) modes and a set of longitudinal modes (modes with zero curl). The former are designated by the symbols $\mathbf{M}_{nm}(\mathbf{r})$ and $\mathbf{N}_{nm}(\mathbf{r})$ while the latter are represented by the symbol $\mathbf{L}_{nm}(\mathbf{r})$. In regions external to the source region, only the transverse modes are required in the expansion of an arbitrary electromagnetic field. Thus, in the region external to a sphere of radius a that completely encloses the small antenna, the electric and magnetic fields can be represented in the following form (Stratton, 1941; Collin, 1990):

$$\mathbf{E}(\mathbf{r}) = \sum_{n,m} C^e_{nm}\mathbf{M}_{nm}(\mathbf{r}) + \sum_{n,m} D^e_{nm}\mathbf{N}_{nm}(\mathbf{r}) \tag{1.3a}$$

$$\mathbf{H}(\mathbf{r}) = \sum_{n,m} C^h_{nm}\mathbf{M}_{nm}(\mathbf{r}) + \sum_{n,m} D^h_{nm}\mathbf{N}_{nm}(\mathbf{r}) \tag{1.3b}$$

where

$$\mathbf{M}_{nm}(\mathbf{r}) = \nabla \times \mathbf{a}_r\left[P^m_n(\cos\theta)k_0 r h^2_n(k_0 r)^{\cos}_{\sin}\,\phi\right] \tag{1.3c}$$

$$\mathbf{N}_{nm}(\mathbf{r}) = \nabla \times \nabla \times \mathbf{a}_r\left[P^m_n(\cos\theta)k_0 r h^2_n(k_0 r)^{\cos}_{\sin}\,\phi\right] \tag{1.3d}$$

In these expressions, $P^m_n(\cos\theta)$ are the associated Legendre polynomials, $h^2_n(k_0 r)$ is the spherical Hankel function of order n, and $k_0 = 2\pi/\lambda_0$ is the free space wave number. C^e_{nm}, C^h_{nm}, D^e_{nm}, and D^h_{nm} are amplitude constants. For each individual mode, the stored electric and magnetic energy divided by the radiated power is independent of the azimuthal index m. Also, the transverse electric or TE_{nm} and the transverse magnetic or TM_{nm} modes are duals of each other with the electric and magnetic fields interchanged, so their Q's are the same. Thus, it is sufficient to consider only the TM_{n0} mode, which is what Chu did.

For the TM_{n0} modes, the electric and magnetic fields are given by

$$E_\theta = C^h_n \frac{-k_0 \sin\theta}{j\omega\varepsilon_0}\frac{dP_n(\cos\theta)}{d(\cos\theta)}\frac{d[k_0 r h^2_n(k_0 r)]}{d(k_0 r)} \tag{1.4a}$$

$$E_r = C_n^h \frac{n(n+1)}{j\omega\varepsilon_0 r^2} P_n(\cos\theta)\left[k_0 r h_n^2(k_0 r)\right] \tag{1.4b}$$

$$H_\phi = C_n^h \frac{\sin\theta}{r} \frac{dP_n(\cos\theta)}{d(\cos\theta)} \left[k_0 r h_n^2(k_0 r)\right] \tag{1.4c}$$

where C_n^h is an amplitude constant. The mode wave impedance at $r = a$ is given by

$$Z_{w,n} = \frac{E_\theta}{H_\phi} = jZ_0 \frac{d\left[k_0 a h_n^2(k_0 a)\right]/d(k_0 a)}{k_0 a h_n^2(k_0 a)} \tag{1.5}$$

Instead of evaluating the stored reactive electric and magnetic energy and the radiated power from the electric and magnetic fields, Chu expanded the mode wave impedance into a continued fraction that could be interpreted as a ladder network with $2n$ elements consisting of alternating capacitors and inductors, and terminated in a normalized resistance of 1 Ω. Conventional circuit analysis could then be used to determine the stored electric energy in the capacitors and the stored magnetic energy in the inductors, as well as the radiated power, which equals the power dissipated in the terminating resistor.

The mode normalized wave impedance can be expressed in the form

$$\frac{Z_{w,n}}{Z_0} = \left[\frac{j}{\rho} + \frac{j}{h_n^2}\frac{dh_n^2}{d\rho}\right] \tag{1.6}$$

where $\rho = k_0 a$. We now use the following recurrence relation for spherical Bessel functions $f_n(\rho)$:

$$f_n = \frac{2n-1}{\rho} f_{n-1} - f_{n-2}$$

and the relation

$$\frac{df_n}{d\rho} = f_{n-1} - \frac{n+1}{\rho} f_n$$

where $\rho = k_0 a$, to get

$$\frac{Z_{w,n}}{Z_0} = \frac{j}{\rho} + \frac{j}{h_n^2}\left[h_{n-1}^2 - \frac{n+1}{\rho}h_n^2\right] = \frac{n}{j\rho} + \cfrac{1}{\cfrac{h_n^2}{jh_{n-1}^2}} = \frac{n}{j\rho} + \cfrac{1}{\cfrac{1}{jh_{n-1}^2}\left[\cfrac{2n-1}{\rho}h_{n-1}^2 - h_{n-2}^2\right]}$$

$$= \frac{n}{j\rho} + \cfrac{1}{\cfrac{2n-1}{j\rho} + \cfrac{1}{\cfrac{h_{n-1}^2}{jh_{n-2}^2}}} = \frac{n}{j\rho} + \cfrac{1}{\cfrac{2n-1}{j\rho} + \cfrac{1}{\cfrac{2n-3}{j\rho} + \ddots \cfrac{1}{\cfrac{3}{j\rho} + \cfrac{1}{\cfrac{1}{j\rho} + 1}}}} \tag{1.7}$$

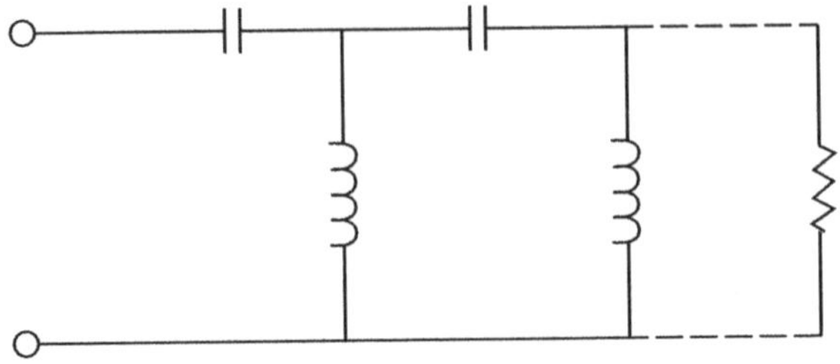

FIGURE 1.1 Chu ladder network.

This expansion can be interpreted as the impedance of the ladder network illustrated in Figure 1.1 for TM modes. L and C values decrease with each step.

The TM_{n0} modes store more electric energy than magnetic energy and hence the mode must be tuned to resonance by adding some additional magnetic energy, that is, by an inductive reactance. If we assume that this is done, then the total stored reactive energy will be twice the electric energy W_e, which equals the energy stored in the capacitors in the equivalent electric circuit. The stored electric energy and the power dissipated in the $1\,\Omega$ terminating resistance can be determined by conventional circuit analysis, but it becomes very tedious to carry out for $n > 3$. For the $n = 1$ mode, the Q was found to be given by

$$Q_1 = \frac{2\omega W_e}{P_r} = \frac{1}{(k_0 a)^3} + \frac{1}{k_0 a} \tag{1.8}$$

Note that the Q in Hansen (1981) has a typo error.

Chu evaluated the Q of the higher order modes by using an approximate equivalent circuit that was obtained as follows. For a series RLC circuit, the input reactance is $X = \omega L - 1/\omega C$ and $dX/d\omega - X = L + 1/\omega^2 C - L = 1/\omega^2 C$. When the input current to the circuit is I, the power dissipated in R is $|I|^2 R/2$ and the electric energy stored in C is $|I|^2/4\omega^2 C$. Hence, the Q is given by

$$Q = \frac{2\omega W_e}{P_r} = \frac{\omega}{\omega^2 CR} = \frac{1}{2R}\left[\frac{dX}{d\omega} - \frac{X}{\omega}\right] \tag{1.9}$$

Chu equated X to the imaginary part of $Z_{w,n}/Z_0$ given by Equation 1.6. Chu also evaluated the Q of the higher order modes using this method. In addition, Chu also evaluated the Q of a combination of a TE_{n1} and a TM_{n1} mode. When the amplitudes of the two modes are equal, a circularly polarized field can be produced. By using his approximate value of the Q, it was found that the Q of the combined modes was one-half that of a single mode.

The ratio of the directive gain to the Q for an antenna radiating a total of N modes was also determined by Chu but he did not find the optimum gain for a given Q, a problem that later authors solved. Any antenna that is contained within a sphere of radius a will have additional energy storage within the enclosing sphere and will consequently have a higher Q. Thus, the Q that Chu found is a lower bound on the Q

of any lossless antenna. Many ESA have a Q that is considerably larger than Chu's lower bound. When the antenna Q is large, one can infer that the bandwidth of the antenna will be small but one cannot always assume that it will be equal to $1/Q$ since the tuning circuit and losses may provide a larger bandwidth. Harrington (1958, 1960, 1961) expanded on the work of Chu, but followed Chu's approximate method to obtain the Q's of TE_{n1} and TM_{n1} modes. Harrington showed that the maximum gain of an antenna, obtained by using only a finite number of TE_{n1} and TM_{n1} modes, was given by

$$G_{\max} = \sum_{n=1}^{N}(2n+1) = N^2 + 2N \tag{1.10}$$

If there was no constraint on the mode amplitudes, an arbitrarily large gain would theoretically be possible. However, high-order modes are very difficult to excite because their wave impedances (wave admittances for the TE_{n0} modes) are very large, so in practice unusually large gains cannot be achieved. The high-order modes also store large amounts of reactive energy, so high gain implies a large antenna Q and a narrow bandwidth. Other investigators have considered optimizing the ratio of gain divided by the antenna Q.

A different approach to Q was taken by Thiele et al. (2003), based on the far-field pattern of a small source. A "pattern" Q is based on the integral of pattern over visible space and that integral over all space, including invisible. Their Q values are higher than those of Chu. An electrically small dipole with sinusoidal current distribution was used to provide the fields for the integrations. These pattern Q values are eight times larger than the Chu results. Dipole bandwidth calculations were also made by Hujanen and Sten (2005). Kalafus (1969) calculates Q for higher modes as well, using series expansions for the integrals of energy. Then coefficients of polynomials representing Q are given. Another calculation of Q due to higher modes is done by Harrington (1960).

An egregious example of claiming antennas that violate the fundamental limitations on small antennas is provided by Underhill and Harper (2002, 2003). For a short folded dipole of length L, they have reactance proportional to kL. However, it is well known that the folded dipole reactance is four times that of the constituent dipole, which is proportional to $1/kL$. For a small loop of diameter D, they have radiation resistance proportional to kD^2, when it is widely accepted that it is proportional to k^4D^4. These errors appear to be due to applying static formulations to electromagnetic problems.

Another paper claims that orthogonal TE and TM modes produce a gain of 3 and a Q half that of either mode (Kwon, 2005). Both are in error; the input power and the peak power density are both doubled, leaving the gain at 1.5 and the Q that of one mode.

An erroneous calculation of bandwidth limitations occurred because of confusion between total energy, stored energy, and radiated energy. This resulted in a bandwidth of 16π times the fundamental limit for VSWR ≤ 2 (Chaloupka, 1992).

Geyi (2003a, 2003b) reexamines the task of maximizing the ratio D/Q, directivity divided by Q. He corrects some inconsistencies in Fante (1969), with the result that the maximum D/Q for a directive antenna (with both TE and TM modes) is $3/Q$, whereas that for an omnidirectional antenna is, as expected, $3/2Q$.

1.3 COLLIN AND ROTHSCHILD *Q* ANALYSIS

The next contribution to evaluating antenna Q was the paper by Collin and Rothschild (1964), where the stored energy was evaluated in terms of the electromagnetic fields. This work provided convenient closed-form formulas for the Q of any mode and was expanded upon by Fante (1969) and also by McLean (1996).

The configuration that will be analyzed consists of a spherical core of radius a with a current sheet located on the surface. The current sheet is chosen so as to excite only a single TE_{n0} mode for which the electric and magnetic fields in the region $r > a$ can be chosen to be

$$E_\phi = C_n^e \frac{\sin\theta}{r} \frac{dP_n(\cos\theta)}{d(\cos\theta)} \left[k_0 r h_n^2(k_0 r)\right] \tag{1.11a}$$

$$H_\theta = C_n^e \frac{k_0 \sin\theta}{j\omega\mu_0 r} \frac{dP_n(\cos\theta)}{d(\cos\theta)} \frac{d\left[k_0 r h_n^2(k_0 r)\right]}{d(k_0 r)} \tag{1.11b}$$

$$H_r = -C_n^e \frac{n(n+1)}{j\omega\mu_0 r^2} P_n(\cos\theta)\left[k_0 r h_n^2(k_0 r)\right] \tag{1.11c}$$

An electric current sheet proportional to $\mathbf{a}_\phi H_\theta(a)$ will support this mode. For simplicity, we will assume that the amplitude constants C_n^e are equal to unity. The TE_{nm} modes store more magnetic energy than electric energy. The Q can be expressed in terms of the total average stored reactive magnetic energy both inside and outside the spherical surface $r = a$ since we will assume that the antenna is tuned to resonance with an additional capacitive reactance. Thus, the Q is given by

$$Q_n = \frac{2\omega W_{\mathrm{m}}}{P_{\mathrm{r}}} \tag{1.12}$$

where P_{r} is the total radiated power. The total energy in the electromagnetic field is infinite since it includes the energy associated with the far-zone radiation field, which is energy that is being transmitted to infinity as radiated power. The radiated power is obtained by integrating the real part of the complex Poynting vector over the surface of a sphere with very large radius and is readily found to be given by

$$P_{\mathrm{r}} = \frac{k_0 \pi}{\omega\mu_0} \frac{2n(n+1)}{2n+1} \tag{1.13}$$

The power flow at infinity is equal to the energy density in the electromagnetic field multiplied by the velocity $c = 1/\sqrt{\mu_0 \varepsilon_0}$ integrated over θ and ϕ. The energy density in the radiation field is split equally between that in the electric field and that in the magnetic field. By using the asymptotic value of the Hankel function, the energy density for the electric and magnetic fields for very large values of r is found to be

$$w_\mathrm{e} = w_\mathrm{m} = \frac{\varepsilon_0 \sin^2\theta}{4r^2} \left[\frac{dP_n(\cos\theta)}{d(\cos\theta)} \right]^2 \tag{1.14}$$

After multiplying by $r^2 \sin\theta$ and integrating over θ and ϕ, we will denote these energy densities by $W_{\mathrm{e,Rad}}$ and $W_{\mathrm{m,Rad}}$. It is found that

$$W_{\mathrm{e,Rad}} = W_{\mathrm{m,Rad}} = \frac{\pi\varepsilon_0 n(n+1)}{2n+1} \tag{1.15}$$

It is easy to verify that $c(W_{\mathrm{e,Rad}} + W_{\mathrm{m,Rad}}) = 2cW_{\mathrm{m,Rad}} = P_\mathrm{r}$. In the original paper by Collin and Rothschild, they made the hypothesis that the energy density $w_\mathrm{e} + w_\mathrm{m}$ should be subtracted from the total energy density $(\varepsilon_0/4)|\mathbf{E}|^2 + (\mu_0/4)|\mathbf{H}|^2$ before integrating over the total volume in order to obtain the average stored reactive energy. After evaluating the total reactive energy, they used the integral of the complex Poynting vector over the surface $r = a$ to obtain an expression for $W_\mathrm{m} - W_\mathrm{e}$, which does not contain the energy associated with the radiation field. By this means, they obtained separate expressions for the average stored electric and magnetic reactive energy. McLean (1996) simplified this procedure by simply subtracting the energy densities w_e and w_m, respectively, from the total electric and magnetic field energy densities. We will follow McLean's procedure in the derivation given below.

By using the expressions for the magnetic field given in Equations 1.11b and 1.11c, the stored magnetic reactive energy in the volume outside the surface $r = a$ is found to be given by

$$W_\mathrm{m} = \int_a^\infty \int_0^\pi \left\{ \frac{\mu_0}{4} \left[|H_\theta|^2 + |H_r|^2 \right] - \frac{\varepsilon_0 \sin^2\theta}{4r^2} \left[\frac{dP_n(\cos\theta)}{d(\cos\theta)} \right]^2 \right\} 2\pi r^2 \sin\theta \, d\theta \, dr \tag{1.16}$$

When r becomes very large, $r^2|H_r|^2$ is asymptotic to $1/r^2$, so the integral of this term vanishes at the upper limit $r = \infty$. The asymptotic limit of the term $\mu_0|H_\theta|^2/4$ is $(\varepsilon_0 \sin^2\theta/4r^2)[dP_n(\cos\theta)/d(\cos\theta)]^2$ and is cancelled by the last term in the above integral. Hence, the integral over r converges as r tends to infinity. If the magnetic energy density in the radiation field had not been subtracted out, the integral would have diverged as r tended toward infinity.

The evaluation of W_m requires the following two integrals:

$$\int_0^\pi [P_n(\cos\theta)]^2 \sin\theta \, d\theta = \frac{2}{2n+1} \tag{1.17a}$$

$$\int_0^\pi \left[\frac{dP_n(\cos\theta)}{d\theta}\right]^2 \sin\theta \, d\theta = \frac{2n(n+1)}{2n+1} \tag{1.17b}$$

After completing the integrations over θ, the expression for W_m reduces to

$$W_m = \frac{\pi\varepsilon_0 n(n+1)}{k_0(2n+1)} \int_{k_0 a}^{\infty} \left\{ \left[\frac{d\rho j_n(\rho)}{d\rho}\right]^2 + \left[\frac{d\rho y_n(\rho)}{d\rho}\right]^2 + n(n+1)\left[j_n^2(\rho) + y_n^2(\rho)\right] - 1 \right\} d\rho \tag{1.18}$$

where j_n and y_n are the spherical Bessel functions of the first and second kinds. In order to carry out the integrations in Equation 1.18, the following integral is used (Morse and Feshbach, 1953):

$$\int \rho^2 [f_n(\rho)]^2 d\rho = \frac{\rho^2}{2}(f_n^2 - f_{n-1} f_{n+1}) \tag{1.19a}$$

where f_n can be j_n or y_n. In order to transform all of the integrals into this form, the following relations are needed:

$$j_n = \frac{\rho}{2n+1}(j_{n-1} + j_{n+1}) \tag{1.19b}$$

$$\frac{d(\rho j_n)}{d\rho} = \frac{\rho}{2n+1}\left[(n+1)j_{n-1} - n j_{n+1}\right] \tag{1.19c}$$

From Equations 1.19b and 1.19c, we can derive the relations

$$j_{n+1} = \frac{n+1}{\rho} j_n - \frac{1}{\rho}\frac{dj_n(\rho)}{d\rho} \tag{1.19d}$$

$$j_{n-1} = \frac{1}{\rho}\left[n j_n + \frac{1}{\rho}\frac{dj_n(\rho)}{d\rho}\right] \tag{1.19e}$$

By using these relations, we can show that

$$\left[\frac{d(\rho j_n)}{d\rho}\right]^2 = \frac{\rho^2}{2n+1}\left[(n+1)j_{n-1}^2 + n j_{n+1}^2\right] - n(n+1)j_n^2 \tag{1.19f}$$

The integral in Equation 1.18 now becomes

$$W_{\mathrm{m}} = \frac{\pi\varepsilon_0 n(n+1)}{k_0(2n+1)} \int_{k_0 a}^{\infty} \left\{ \frac{(n+1)\rho^2}{2n+1} \left[j_{n-1}^2(\rho)+y_{n-1}^2(\rho)\right] + \frac{n\rho^2}{2n+1}\left[j_{n+1}^2(\rho)+y_{n+1}^2(\rho)\right] - 1 \right\} d\rho$$

$$= \frac{\pi\varepsilon_0 n(n+1)}{k_0(2n+1)} \left\{ k_0 a - \frac{(k_0 a)^3 (n+1)}{2(2n+1)} \left[j_{n-1}^2(k_0 a) - j_{n-2}(k_0 a) j_n(k_0 a) \right.\right.$$

$$\left.\left. + \frac{n}{n+1} j_{n+1}^2(k_0 a) - \frac{n}{n+1} j_n(k_0 a) j_{n+2}(k_0 a)\right] \right\} + \text{corresponding terms in}$$

$$y_n, y_{n-1}, y_{n-2}, y_{n+1}, \text{and } y_{n+2} \qquad (1.20)$$

The quality factor Q for the nth mode, which we will designate by the symbol Q_n^{TE}, is obtained by multiplying by $2\omega/P_{\mathrm{r}}$ and is given by

$$Q_n^{\mathrm{TE}} = k_0 a - \frac{(k_0 a)^3 (n+1)}{2(2n+1)} \left[j_{n-1}^2(k_0 a) - j_{n-2}(k_0 a) j_n(k_0 a) + \frac{n}{n+1} j_{n+1}^2(k_0 a) \right.$$

$$\left. - \frac{n}{n+1} j_n(k_0 a) j_{n+2}(k_0 a) \right] + \text{corresponding terms in}$$

$$y_n,\ y_{n-1},\ y_{n-2},\ y_{n+1},\ \text{and}\ y_{n+2}$$

This expression can be simplified by using the recurrence relation (Equation 1.19b) to eliminate the Bessel functions of order $n-1$, $n-2$, and $n+2$. When this is done, we obtain

$$Q_n^{\mathrm{TE}} = k_0 a - \left[\frac{(k_0 a)^2}{2} + (n+1) k_0 a\right] \left[j_n^2(k_0 a) + y_n^2(k_0 a)\right] - \frac{(k_0 a)^3}{2} \left[j_{n+1}^2(k_0 a) + y_{n+1}^2(k_0 a)\right]$$

$$+ \frac{2n+3}{2} (k_0 a)^2 \left[j_n(k_0 a) j_{n+1}(k_0 a) + y_n(k_0 a) y_{n+1}(k_0 a)\right] \qquad (1.21)$$

which is the same expression as that given in the Collin and Rothschild's (1964) paper. The Q_n^{TE} can be expressed as a power series in inverse powers of $k_0 a$ by using

$$(k_0 a)^2 \left[j_n^2(k_0 a) + y_n^2(k_0 a)\right] = C_n^2 + D_n^2 \qquad (1.22a)$$

$$(k_0 a)^2 \left[j_n(k_0 a) j_{n+1}(k_0 a) + y_n(k_0 a) y_{n+1}(k_0 a)\right] = C_n D_{n+1} - D_n C_{n+1} \qquad (1.22b)$$

where

$$C_n = \sum_{m=0}^{2m \le n} \frac{(-1)^m (n+2m)!}{(2m)!(n-2m)!2^{2m}(k_0a)^{2m}}$$

and

$$D_n = \sum_{m=0}^{2m \le n-1} \frac{(-1)^m (n+2m+1)!}{(2m+1)!(n-2m-1)!2^{2m+1}(k_0a)^{2m+1}}$$

For the first three modes, the results are

$$Q_1 = \frac{1}{k_0a} + \frac{1}{(k_0a)^3} \tag{1.23a}$$

$$Q_2 = \frac{3}{k_0a} + \frac{6}{(k_0a)^3} + \frac{18}{(k_0a)^5} \tag{1.23b}$$

$$Q_3 = \frac{6}{k_0a} + \frac{21}{(k_0a)^3} + \frac{135}{(k_0a)^5} + \frac{675}{(k_0a)^7} \tag{1.23c}$$

The Q of the first three modes is shown in Figure 1.2. Note that the Q rapidly becomes very large as soon as the parameter k_0a becomes less than unity, and that Equations 1.23a–1.23c are exact.

Collin and Rothschild (1964) applied the same method to calculate the Q of cylindrical modes excited outside the surface of a cylinder with radius a. The Q of cylindrical modes was found to have a similar dependence on the radius of the cylinder as the spherical modes have on the radius of the circumscribing sphere.

The TM_{n1} modes are the dual of the TE_{n1} modes and have the same value for the Q_n. The expressions for the electric and magnetic reactive energies are interchanged, so $W_e > W_m$. If the TE_{10} and TM_{10} modes are excited with equal amplitude from a one-port input network, then the electromagnetic field outside the sphere with radius a will contain equal amounts of reactive electric and magnetic energies. Since the TE_{n1} and TM_{n1} modes are orthogonal for energy storage and radiated power, these quantities can be summed when both sets of modes are excited. In general, there will also be some energy stored within the circumscribing sphere. We will let these be represented by W_e' and W_m'. For the purpose of discussion, we will assume that $W_m > W_e$. The following situations can occur:

$$W_m + W_m' > W_e + W_e'$$

$$W_m + W_m' < W_e + W_e'$$

For the first case, we would need to add some stored electric energy ΔW_e in order to tune the system to resonance, that is,

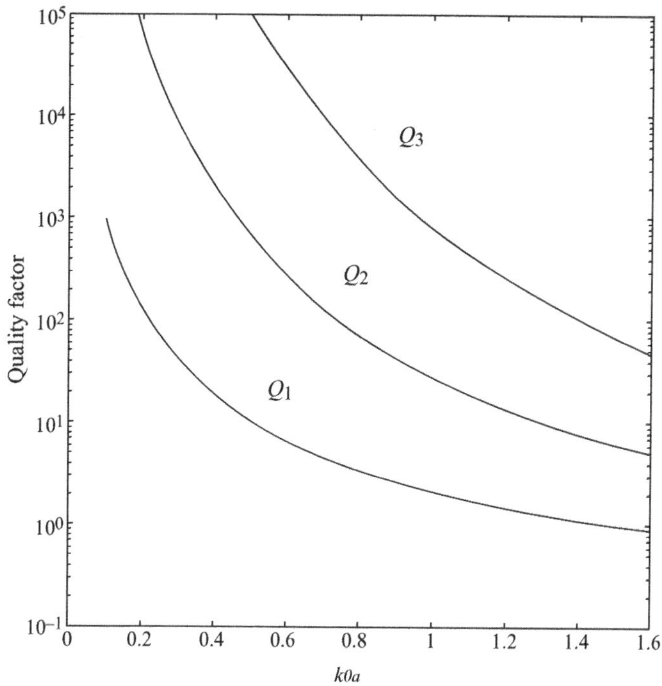

FIGURE 1.2 The quality factors (or the first three TE_{n0} and TM_{n0} modes. Only the stored reactive energy outside the circumscribing sphere of radius a is included.

$$W_m + W_m' = W_e + W_e' + \Delta W_e$$

Clearly, if the Q is calculated using W_m for the time-averaged stored magnetic energy, this will give a lower bound on the antenna Q since the total stored magnetic energy is larger because it includes the internal stored magnetic energy W_m'. For the second case, we would need to add additional magnetic energy ΔW_m such that

$$W_m + W_m' + \Delta W_m = W_e + W_e'$$

Again it is clear that using only the external stored magnetic energy will give a lower bound on the antenna Q.

If the TE_{10} mode is excited with a phase angle $\pi/2$ relative to that for the TM_{10} mode but with an equal amplitude, then the radiated field everywhere will be circularly polarized. The power across any spherical surface will be independent of time. If the internal stored electric and magnetic energies are also balanced, then the system will be resonant at all frequencies. The radiated power will be twice that of a single mode. The stored energy will be the sum of W_m from the TE_{10} mode and that from the TM_{10} mode, which is proportional to $1/k_0a$ and thus leads to the following lower bound on the Q (McLean, 1996; Collin, 1998):

$$Q_{\mathrm{TE+TM}} = \frac{1}{2}\left\{\left[\frac{1}{(k_0a)^3} + \frac{1}{k_0a}\right] + \frac{1}{k_0a}\right\} = \frac{1}{2(k_0a)^3} + \frac{1}{k_0a} \tag{1.24}$$

1.4 THAL ANTENNA Q

Any antenna that is contained within a sphere of radius a will have additional energy storage within the enclosing sphere and will consequently have a higher Q. Thus, the Q that Chu found is a lower bound on the Q of any lossless antenna. Many ESA have a Q that is considerably larger than Chu's lower bound. When the antenna Q is large, one can infer that the bandwidth of the antenna will be small but one cannot always assume that it will be equal to $1/Q$ since the tuning circuit and losses may provide for a larger bandwidth.

In order to complete the derivation of the new lower bound on antenna Q, we need to consider the effects of energy stored within the sphere of radius a. In two recent papers, Thal (1978, 2006) reevaluated the Q of TE_{n1} and TM_{n1} modes by assuming that the antenna consisted of a suitable current sheet on the surface of the sphere of radius a. This allowed the modes excited in the interior of the sphere to be included in the energy storage and hence led to larger values for the minimum achievable Q. This work was based on the use of continued fraction expansions for the mode impedances in both the internal and external regions. This current sheet can be chosen so as to excite a single TE_{n0} or TM_{n0} mode. The only boundary condition that needs to be applied is the continuity of the tangential electric field across the current sheet. Thal extended the circuit analysis of Chu by developing a ladder network that included the energy inside the enclosing sphere.

Hansen and Collin (2009) extended the exact formulation in terms of spherical modes to include the energy stored inside the sphere. The result is a quotient of spherical Bessel and Hankel functions. Numerical values are shown in Table 1.1, and as expected these agree with those published by Thal. Figure 1.3 shows the Chu-Q for TE_1 and TM_1, and the Thal-Q for TE_1 and Thal-Q for TM_1. Exact formulas are those in Section 1.5 for $\mu = 1$ and $\varepsilon = 1$.

TABLE 1.1 New Q Values

ka	Chu-Q (TM or TE)	Thal-Q (TM)	Thal-Q (TE)
0.1	1010.0	1506.0	3030.0
0.15	302.96	448.51	908.90
0.2	130.00	190.58	390.00
0.25	68.000	98.506	204.00
0.30	40.370	57.684	121.11
0.35	26.181	36.850	78.540
0.40	18.125	25.111	54.380
0.45	13.196	17.991	39.590
0.50	10.000	13.421	30.004

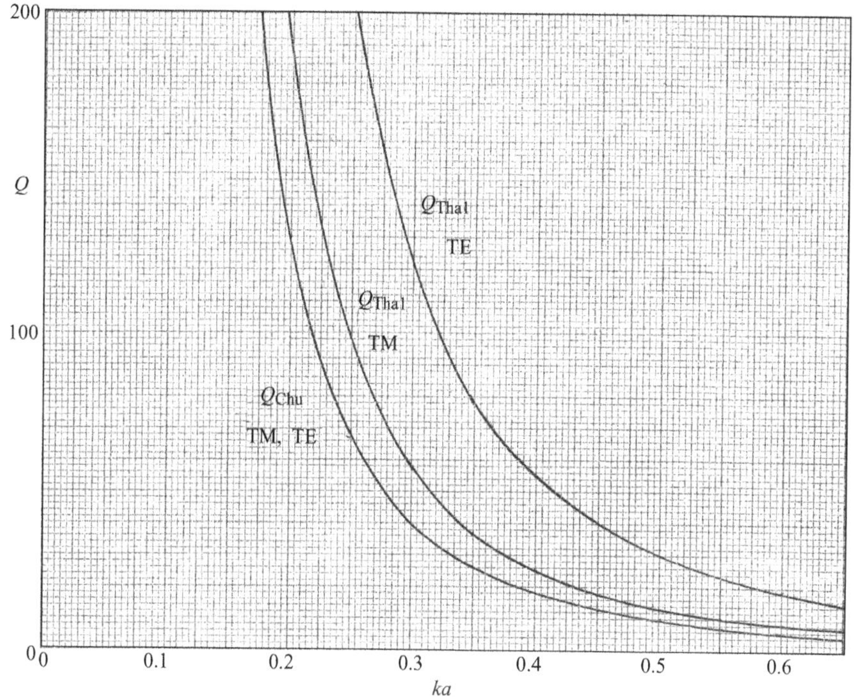

FIGURE 1.3 Q for lowest order modes.

The increase in Q due to the energy stored inside the sphere is (Hansen and Collin, 2009)

$$\Delta Q = \mathrm{SF}\left\langle \frac{(ka)^3}{2}\left[j_n^{\,2} - j_{n-1}j_{n+1}\right] + \frac{(ka)^2}{2n+1}\left[(n+1)\,j_n\,j_{n-1} - nj_n\,j_{n+1}\right]\right\rangle \quad (1.25)$$

The scale factors are

$$\mathrm{SF_{TM}} = 1 + \frac{\left[(n+1)y_{n-1} - ny_{n+1}\right]^2}{\left[(n+1)j_{n-1} - nj_{n+1}\right]^2}$$
$$\mathrm{SF_{TE}} = \frac{[j_n(ka)]^2 + [y_n(ka)]^2}{[j_n(ka)]^2} \quad (1.26)$$

Unlike the Chu-Q case, the new formulas do not have Q expressed as a two- or three-term formula. This was remedied by Hansen and Collin (2009) who performed a least pth fit to the exact values for both the TM_1 and TE_1 modes for two terms. Least-squares fitting has a limitation in that the errors at the interval ends are different from those in the middle of the interval. This was corrected by Bandler and Charalambous (1972) with the least pth fit. It takes the pth root of the sum of the

TABLE 1.2 TM Q Formula Errors

ka	Q_{new}	Q_{approx}	% Error
0.10	1506.0	1507.1	0.07
0.15	448.51	449.20	0.15
0.20	190.58	191.00	0.22
0.25	98.506	98.830	0.33
0.30	57.684	57.910	0.39
0.35	36.850	37.010	0.43
0.40	25.111	25.210	0.39
0.45	17.991	18.030	0.22
0.50	13.421	13.410	−0.08

function values each raised to the pth power. For $p \geq 10$, the errors are evenly distributed. The calculated values were fitted with $p = 20$, for both TM_1 and TE_1 modes. The TM coefficients were close to 0.707 and 1.5, so these were used; TE coefficients were close to 3.

$$\begin{aligned} TM_1: \quad & Q = \frac{1}{\sqrt{2}ka} + \frac{1.5}{(ka)^3} \\ TE_1: \quad & Q = \frac{3}{ka} + \frac{3}{(ka)^3} \end{aligned} \tag{1.27}$$

Table 1.2 shows the errors versus ka for the TM_1 case; the maximum error is 0.43% over the important range of $ka \leq 0.5$; for the TE_1 formula, the errors are even smaller. These useful formulas will be easy to remember!

1.5 RADIAN SPHERE WITH MU AND/OR EPSILON: TE MODES

One could assume that the spherical core was a material with permittivity and permeability greater (or less) than those of free space. Indeed, Wheeler (1958) did evaluate a spherical antenna consisting of a coil wound on the surface of a sphere, with a permeability greater than that of free space. Two recent papers by Kim et al. (2010) and Kim and Breinbjerg (2011) also address this problem. See also McLean, Foltz, and Sutton, 2011. The analysis starts with the Collin and Rothschild's (1964) and Hansen and Collin's (2009) papers, modifying the analysis to allow a dielectric–magnetic core. The papers by Kim's group show that for TE modes and a magnetic core, a Q approaching the Chu lower bound can be realized when the radius of the core tends toward zero.

Now consider the case of the excitation of TE_{n0} modes and assume that the interior of the sphere is filled with lossless material having a relative permittivity ε_r and a relative permeability μ_r. The only case that can be analyzed without having to specify the details of the system of currents used to excite the electromagnetic field is the use of a current sheet located on the surface $r = a$. This current sheet can be chosen so as to excite a single TE_{n0} or TM_{n0} mode. The only boundary condition that needs to be applied is the continuity of the tangential electric field across the current sheet. For the case of TE_{n0} modes, the fields are given by Equations 1.11a–1.11c but with the Hankel functions replaced by the spherical Bessel functions $j_n(ka)$ and k_0 replaced by $k = \sqrt{\varepsilon_r \mu_r} k_0$, thus

$$E_\phi = C_n^e \frac{\sin\theta}{r} \frac{dP_n(\cos\theta)}{d(\cos\theta)} [krj_n(kr)] \tag{1.28a}$$

$$H_\theta = C_n^e \frac{k\sin\theta}{j\omega\mu_r\mu_0 r} \frac{dP_n(\cos\theta)}{d(\cos\theta)} \frac{d[krj_n(kr)]}{d(kr)} \tag{1.28b}$$

$$H_r = -C_n^e \frac{n(n+1)}{j\omega\mu_r\mu_0 r^2} P_n(\cos\theta)[krj_n(kr)] \tag{1.28c}$$

The stored magnetic energy within the sphere is given by an integral similar to Equation 1.18 but with the terms involving y_n dropped, changing k_0 to k, and ε_0 to $\varepsilon_r\varepsilon_0$, which gives

$$W_m = |C_n^e|^2 \frac{\pi\varepsilon_r\varepsilon_0 n(n+1)}{k(2n+1)} \int_0^{ka} \left\{ \left[\frac{d\rho j_n(\rho)}{d\rho}\right]^2 + n(n+1) j_n^2(\rho) \right\} d\rho \tag{1.29a}$$

Note that $\rho = \sqrt{\varepsilon_r \mu_r} k_0 a$ and there is no propagating energy density subtracted at infinity. In addition, the limits on the integral are now from 0 to ka. The evaluation of this integral is similar to that for Equation 1.18 and can be inferred to be

$$W_m = |C_n^e|^2 \frac{\pi\varepsilon_r\varepsilon_0 n(n+1)}{k(2n+1)} \left\{ \frac{(ka)^3(n+1)}{2(2n+1)} \left[j_{n-1}^2(ka) - j_{n-2}(ka) j_n(ka) + \frac{n}{n+1} j_{n+1}^2(ka) - \frac{n}{n+1} j_n(ka) j_{n+2}(ka) \right] \right\} \tag{1.29b}$$

We can simplify this expression by using the Bessel function recurrence relations to eliminate the Bessel functions of order $n-2$ and $n+2$. This gives the result

$$W_m = |C_n^e|^2 \frac{\pi\varepsilon_r\varepsilon_0 n(n+1)}{k(2n+1)} \left\{ \frac{\rho^3}{2} \left[j_n^2(\rho) - j_{n-1}(\rho) j_{n+1}(\rho) \right] + \frac{\rho^2}{2n+1} \left[(n+1) j_n(\rho) j_{n-1}(\rho) - n j_n(\rho) j_{n+1}(\rho) \right] \right\} \tag{1.29c}$$

When we multiply by 2ω and divide by the radiated power given by Equation 1.13, we obtain the change in the antenna Q due to the energy stored internal to the sphere. We will denote this change by ΔQ_n^{TE}, which is given by

$$\Delta Q_n^{\mathrm{TE}} = |C_n^e|^2 \sqrt{\frac{\varepsilon_{\mathrm{r}}}{\mu_r}} \left\{ \frac{\rho^3}{2} \left[j_n^2(\rho) - j_{n-1}(\rho) j_{n+1}(\rho) \right] \right.$$

$$\left. + \frac{\rho^2}{2n+1} \left[(n+1) j_n(\rho) j_{n-1}(\rho) - n j_n(\rho) j_{n+1}(\rho) \right] \right\} \tag{1.30}$$

The last step is to find the value of $|C_n^e|^2$, which is the scale factor that the interior field must be multiplied by, from the condition that the tangential electric field must be continuous across the current sheet. This condition gives

$$|C_n^e|^2 = \mathrm{SF}_{\mathrm{TE}} = \frac{k_0^2 \left[j_n^2(k_0 a) + y_n^2(k_0 a) \right]}{k^2 j_n^2(ka)} \tag{1.31}$$

The final result for the change in the Q for the TE_{n0} mode due to energy stored within the circumscribing sphere is

$$\Delta Q_n^{\mathrm{TE}} = \frac{k_0}{\mu_r k} \frac{j_n^2(k_0 a) + y_n^2(k_0 a)}{j_n^2(ka)} \left\{ \frac{\rho^3}{2} \left[j_n^2(\rho) - j_{n-1}(\rho) j_{n+1}(\rho) \right] \right.$$

$$\left. + \frac{\rho^2}{2n+1} \left[(n+1) j_n(\rho) j_{n-1}(\rho) - n j_n(\rho) j_{n+1}(\rho) \right] \right\} \tag{1.32a}$$

When $\varepsilon_{\mathrm{r}} = \mu_{\mathrm{r}} = 1$, the above result agrees with that given by Hansen and Collin (2009).

For the TE_{10} mode, a simplified expression is easily derived by using the sine and cosine expressions for the spherical Bessel functions, thus

$$\Delta Q_1^{\mathrm{TE}} = \sqrt{\frac{\varepsilon_{\mathrm{r}}}{\mu_{\mathrm{r}}}} \frac{(ka)^2}{(k_0 a)^2} \frac{1 + (k_0 a)^2}{\left[(ka)^2 \cos^2 ka - (ka) \sin 2ka + \sin^2 ka \right]}$$

$$\times \left[\frac{ka}{2} - \frac{\sin 2ka}{4} - \frac{\cos^2 ka}{ka} + \frac{\sin 2ka}{(ka)^2} - \frac{\sin^2 ka}{(ka)^3} \right] \tag{1.32b}$$

We can obtain an alternative expression by using

$$1 + (k_0 a)^2 = (k_0 a)^3 \left[\frac{1}{(k_0 a)^3} + \frac{1}{k_0 a} \right] = (k_0 a)^3 Q_{1,\mathrm{Chu}}$$

where $Q_{1,\mathrm{Chu}}$ is given by Equation 1.23a and is the contribution to the Q due to the external stored energy. We now find that the new lower bound on the total Q for the TE_{10} mode can be expressed in the form

$$Q_1^{\mathrm{TE}} = Q_{1,\mathrm{Chu}} + \Delta Q_1^{\mathrm{TE}}$$

$$= \left\{ 1 + \frac{2}{\mu_r} \frac{\left[(ka)^4/4 - \left((ka)^3 \sin 2ka\right)/8 - \left((ka)^2 \cos^2(ka)\right)/2 + (ka \sin 2ka)/2 - \left(\sin^2(ka)\right)/2\right]}{\left[(ka)^2 \cos^2(ka) - ka \sin 2ka + \sin^2(ka)\right]} \right\} Q_{1,\mathrm{Chu}} \tag{1.33}$$

This result is the same as that given by Kim et al. (2010). However, the above authors do not give any formulas for the Q for the higher order TE_{n0} modes. The above results also support the result $Q_1 = (1 + 2/\mu_r)(k_0 a)^{-3}$ given many years ago by Wheeler (1958).

From Equation 1.33, it can be seen that the new lower bound for the total Q depends only on the permeability parameter μ_r and the size of the core through the parameter $ka = \sqrt{\varepsilon_r \mu_r}(k_0 a)$. For very small values of ka, the contribution ΔQ_n^{TE} to the total Q of the antenna is very small provided that the permeability is very large. Hence, for a core with a small value of ka, the lower bound is very close to the value of $Q_{1,\mathrm{Chu}}$. As ka increases in value, the ΔQ_n^{TE} increases but the total Q decreases because $Q_{1,\mathrm{Chu}}$ is decreasing at a rapid rate. After Q reaches a minimum value, it begins to increase without limit as the resonant frequency of the internal spherical core is approached. The resonant frequency occurs when the denominator term $j_n(ka)$ in Equation 1.31 equals zero. At the resonant frequency, the tangential electric field equals zero on the interior side of the surface of the circumscribing sphere. Since the tangential electric field is continuous across the current sheet, it must also be zero on the exterior side of the current sheet and consequently there is no external radiation at the resonant frequencies. This is a manifestation of the nonuniqueness of the external scattering problem for a sphere at the internal resonances.

When ε_r and μ_r are equal to unity, ΔQ_1^{TE} for the TE_{10} mode is very close to $2Q_{1,\mathrm{Chu}}$ for $k_0 a < 0.7$, which makes the new lower bound on the antenna Q for this mode approximately three times as large as that obtained by considering only the energy stored outside of the circumscribing sphere. In Figure 1.4a, we show some typical results for the new lower bound on the Q as a function of $k_0 a$ for $\mu_r = 4$, 16, and 64 and $\varepsilon_r = 4$. The curve for $\mu_r = 64$ shows a large increase in Q when the frequency approaches the resonance value. These resonances are also encountered for the lower values of μ_r, and also for the case of an air core, but at larger values of $k_0 a$. For μ_r equal to 16 and larger, the curves of Q versus $k_0 a$ are almost identical except in the near vicinity of the resonances, which are different for each case because these occur when $ka = \tan ka$ for the $k_0 a$ mode and thus depend on both core permittivity and permeability. From Equation 1.33, we can see that the new lower bound Q_1^{TE} for the TM_{10} mode depends only on ka and the relative permeability μ_r. For this reason, the same results shown in Figure 1.4a are shown in Figure 1.4b plotted as a function of ka. These curves show additional resonance points and illustrate the minimum values of Q_1^{TE} that can be achieved. Note that when Q_1^{TE} is plotted as a factor of $ka\,Q_1^{\mathrm{TE}}$ decreases with an increase in μ_r, but when plotted as a function of ka the quality factor increases with an increase in μ_r.

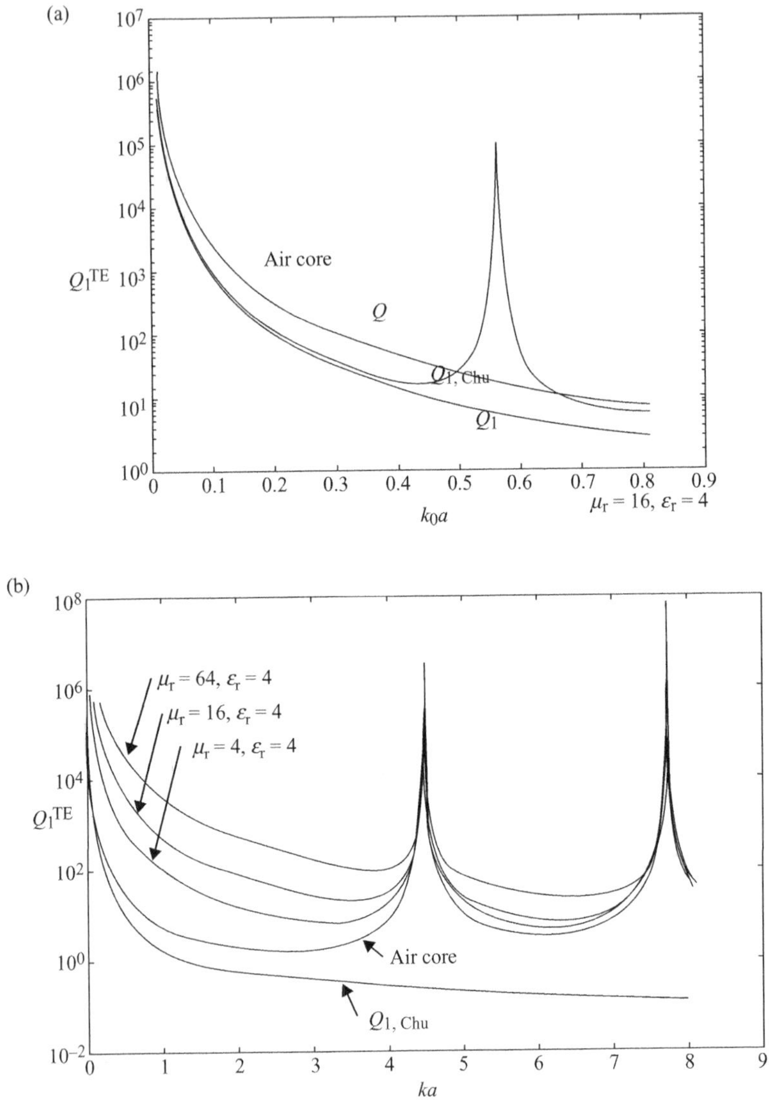

FIGURE 1.4 (a) A plot of lower bound Q_1^{TE} for TF_{10} mode as a function of ka for various electrical parameters for the spherical core. (b) Comparison of the new lower bound for Q_1^{TE} for the TE_{10} mode with the new lower bound taking into account internal energy stored in a core with relative permittivity of 16. Also shown is the Q for an air core and the original $Q_{1,\text{Chu}}$ that is based only on the external stored energy.

In the paper by Kim and Breinbjerg (2011), the ratio of the total stored magnetic field energy to the total stored electric field energy is plotted as a function of k_0a for μ_r equal to 1, 2, 8, and 100, with $\varepsilon_r = 1$. It was found that the stored electric field energy became equal to the stored magnetic field energy at the cavity resonant frequencies, but never exceeded the stored magnetic field energy. We have verified these calculations for ka up to 100 and also included ε_r values of 2, 4, 16, 64, and 100. The same property that the total stored electric field energy did not exceed the total stored magnetic field energy continued to hold. Thus, the formula given by Equation 1.33 for the Q of the TE_{10} will hold for all values of k_0a, which is contrary to a conclusion given by Kim and Breinbjerg (2011).

The next issue we wish to explore is whether or not the frequency dependence of the tuned antenna admittance will result in a 3 dB bandwidth that is equal to $2/Q_1^{TE}$, where Q_1^{TE} is the Q of the TE_{10} mode with the dielectric–magnetic core and no external conductive loading; that is, Q_1^{TE} is the unloaded antenna Q. The antenna configuration analyzed above consists of a dielectric–magnetic core of radius a and wound with a current sheet in the ϕ direction. The admittance presented to the current sheet source is the parallel combination of the wave admittance Y_e looking in the outward direction with the wave admittance Y_i looking inward from the surface at $r = a$. These wave admittances can be obtained from the expressions for the fields of the TE_{10} mode given in Equations 1.11a, 1.11b, 1.28a, and 1.28b and are, after normalization with respect to the characteristic admittance $Y_0 = (\varepsilon_0/\mu_0)^{1/2}$ of free space,

$$Y_e = -\frac{H_\theta}{E_\phi} = j\frac{d\left[(k_0a)h_1^2(k_0a)\right]/(dk_0a)}{(k_0a)h_1^2(k_0a)} \tag{1.34a}$$

$$Y_i = -j\sqrt{\frac{\varepsilon_r}{\mu_r}}\frac{d[(ka)j_1(ka)]/d(ka)}{(ka)j_1(ka)} \tag{1.34b}$$

We now use

$$xj_1(x) = \frac{\sin(x)}{x} - \cos(x)$$

and

$$xh_1^2(x) = \frac{\sin(x)}{x} - \cos(x) + j\left[\frac{\cos(x)}{x} + \sin(x)\right]$$

to obtain

$$Y_e = \frac{j\left\{k_0a\cos(k_0a) - \sin(k_0a) + (k_0a)^2\sin(k_0a) + j\left[(k_0a)^2\cos(k_0a) - \cos(k_0a) - k_0a\sin(k_0a)\right]\right\}}{k_0a\sin(k_0a) - (k_0a)^2\cos(k_0a) + j\left[k_0a\cos(k_0a) + (k_0a)^2\sin(k_0a)\right]}$$

$$= \frac{(k_0a)^3 - j}{k_0a\left[1 + (k_0a)^2\right]} \tag{1.35a}$$

$$Y_{\mathrm{i}} = j\sqrt{\frac{\varepsilon_{\mathrm{r}}}{\mu_{\mathrm{r}}}}\left[\frac{\sin(ka) - ka\cos(ka) - (ka)^2\sin(ka)}{ka\sin(ka) - (ka)^2\cos(ka)}\right] \tag{1.35b}$$

The external admittance Y_{e} is that of a lumped element circuit consisting of an inductive reactance jk_0a in parallel with the reactance of a series capacitor and resistor $1 + 1/jk_0a$ (Harrington, 1961). The internal admittance cannot be represented by a lumped element circuit, except when ka is small so that a first-order power series expansion of Y_{i} can be used. In the low-frequency range, Y_{i} can be approximated as an inductive reactance $jka = j\omega\mu_{\mathrm{r}}\mu_0 a\sqrt{\varepsilon/\mu}$, which is connected in parallel with $jk_0a = j\omega\mu_0\sqrt{\varepsilon_0/\mu_0}$. It can be seen that the parallel combination of these two inductive reactances will be dominated by the jk_0a one when μ_{r} and ε_{r} are large. The admittance seen by the current source is $Y_{\mathrm{e}} + Y_{\mathrm{i}}$. This admittance is inductive. The antenna can be tuned to resonance by connecting a capacitive admittance $jB_{\mathrm{c}} = j\,\mathrm{Im}(Y_{\mathrm{e}} + Y_{\mathrm{i}})$ in parallel. We will assume that $B(\omega) = B(\omega_{\mathrm{c}})\omega/\omega_{\mathrm{c}}$ so that at the center frequency ω_{c} the input admittance is a pure conductance equal to the radiation conductance of the antenna. Away from the center frequency, the antenna input admittance will no longer be a pure conductance. The issue we wish to explore is whether or not the frequency dependence of the tuned antenna admittance will result in a 3 dB bandwidth that is equal to $2/Q_1^{\mathrm{TE}}$, where Q_1^{TE} is the Q of the TE_{10} mode with the dielectric–magnetic core and no external conductive loading; that is, Q_1^{TE} is the unloaded antenna Q. A sample evaluation of the VSWR as a function of k_0a for the loaded antenna has been calculated for the case of an antenna with $k_0a = 0.25$ at the center frequency, with a core having $\varepsilon_{\mathrm{r}} = 4$ and $\mu_{\mathrm{r}} = 16$, and having a load conductance equal to that of the conductance of Y_{e} at the center frequency. For this antenna, the Q is very nearly equal to $(1 + 2/\mu_{\mathrm{r}})Q_{1,\mathrm{Chu}}$, which equals 76.5. The VSWR is shown in Figure 1.5. From the figure, it was estimated that the 3 dB bandwidth was 0.0065 in units of k_0a, which gives a fractional bandwidth of $0.0065/0.25 = 0.026$. From this bandwidth, we can determine the circuit loaded Q, which is given by $1/0.026 = 38.45$. The unloaded antenna Q is equal to twice this value and is 76.9, which is slightly greater than the theoretical lower bound given by Equation 1.33. This exmple suggests that an antenna using a core with a large value of permeability and with the excitation in the form of a surface winding in the azimuthal direction is an optimum design.

1.6 RADIAN SPHERE WITH MU AND/OR EPSILON: TM MODES

We will now analyze an antenna consisting of a dielectric–magnetic core that is excited by a current sheet in the θ direction so as to radiate only TM_{n1} modes. The external fields are given by Equations 1.4a–1.4c. The internal fields have the same form but with the Hankel functions replaced by Bessel functions, thus

$$E_\theta = D_n^h \frac{-k\sin\theta}{j\omega\varepsilon_{\mathrm{r}}\varepsilon_0 r}\frac{dP_n(\cos\theta)}{d(\cos\theta)}\frac{d[krj_n(kr)]}{d(kr)} \tag{1.36a}$$

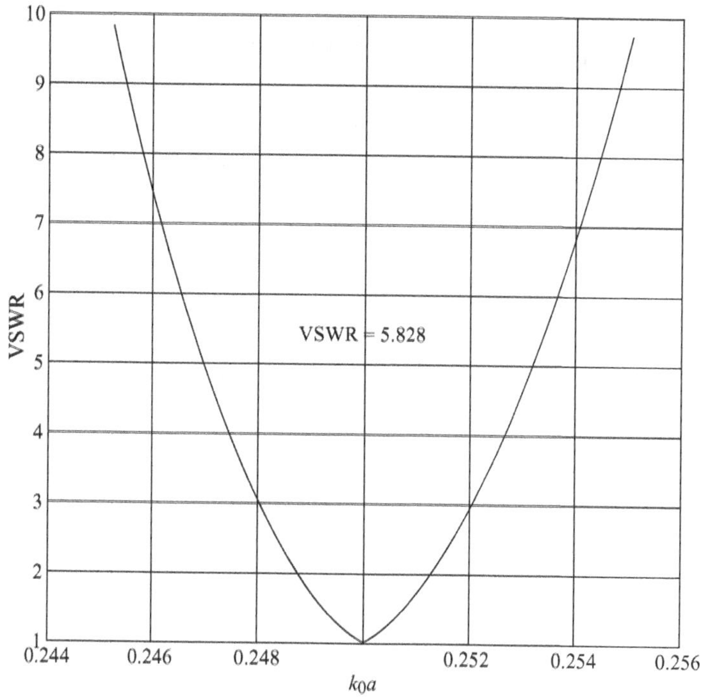

FIGURE 1.5 VSWR as a function of k_0a for an antenna tuned to resonance at the frequency corresponding to $k_0a = 0.25$. The core has a relative permittivity of 4 and a relative permeability of 16. The 3 dB loaded antenna Q is equal to 38.45. The unloaded antenna Q equals 76.9.

$$E_r = D_n^h \frac{n(n+1)}{j\omega\varepsilon_r\varepsilon_0 r^2} P_n(\cos\theta)[krj_n(kr)] \tag{1.36b}$$

$$H_\phi = D_n^h \frac{\sin\theta}{r} \frac{dP_n(\cos\theta)}{d(\cos\theta)} [krj_n(kr)] \tag{1.36c}$$

For these modes, the stored electric energy is greater than the stored magnetic energy. For the external fields, the stored electric energy is given by an expression similar to that in Equation 1.20 since the TM_{n1} modes are the dual of the TE_{n0} modes. Thus, the external stored energy leads to the same contribution to the antenna Q that is given by Equation 1.21. Consequently, we need to evaluate only the additional contribution to the antenna Q arising from the internal stored energy. The integral for the stored electric energy will be similar to that for the stored magnetic energy as given by Equation 1.29a. By using the expressions from Equations 1.36a and 1.36c, we obtain

$$
\begin{aligned}
W_{\mathrm{e}} &= \int_0^a\int_0^\pi \frac{\varepsilon_r\varepsilon_0}{4}\left[|E_\theta|^2+|E_r|^2\right]2\pi r^2\sin\theta\, d\theta\, dr \\
&= \int_0^a\int_0^\pi |D_n^h|^2\frac{\mu_r\mu_0}{4}\left\{\left\{\frac{dP_n(\cos\theta)}{d(\cos\theta)}\frac{d[krj_n(kr)]}{d(kr)}\right\}^2\right. \\
&\qquad \left. +\left[n(n+1)P_n(\cos\theta)j_n(kr)\right]^2\right\}2\pi\sin^3\theta\, d\theta\, dr \\
&= \frac{\mu_{\mathrm{r}}\mu_0\pi}{k}\frac{n(n+1)}{2n+1}\int_0^{ka}|D_n^h|^2\left\{\left[\frac{d\rho j_n(\rho)}{d\rho}\right]^2+n(n+1)j_n^2(\rho)\right\}d\rho
\end{aligned}
\tag{1.37a}
$$

where $\rho = ka$. This integral is the same as that in Equation 1.26 and the result will be the same as that given by Equation 1.27. Thus, we have

$$
\begin{aligned}
W_{\mathrm{e}} = |D_n^h|^2\frac{\pi\mu_{\mathrm{r}}\mu_0 n(n+1)}{k(2n+1)}&\left\{\frac{(ka)^3(n+1)}{2(2n+1)}\left[j_{n-1}^2(ka)-j_{n-2}(ka)j_n(ka)\right.\right. \\
&\left.\left.+\frac{n}{n+1}j_{n+1}^2(ka)-\frac{n}{n+1}j_n(ka)j_{n+2}(ka)\right]\right\}
\end{aligned}
\tag{1.37b}
$$

The total radiated power is given by Equation 1.13 multiplied by μ_0/ε_0 and is $[2n(n+1)/(2n+1)]\pi\sqrt{\mu_0/\varepsilon_0}$, so the contribution ΔQ_n^{TM} to the antenna Q from the internal stored energy will be similar to that in Equation 1.29:

$$
\begin{aligned}
\Delta Q_n^{\mathrm{TM}} = \frac{2\omega W_{\mathrm{e}}}{P_{\mathrm{r}}} = |D_n^h|^2\sqrt{\frac{\mu_{\mathrm{r}}}{\varepsilon_{\mathrm{r}}}}&\left\{\frac{\rho^3}{2}\left[j_n^2(\rho)-j_{n-1}(\rho)j_{n+1}(\rho)\right]\right. \\
&\left.+\frac{\rho^2}{2n+1}\left[(n+1)j_n(\rho)j_{n-1}(\rho)-nj_n(\rho)j_{n+1}(\rho)\right]\right\}
\end{aligned}
\tag{1.38}
$$

We must also choose the scale factor $\mathrm{SF_{TM}} = |D_n^h|^2$ so that the tangential electric field is continuous across the current sheet. This requires the constant D_n^h to be given by

$$
D_n^h = \sqrt{\frac{\varepsilon_{\mathrm{r}}}{\mu_{\mathrm{r}}}}\frac{d\left[\rho_0 h_n^2(\rho_0)/d\rho_0\right]}{d\left[\rho j_n(\rho)/d\rho\right]}
$$

which gives

$$
\mathrm{SF_{TM}} = |D_n^h|^2 = \frac{\varepsilon_{\mathrm{r}}}{\mu_{\mathrm{r}}}\frac{\left[d\rho_0 j_n(\rho_0)/d\rho_0\right]^2+\left[d\rho_0 y_n(\rho_0)/d\rho_0\right]^2}{\left[d\rho j_n(\rho)/d\rho\right]^2}
\tag{1.39}
$$

The final expression obtained for ΔQ_n^{TM} is

$$\Delta Q_n^{\mathrm{TM}} = \mathrm{SF}_{\mathrm{TM}}\sqrt{\frac{\mu_{\mathrm{r}}}{\varepsilon_{\mathrm{r}}}}\left\{\frac{\rho^3}{2}\left[j_n^2(\rho)-j_{n-1}(\rho)j_{n+1}(\rho)\right]+\frac{\rho^2}{2n+1}\left[(n+1)j_n(\rho)j_{n-1}(\rho)\right.\right.$$
$$\left.\left.-nj_n(\rho)j_{n+1}(\rho)\right]\right\} \tag{1.40}$$

Hence, for the TM_{n0} modes excited by a current sheet on a dielectric–magnetic spherical core, the new lower bound for the Q is

$$Q_n^{\mathrm{TM}} = \Delta Q_n^{\mathrm{TM}} + \left\{k_0a - \left[\frac{(k_0a)^2}{2}+(n+1)k_0a\right]\left[j_n^2(k_0a)+y_n^2(k_0a)\right] - \frac{(k_0a)^3}{2}\left[j_{n+1}^2(k_0a)\right.\right.$$
$$\left.\left.+y_{n+1}^2(k_0a)\right]+\frac{2n+3}{2}(k_0a)^2\left[j_n(k_0a)j_{n+1}(k_0a)+y_n(k_0a)y_{n+1}(k_0a)\right]\right\} \tag{1.41}$$

For $n = 1$, this expression can be simplified to the form

$$Q_1^{\mathrm{TM}} = \Delta Q_1^{\mathrm{TM}} + \left[\frac{1}{(k_0a)^3}+\frac{1}{k_0a}\right] \tag{1.42}$$

where

$$\Delta Q_1^{\mathrm{TM}} = \sqrt{\frac{\varepsilon_{\mathrm{r}}}{\mu_{\mathrm{r}}}}\left[\frac{\rho^4(1+\rho_0^4-\rho_0^2)}{\rho_0^4\left[\rho^2\cos^2(\rho)+(1-\rho^2)^2\sin^2(\rho)-\rho(1-\rho^2)\sin(2\rho)\right]}\right]$$
$$\times\left[\frac{\rho}{2}-\frac{\sin^2(\rho)}{\rho^3}+\frac{(4\rho-\rho^3)\sin(2\rho)}{4\rho^3}-\frac{\cos^2(\rho)}{\rho}\right] \tag{1.43}$$

Typical values of Q_1^{TM} are shown in Figure 1.6a for representative values of ε_{r} and μ_{r}. For an air core, the above formula shows that for small values of $k_0a = \rho_0$ the new lower bound on Q_1^{TM} is very nearly equal to 1.5 $Q_{1,\mathrm{Chu}}$ and decreasing to 1.34 $Q_{1,\mathrm{Chu}}$ for $k_0a = 0.5$, where $Q_{1,\mathrm{Chu}}$ is given by the term appearing after the term ΔQ_1^{TM} in the above equation. For the case of a relative permeability greater than 1, or a relative permittivity greater than 1, or when both are greater than 1, the Q is increased.

Overall, the relative permeability has very little effect on the Q as long as it does not result in a value for ka that is close to the resonant frequency for the core. Thus, an air core is the best choice for TM_{n1} modes. Resonances will occur when ε_{r} and μ_{r} are large enough to make $d[\rho j_n(\rho)]/d\rho = 0$ and these will occur for smaller values of k_0a. At the

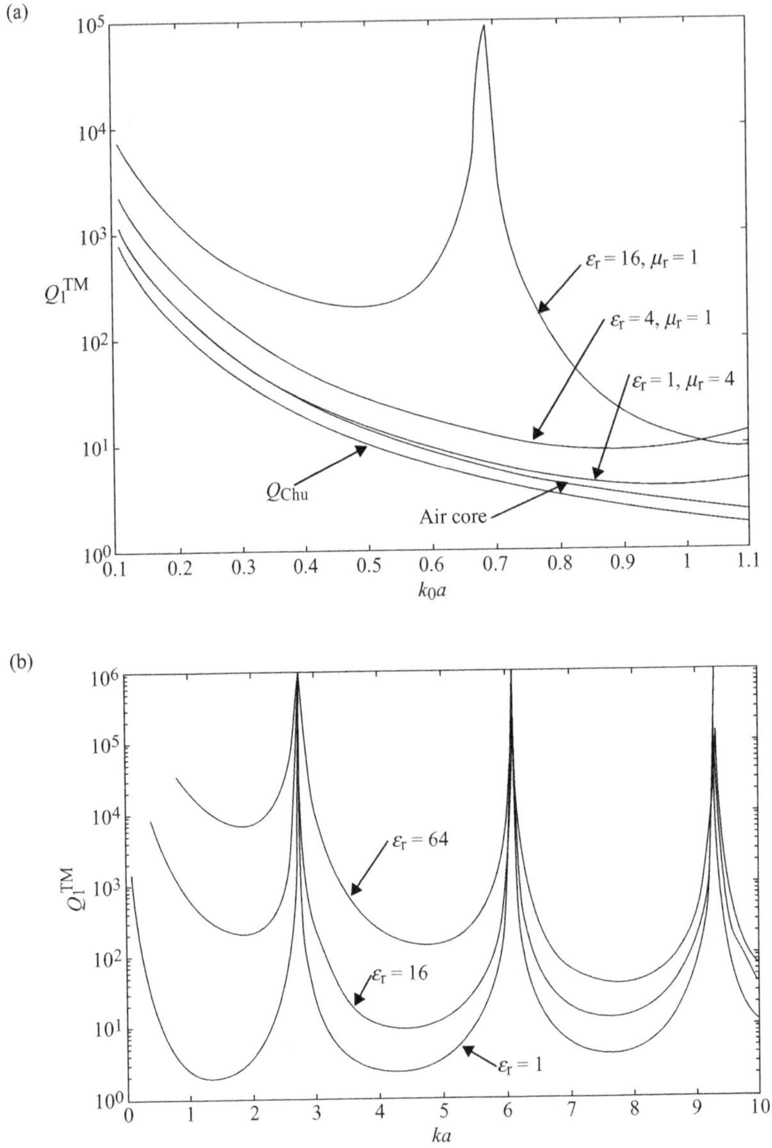

FIGURE 1.6 (a) Q_1^{TM} for TM_{10} mode as a function of k_0a for various electrical parameters of the spherical core. For an air core, the Q_1^{TM} is approximately 1.5 times the Q_{Chu} value. (b) Q_1^{TM} plotted as a function of ka for ε_r equal to 1, 16, and 64 with μ_r equal to 1.

resonant frequencies, the external radiation vanishes and since the core has been assumed to be lossless Q_1^{TM} becomes infinite. In Figure 1.6b, the value of Q_1^{TM} as a function of ka for ε_r equal to 16 and 64, with $\mu_r = 1$, is shown in order to illustrate the multiple resonances that occur.

The final topic we will examine is the evaluation of the Q of the antenna system, consisting of the dielectric–magnetic core wound with a current sheet in the θ direction, in terms of the frequency behavior of its equivalent circuit. It will be shown that the loaded 6 dB fractional bandwidth calculated from the equivalent circuit, when tuned to resonance by a shunt inductive reactance, is equal to twice the Q given by Equation 1.42, to a high degree of accuracy. This result is similar to that found for the TE_{10} mode driven by a current sheet. It is a verification of the generally held assumption that for simple antennas the fractional bandwidth is inversely proportional to the antenna Q.

For the TM_{10} mode, the wave admittance seen looking outward from the current sheet at $r = a$ may be found using Equations 1.4a and 1.4c and is given by

$$
\begin{aligned}
Y_e &= \frac{H_\phi}{E_\theta} = \frac{-j(k_0a)h_1^2(k_0a)}{d\left[k_0ah_1^2(k_0a)\right]/d(k_0a)} \\
&= \frac{\left[\rho_0\cos(\rho_0)+\rho_0^2\sin(\rho_0)-j\rho_0\sin(\rho_0)+j\rho_0^2\cos(\rho_0)\right]}{\left[\rho_0\cos(\rho_0)+\rho_0^2\sin(\rho_0)-\sin(\rho_0)\right]+j\left[\rho_0^2\cos(\rho_0)-\cos(\rho_0)-\rho_0\sin(\rho_0)\right]} \\
&= \frac{\rho_0^4+j\rho_0}{1+\rho_0^4-\rho_0^2}
\end{aligned}
\tag{1.44}
$$

after normalization with respect to the characteristic admittance of free space. Note that $\rho_0 = k_0a$. The equivalent circuit for the external admittance consists of a capacitive admittance jk_0a connected in series to an inductive reactance jk_0a in parallel with a $1\,\Omega$ resistance (Harrington, 1961).

The normalized admittance seen looking inward from the current sheet can be found by using Equations 1.36a and 1.36c and is given by

$$
Y_i = j\sqrt{\frac{\varepsilon_r}{\mu_r}}\frac{\rho j_1(\rho)}{d[\rho j_1(\rho)]/d\rho} = j\sqrt{\frac{\varepsilon_r}{\mu_r}}\frac{\rho\sin(\rho)-\rho^2\cos(\rho)}{\rho^2\sin(\rho)-\sin(\rho)+\rho\cos(\rho)}
\tag{1.45}
$$

where $\rho = ka$. Both Y_e and Y_i are capacitive for TM modes. The antenna can be tuned to resonance by connecting a parallel inductive admittance in parallel with $Y_e + Y_i$ such that $-jB_L + j\mathrm{Im}(Y_e + Y_i) = 0$ at the center frequency, where we have assumed that at the center frequency (resonant frequency) the net susceptance of the tuned circuit is zero. Away from the resonant frequency, we will assume that B_L is given by $B_L(\omega) = B_L(\omega_c)\omega_c/\omega$, where ω_c is the resonant frequency of the circuit. The last assumption will be that the circuit is loaded by a conductance equal to the radiation conductance $\mathrm{Re}(Y_e + Y_i)$ at the resonant frequency. We can now evaluate the input reflection coefficient seen by the current source and the resultant VSWR. From the fractional 6 dB bandwidth, the Q_1^{TM} for the antenna system described can be found from the relationship $Q_1^{TM} = 2/\mathrm{BW}$, where BW is the 6 dB fractional bandwidth of

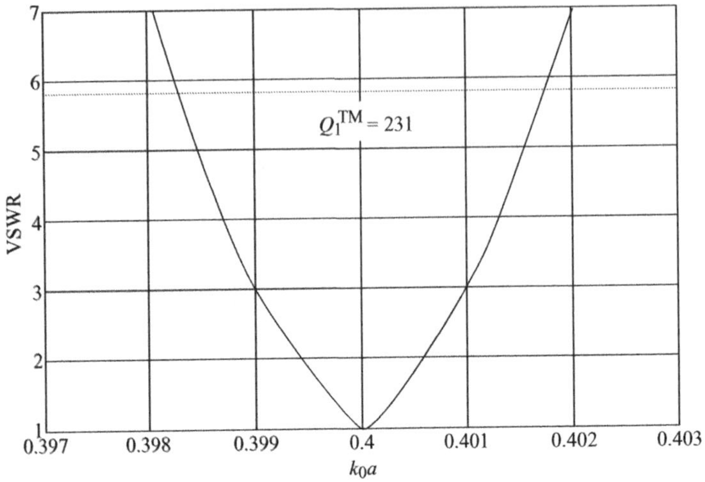

FIGURE 1.7 VSWR as a function of k_0a for the TM_{10} mode with a dielectric core with a relative permittivity of 16. The quality factor Q_1^{TM} is equal to 231 for the unloaded antenna.

the loaded tuned circuit. In Figure 1.7, we show the VSWR curve for the case where $\varepsilon_r = 16$ and the center frequency corresponds to k_0a equal to 0.4. From the VSWR curve, the estimate of the quality factor Q_1^{TM} is 231. The calculated value from Equation 1.42 is 230.2. Several other cases were also checked and gave similar very close agreements between the circuit-based Q and the value based on stored energy.

The preceding work assumed a spherical core of permeability material. Stuart and Yaghjian (2010) have shown that a solid core is not necessary. They used a monopole with a circular plate top hat; a high-μ cylindrical shell of diameter equal to that of the top hat, and height equal to that of the monopole, was added to the antenna. Results using $\mu > 100$ were close to the Chu limit. Figure 1.8 shows the decrease of Q as μ increases; also shown is the importance of an ε close to unity. See also Kim and Breinbjerg (2011).

1.7 EFFECTS OF CORE LOSSES

The results derived above showed that for TE_{n0} and TM_{n0} modes excited by means of current sheets located on the surface of a sphere of solid dielectric or magnetic material, the resultant Q would be dependent on the constituent parameters of the core material. The use of dielectric material with a relative permittivity greater than unity increased the quality factor of TM_{n0} modes and thus would not be a useful way to increase the bandwidth of the antenna. On the other hand, the use of magnetic material with a relative permittivity greater than unity had the effect of significantly reducing the Q of excited TE_{n0} modes. Since physical materials will have some loss,

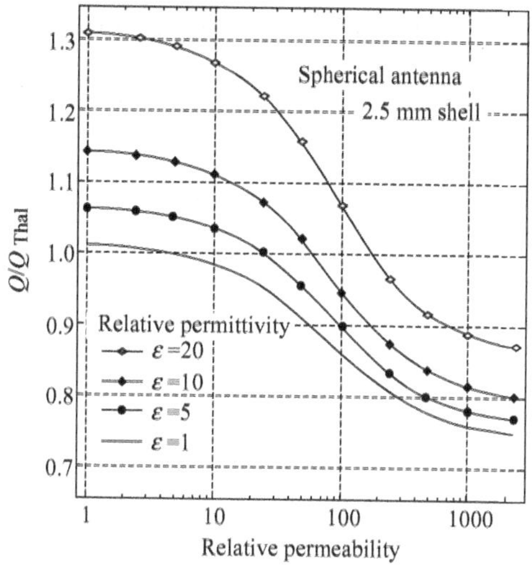

FIGURE 1.8 Top hat monopole with cylindrical permeable shell. Courtesy of Stuart and Yaghjian (2010).

an inevitable penalty to be paid by the use of a solid core is the increase in the loss and a decrease in the efficiency of the antenna. The reduction in the Q results in an increase in the bandwidth, which is normally a desirable effect, but it is offset by the decrease in efficiency.

For isotropic lossy material, the power loss is related to the average stored electric and magnetic energy within the core through the imaginary parts of the constitutive parameters. Consider the integral of the inward directed complex Poynting vector over the surface of the core at $r = a$,

$$-\frac{1}{2}\int_0^{2\pi}\int_0^{\pi} \mathbf{E}\times\mathbf{H}^*\cdot\mathbf{a}_r \sin\theta\, d\theta\, d\phi = \frac{j\omega}{2}\int_0^{a}\int_0^{2\pi}\int_0^{\pi}\left[\mu_0(\mu_r' - j\mu_r'')\mathbf{H}\cdot\mathbf{H}^* - \varepsilon_0(\varepsilon_r' + j\varepsilon_r'')\right]\sin\theta\, d\theta\, d\phi\, dr$$

The real part of the right-hand side represents the average power loss within the core. The imaginary part represents $2j\omega(W_\mathrm{m} - W_\mathrm{e})$, where W_m and W_e are the time-averaged stored magnetic and electric energies in the core material, respectively. When the core losses are small, the stored energy may be evaluated by assuming that the losses are zero. In terms of the stored energy, we can express the power loss in the following form:

$$P_\mathrm{L} = \frac{2\omega\mu_\mathrm{r}''}{\mu_\mathrm{r}'}W_\mathrm{m} + \frac{2\omega\varepsilon_\mathrm{r}''}{\varepsilon_\mathrm{r}'}W_\mathrm{e} \qquad (1.46)$$

The new reduced Q of the system is given by Q', where

$$Q' = \frac{P_r}{P_r + P_L} Q \tag{1.47}$$

and Q is the quality factor in the absence of core losses. P_r is the average radiated power from the excited TE_{n0} or TM_{n0} mode. The reduced efficiency of the antenna is given by the factor $P_r/(P_r + P_L)$. For the TE_{n0} and TM_{n0} modes, we have already obtained expressions for W_m and W_e, respectively. The other stored energy function can be obtained by direct evaluation of the volume integral of the corresponding energy density.

For the TE_{n0} modes, the stored energy in the magnetic field in the core is given by Equation 1.29c, which is repeted below:

$$W_m = |C_n^e|^2 \frac{\pi \varepsilon_r' \varepsilon_0 n(n+1)}{k(2n+1)} \left\{ \frac{\rho^3}{2} \left[j_n^2(\rho) - j_{n-1}(\rho) j_{n+1}(\rho) \right] \right.$$

$$\left. + \frac{\rho^2}{2n+1} \left[(n+1) j_n(\rho) j_{n-1}(\rho) - n j_n(\rho) j_{n+1}(\rho) \right] \right\} \tag{1.48}$$

The stored electric field energy is given by (see Equations 1.17b and 1.19a)

$$\begin{aligned} W_e &= \frac{\varepsilon_r' \varepsilon_0}{4} \int_0^{\pi} \int_0^{a} |E_\phi|^2 2\pi r^2 \sin\theta \, d\theta \, dr \\ &= \frac{\varepsilon_r' \varepsilon_0}{4k} \int_0^{\pi} \int_0^{a} |C_{ne}|^2 \left[\frac{dP_n(\cos\theta)}{d\theta} \right]^2 k^2 r^2 j_n^2(kr) 2\pi \sin\theta \, d\theta \, d(kr) \\ &= \frac{\varepsilon_r' \varepsilon_0}{2k} \pi |C_{ne}|^2 \frac{n(n+1)}{2n+1} (kr)^3 \left[j_n^2(kr) - j_{n-1}(kr) j_{n+1}(kr) \right] \end{aligned} \tag{1.49}$$

The loss arising from the magnetic material is obtained by multiplying W_m given by Equation 1.48 by the factor $2\omega\mu_r''/\mu_r'$. Similarly, the power loss in the core due to the lossy dielectric material is given by multiplying Equation 1.49 by the factor $2\omega\varepsilon_r''/\varepsilon_r'$. For the TE_{10} mode, the expressions for the power loss in the core are given by

$$P_{Lm} = \frac{4\pi}{3} \sqrt{\frac{\varepsilon_r' \varepsilon_0}{\mu_r' \mu_0}} \frac{\mu_r''}{\mu_r'} |C_{ne}|^2 \left[\frac{\rho^3}{2} (j_1^2 - j_0 j_2) + \frac{\rho^2}{3} (2 j_1 j_0 - j_1 j_2) \right] \tag{1.50}$$

$$P_{Le} = \frac{4\pi}{3} \sqrt{\frac{\varepsilon_r' \varepsilon_0}{\mu_r' \mu_0}} \frac{\varepsilon_r''}{\varepsilon_r'} |C_{ne}|^2 \left[\frac{\rho^3}{2} (j_1^2 - j_0 j_2) \right] \tag{1.51}$$

where $\rho = ka$ is the argument for the Bessel functions.

A useful parameter is the ratio of power loss to power radiated

$$\eta_{\mathrm{TE}} = \frac{P_{\mathrm{Lm}} + P_{\mathrm{Le}}}{P_{\mathrm{r}}} \tag{1.52}$$

in terms of which the new lower Q of the antenna with a lossy core is given by

$$Q' = \frac{Q'_{\mathrm{TE}}}{1 + \eta_{\mathrm{TE}}} \tag{1.53}$$

For the special case of the TE_{10} mode when $n = 1$, we obtain

$$\eta_{\mathrm{TE}} = \frac{P_{\mathrm{L}}}{P_{\mathrm{r}}} = \left|C_1^e\right|^2 \sqrt{\frac{\varepsilon'_{\mathrm{r}}}{\mu'_{\mathrm{r}}}} \left\{ \frac{\mu''_{\mathrm{r}}}{\mu'_{\mathrm{r}}} \left[\frac{\rho}{2} - \frac{\sin(2\rho)}{4} - \frac{\sin^2(\rho)}{\rho^3} + \frac{\sin(2\rho)}{\rho^2} - \frac{\cos^2(\rho)}{\rho} \right] + \frac{\varepsilon''_{\mathrm{r}}}{\varepsilon'_{\mathrm{r}}} \left[\frac{\rho}{2} + \frac{\sin(2\rho)}{4} - \frac{\sin^2(\rho)}{\rho} \right] \right\} \tag{1.54}$$

where

$$\left|C_1^e\right|^2 = \frac{\rho^2}{\rho_0^2} \frac{(1 + \rho_0^2)}{\left[\sin^2\rho + \rho^2\cos^2\rho - \rho\sin(2\rho)\right]}$$

A similar derivation may be carried out to obtain an expression for the total core loss for the TE_{10} mode. The factor η_{TM} for the TE_{10} mode is given by

$$\eta_{\mathrm{TM}} = \frac{P_{\mathrm{L}}}{P_{\mathrm{r}}} = \left|D_1^h\right|^2 \sqrt{\frac{\mu'_{\mathrm{r}}}{\varepsilon'_{\mathrm{r}}}} \left\{ \frac{\varepsilon''_{\mathrm{r}}}{\varepsilon'_{\mathrm{r}}} \left[\frac{\rho}{2} - \frac{\sin(2\rho)}{4} - \frac{\sin^2(\rho)}{\rho^3} + \frac{\sin(2\rho)}{\rho^2} - \frac{\cos^2(\rho)}{\rho} \right] + \frac{\mu''_{\mathrm{r}}}{\mu'_{\mathrm{r}}} \left[\frac{\rho}{2} + \frac{\sin(2\rho)}{4} - \frac{\sin^2(\rho)}{\rho} \right] \right\} \tag{1.55}$$

where

$$\left|D_1^h\right|^2 = \sqrt{\frac{\varepsilon'_{\mathrm{r}}}{\mu'_{\mathrm{r}}}} \frac{\rho^4}{\rho_0^4} \frac{(\rho_0^4 - \rho_0^2 + 1)}{\left[\rho^2\cos^2\rho + (1 - \rho^2)^2\sin^2\rho + (\rho^3 - \rho)\sin(2\rho)\right]}$$

Representative values of Q_1^{TM} and Q_1^{TE} are plotted in Figure 1.9a and b as a function of k_0a for a core with relative permittivity of 4 and a relative permeability of 16, and with loss tangents $\varepsilon'_{\mathrm{r}}/\varepsilon''_{\mathrm{r}} = \mu'_{\mathrm{r}}/\mu''_{\mathrm{r}}$ equal to 0.01, 0.001, and 0.0001. For the TE_{10} mode, the Q is about 5–10 times larger than that for the TE_{10} mode with the same core parameters.

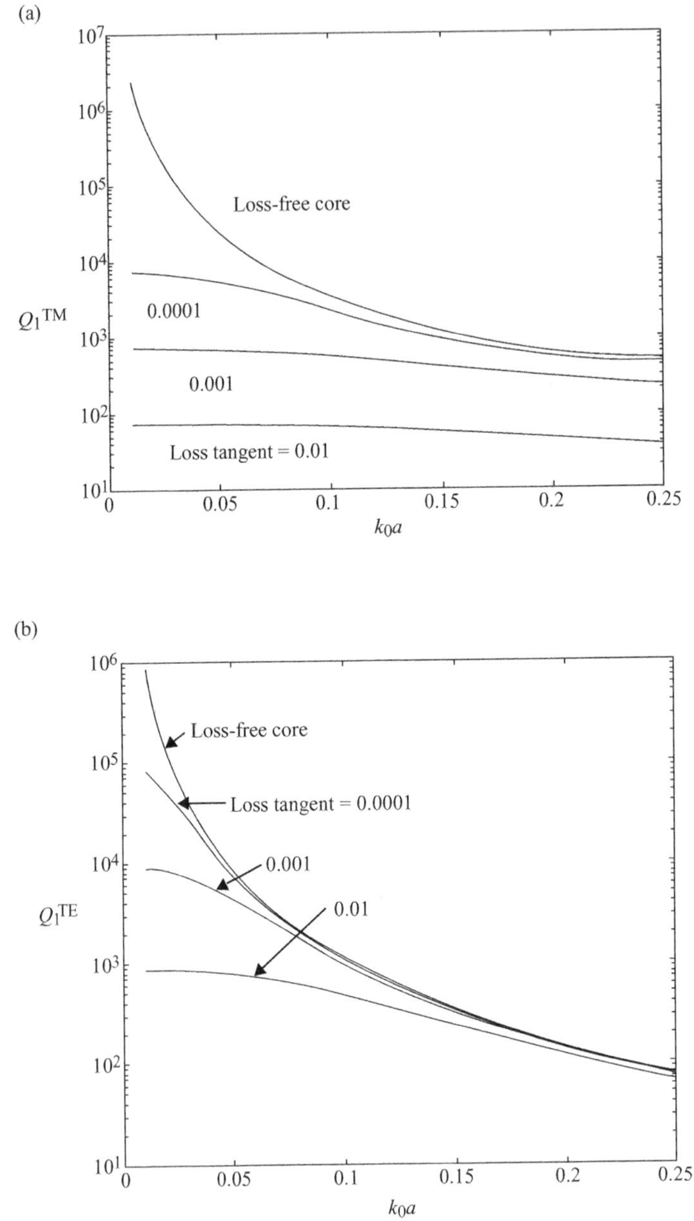

FIGURE 1.9 (a) The Q of the TMio mode for a core with relative permittivity of 4 and relative permeability of 16, for loss tangents $\varepsilon_r'/\varepsilon_r'' = \mu_r'/\mu_r''$ equal to 0.01, 0.001, and 0.0001. (b) The Q of the TE_{10} mode for a core with relative permittivity of 4 and relative permeability of 16, for loss tangents $\varepsilon_r'/\varepsilon_r'' = \mu_r'/\mu_r''$ equal to 0.01, 0.001, and 0.0001.

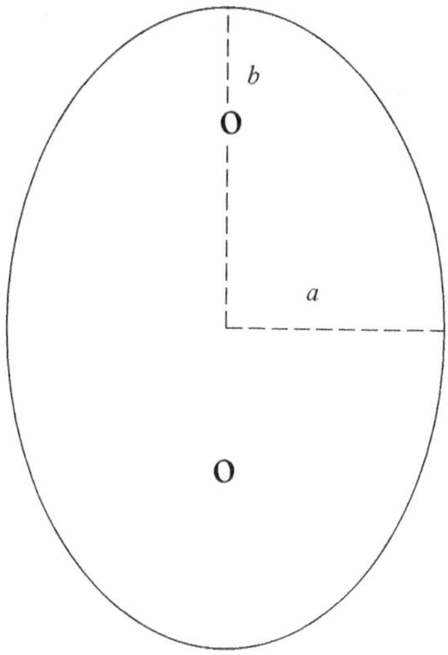

FIGURE 1.10 Prolate spheroid.

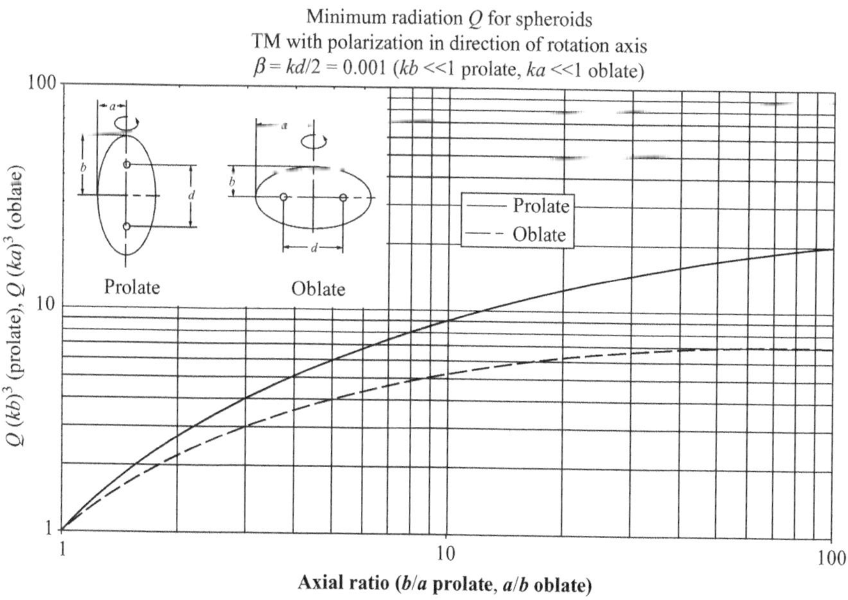

FIGURE 1.11 Q of spheroid in sphere. Courtesy of Peder Hansen, SPAWAR, San Diego, CA.

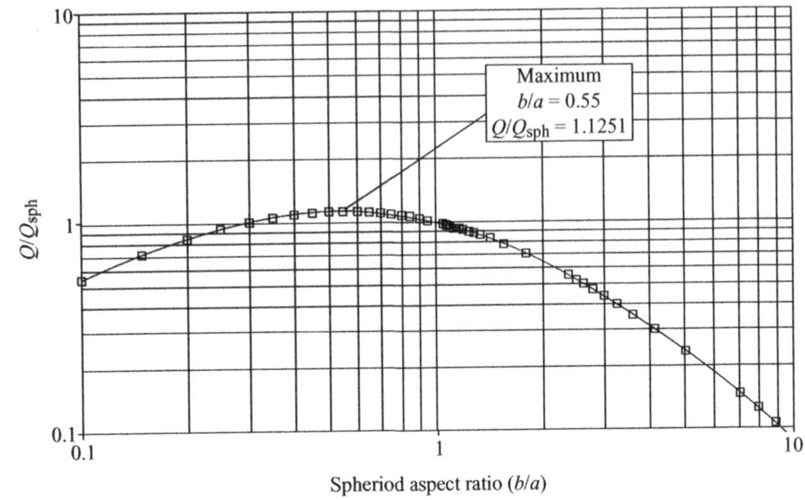

FIGURE 1.12 Q ratio: equal-volume sphere and spheroid. Courtesy of Hansen and Adams (2010).

1.8 *Q* FOR SPHEROIDAL ENCLOSURES

A rigorous Chu-type fundamental limitation-type formulation for nonspherical volumes would require that a low-order mode exists in the enclosure, and that stored and radiated energy can be calculated. The prolate spheroid, which approximates a cylinder, and the oblate spheroid, which approximates a flat disk, are candidates. The first effort was made by Foltz and McLean (1999) who attempted a numerical partial fraction expansion (ala Chu) of the prolate spheroid modal impedance. A Herculean spheroidal mode approach has been underway by Adams and Hansen (2004, 2008) of SPAWAR. Spheroidal functions were described by Hobson (1931), Stratton (1941), and Morse and Feshbach (1953). EM applications have been discussed by Li et al. (2002). Tables of functions are provided by Stratton et al. (1956), Flammer (1957), and Chang and Yeh (1966).

The spherical mode theory developed by Collin utilized closed-form integrals and recursion relationships (Equations 1.17 and 1.19) to develop the energy formulas. The spheroidal mathematical toolbox is much more sparse, so that numerical methods were used by Adams and Hansen to supplement the few available functional relationships. Figure 1.10 shows the prolate spheroid geometry, where the spheroid height is $2b$ and the diameter is $2a$. Because the Q of a spheroid depends upon both the dimensions in wavelengths and the prolateness, data in only two dimensions can be provided by comparing the spheroid with a sphere. Figure 1.11 shows the ratio of spheroid Q to Q of the sphere that just encloses the spheroid. It can be observed that the relative Q increases as the prolate spheroid becomes thinner, or as the oblate spheroid becomes fatter. Figure 1.12 shows the Q ratio for a sphere and a spheroid of equal volume. As shown, the Q peaks slowly at $b/a = 0.55$, with a Q ratio of 1.1251.

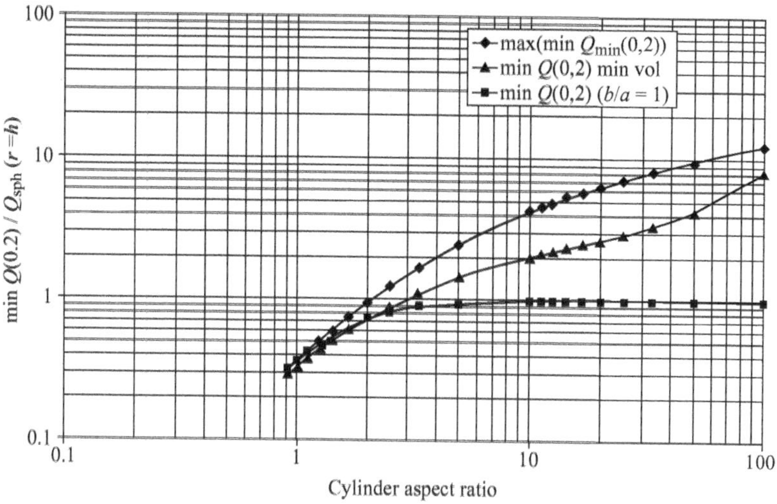

FIGURE 1.13 *Q* ratio: cylinder in prolate spheroid. Courtesy of Adams and Hansen (2008).

The Q/Q_{sphere} ratio versus cylinder aspect ratio for a cylinder enclosed in a prolate spheroid is shown in Figure 1.13. Data are shown for three cases: maximum Q, Q for maximum volume, and Q for the cylinder enclosed in a hemisphere.

All these data are summed up in Figure 1.14, which shows normalized Q versus aspect ratio, when sphere and spheroid have equal Q, for several dimensional equivalences.

Wheeler (1975), as mentioned earlier, used electrostatics and magnetostatics to develop Q boundaries for cylindrical boundaries. This work was continued by Gustafsson and colleagues (Gustafsson and Nordebo, 2006; Gustafsson et al.,

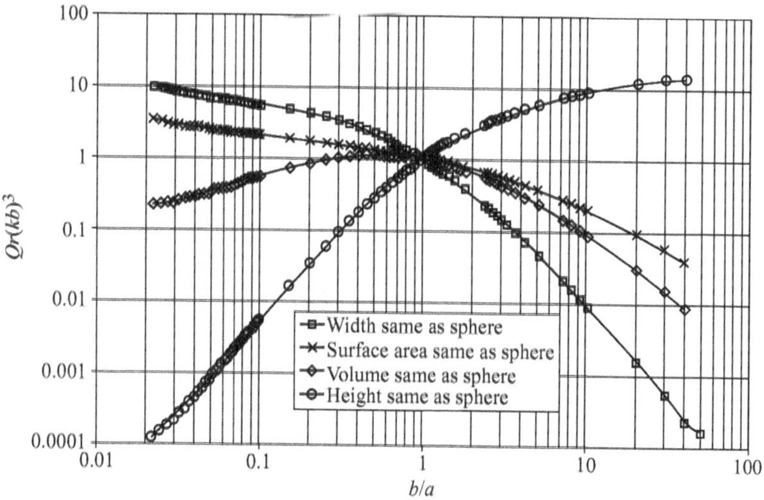

FIGURE 1.14 Sphere and spheroid of equal *Q*. Courtesy of Hansen and Adams (2010).

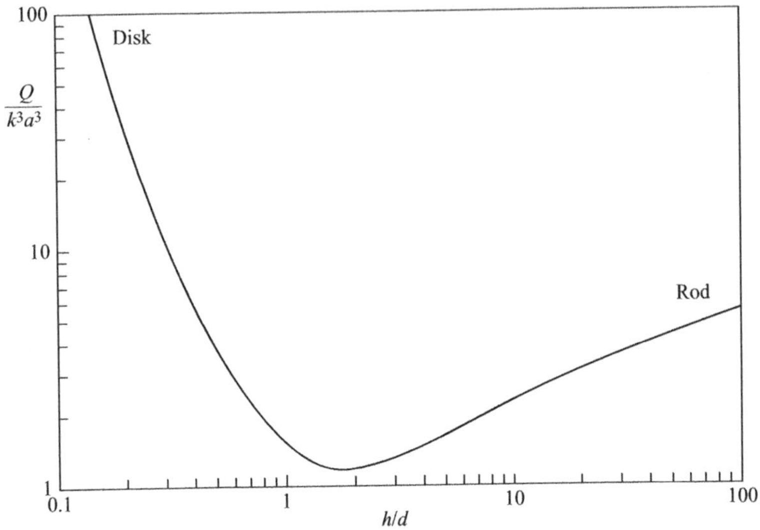

FIGURE 1.15 Static Q ratio for cylinder versus height/diameter. Courtesy of Gustafsson et al. (2009).

2007, 2009). Figure 1.15 shows Q/Q_{Chu} as a function of cylinder height/diameter. See also Yaghjian and Stuart (2010). These data, based on static fields, are accurate only for $ka \ll 1$.

REFERENCES

ADAMS, R. C. AND HANSEN, P. M. *The Evaluation of 'Q' in an Electrically Small Antenna in Prolate Spheroidal Coordinates*, TD 3188, USN, SPARWAR, San Diego, CA, 2004.

ADAMS, R. C. AND HANSEN, P. M. *The Modal Decomposition of the Quality Factor of an Antenna in Prolate Spheroidal Coordinates*, TR 1969, USN, SPAWAR, San Diego, CA, September 2008.

BANDLER, J. W. AND CHARALAMBOUS, C. Practical Least *p*th Optimization of Networks. *Trans. IEEE*, Vol. MTT-20, December 1972, pp. 834–840.

CHALOUPKA, H. On the Frequency Bandwidth of Functionally Small Antennas. *Proc. URSI Symp. EM Theory*, August 1992, pp. 266–268.

CHANG, C. AND YEH, C. *The Radial Prolate Spheroidal Functions*, EE Report 166, USC, Los Angeles, CA, June 1966.

CHU, L. J. Physical Limitations of Omni-Directional Antennas. *J. Appl. Phys.*, Vol. 19, December 1948, pp. 1163–1175.

COLLIN, R. E. *Field Theory of Guided Waves*, IEEE Press, Piscataway, NJ, 1990, Chapter 2.

COLLIN, R. E. Minimum *Q* of Small Antennas. *J. Electromagn. Waves Appl.*, Vol. 12, 1998, pp. 1369–1393.

COLLIN, R. E. AND ROTHSCHILD, S. Evaluation of Antenna *Q*. *Trans. IEEE*, Vol. AP-12, January 1964, pp. 23–27.

FANTE, R. L. Quality Factor of General Ideal Antennas. *Trans IEEE*, Vol. AP-17, March 1969, pp. 151–155.

FLAMMER, C. *Spheroidal Wave Functions*, Stanford University Press, 1957.

FOLTZ, H. D. AND MCLEAN, J. S. Limits on the Radiation *Q* of Electrically Small Antennas Restricted to Oblong Boundary Regions. *IEEE AP Symposium*, 1999, pp. 2702–2705.

GEYI, W. Physical Limitations of Antenna. *Trans. IEEE*, Vol. AP-51, August 2003a, pp. 2116–2123.

GEYI, W. A Method for the Evaluation of Small Antenna *Q*. *Trans. IEEE*, Vol. AP-51, August 2003b, pp. 2124–2129.

GUSTAFSSON, M. AND NORDEBO. S. Bandwidth, *Q* Factor, and Resonance Models of Antennas. *Prog. Electromagn. Res.*, Vol. 62, 2006, pp. 1–20.

GUSTAFSSON, M. ET AL. Physical Limitations on Antennas of Arbitrary Shape. *Proc. R. Soc. A*, Vol. 463, 2007, pp. 2589–2607.

GUSTAFSSON, M. ET AL. Illustrations of New Physical Bounds on Linearly Polarized Antennas. *Trans. IEEE*, Vol. AP-57, May 2009, pp. 1319–1327.

HANSEN, P. M. AND ADAMS, R. C. The Minimum Value for the Quality Factor of an Electrically Small Antenna in Spheroidal Coordinates: TM Case. *IEEE APS Symp.*, Toronto, July 2010.

HANSEN, R. C. Fundamental Limitations in Antennas. *Proc. IEEE*, Vol. 69, February 1981, pp. 170–182.

HANSEN, R. C. *Q* and Bandwidth of Electrically Small Antennas. *Microwave Opt. Technol. Lett.*, Vol. 49, May 2007, pp. 1170–1171.

HANSEN, R. C. AND COLLIN, R. E. A New Chu Formula for *Q*. *IEEE Antenn. Propag. Mag.*, Vol. 51, October 2009, pp. 38–41.

HANSEN, W. W. A New Type of Expansion in Radiation Problems. *Phys. Rev.*, Vol. 47, 1935, pp. 139–143.

HARRINGTON, R. F. On the Gain and Bandwidth of Directional Antennas. *Trans. IEEE*, Vol. AP-6, 1958, pp. 219–225.

HARRINGTON, R. F. Effect of Antenna Size on Gain, Bandwidth, and Efficiency. *J. Res. NBS*, Vol. 64D, January–February 1960, pp. 1–12.

HARRINGTON, R. F. *Time Harmonic Electromagnetic Fields*, McGraw-Hill, New York, 1961.

HARRINGTON, R. F. Antenna Excitation for Maximum Gain. *Trans. IEEE*, Vol. AP-13, November 1965, pp. 896–903.

HOBSON, E. W. *The Theory of Spherical and Ellipsoidal Harmonics*, Cambridge University Press, 1931; Chelsea Publishing Co., 1965.

HUJANEN, A. AND STEN, J. C. E. Bandwidth Limitations of Impedance Matched Ideal Dipoles. *Trans. IEEE*, Vol. AP-53, October 2005, pp. 3236–3239.

KALAFUS, R. M. On the Evaluation of Antenna Quality Factors. *Trans. IEEE*, Vol. AP-17, November 1969, pp. 279–732.

KIM, O. S. AND BREINBJERG, O. Lower Bound for the Radiation *Q* of Electrically Small Magnetic Dipole Antennas with a Solid Magnetodielectric Core. *Trans. IEEE*, Vol. AP-59, February 2011, pp. 679–681.

KIM, O. S. AND BREINBJERG, O. Reaching the Chu Lower Bound on *Q* with Magnetic Dipole Antennas Using a Magnetic-Coated PEC Core. *Trans. IEEE*, Vol. AP-59, August 2011.

Kim, O. S., Breinbjerg, O., and Yaghjian, A. D. Electrically Small Magnetic Dipole Antennas with Quality Factors Approaching the Chu Lower Bound. *Trans. IEEE*, Vol. AP-58, June 2010, pp. 1898–1906.

Kwon, D.-H. On the Radiation *Q* and the Gain of Crossed Electric and Magnetic Dipole Moments. *Trans. IEEE*, Vol. AP-53, May 2005, pp. 1681–1687.

Li, L.-W., Kang, X.-K., and Leong, M.-S. *Spheroidal Wave Functions in Electromagnetic Theory*, Wiley, 2002.

McLean, J. S. A Re-examination of the Fundamental Limits on the Radiation *Q* of Electrically Small Antennas. *Trans. IEEE*, Vol. AP-44, May 1996, pp. 672–676.

McLean, J.S., Foltz. H., and Sutton, R. Broadband, Electrically-Small Spherical-Wire and Generalized Loop Antennas Exploiting Inhomogeneous Magentic Cores. 20th Int. Conf. on Applied Electromagnetics and Communications, September 2011.

Morse, P. M. and Feshbach, H. *Methods of Theoretical Physics*, McGraw-Hill, New York, 1953, Chapter 11.

Rhodes, D. R. On the Stored Energy of Planar Apertures. *Trans. IEEE*, Vol. AP-14, November 1966, pp. 676–683.

Rhodes, D. R. *Synthesis of Planar Antenna Sources*, Oxford Clarendon Press, 1974, Section 5.4.

Stratton, J. A. *Electromagnetic Theory*, McGraw-Hill, New York, 1941, Chapter 7.

Stratton, J. A. et al. *Spheroidal Wave Functions*, Technology Press of MIT–Wiley, 1956.

Stuart, H. R. and Yaghjian, A. D. Approaching the Lower Bounds on *Q* for Electrically Small Electric-Dipole Antennas Using High Permeability Shells. *Trans. IEEE*, Vol. AP-58, December 2010, pp. 3865–3872.

Terman, R. F. *Radio Engineers' Handbook*, McGraw-Hill, New York, 1943.

Thal, H. L., Jr., Exact Circuit Analysis of Spherical Waves. *Trans. IEEE*, Vol. AP-26, March 1978, pp. 282–287.

Thal, H. L., Jr. New Radiation *Q* Limits for Spherical Wave Antennas. *Trans. IEEE*, Vol. AP-54, October 2006, pp. 2757–2763.

Thiele, G. A., Detweiler, P. L., and Penno, R. P. On the Lower Bound of the Radiation *Q* for Electrically Small Antennas. *Trans. IEEE*, Vol. AP-51, June 2003, pp. 1263–1269.

Underhill, M. J. and Harper, M. Simple Circuit Model of Small Tuned Loop Antenna Including Observable Environmental Effects. *Electron. Lett.*, Vol. 38, August 2002, pp. 1006–1008.

Underhill, M. J. and Harper, M. Small Antenna Input Impedances That Contradict Chu–Wheeler *Q* Criterion. *Electron. Lett.*, Vol. 39, May 2003, pp. 828–830.

Wheeler, H. A. Fundamental Limitations of Small Antennas. *Proc. IRE*, Vol. 35, December 1947, pp. 1479–1484.

Wheeler, H. A. The Spherical Coil as an Inductor, Shield, or Antenna. *Proc. IRE*, No. 9, 1958, pp. 1595–1602.

Wheeler, H. A. Small Antennas. *Trans. IEEE*, Vol. AP-23, July 1975, pp. 462–469.

Yaghjian, A. D. and Stuart, H. R. Lower Bounds on the *Q* of Electrically Small Dipole Antennas. *Trans. IEEE*, Vol. AP-58, October 2010, pp. 3114–3121.

CHAPTER 2

BANDWIDTH AND MATCHING

2.1 INTRODUCTION

Chapter 1 was concerned with ESA values of Q. In this chapter, the emphasis is on impedance matching limitations and the relation between Q and bandwidth (BW). Key names of the major contributors are Foster (reactance theorem), Fano (matching limitations), and Moreno (loss magnification). Finally, a comparison is made between a matched short whip antenna and a whip with a high-impedance preamp.

In relating bandwidths for different VSWR, a convenient formula is

$$\mathrm{BW}_x\sqrt{\mathrm{VSWR}_x}/(\mathrm{VSWR}_x - 1) = \mathrm{BW}_y\sqrt{\mathrm{VSWR}_y}/(\mathrm{VSWR}_y - 1) \tag{2.1}$$

For example,

$$\mathrm{BW}_{\mathrm{HP}} = 2\sqrt{2}\,\mathrm{BW}_{\mathrm{VSWR}=2} \tag{2.2}$$

2.2 FOSTER'S REACTANCE THEOREM AND SMITH CHART

In 1924, R. M. Foster published a paper showing that a lossless reactance always has a positive slope of reactance or susceptance with frequency (Foster, 1924). This has become known as Foster's reactance theorem. Such networks have their poles and zeroes on the real axis, and they alternate. See also Bode (1945) and Guillemin (1935). Because all antennas have a virtual loss due to the radiation resistance, Foster's theorem should not apply to antennas. There has been some controversy regarding whether Foster's reactance theorem applies to antennas, and in particular,

Small Antenna Handbook, Robert C. Hansen and Robert E. Collin.

to electrically small antennas. Best (2004) claims that it does not apply near antiresonance, where the reactance is rapidly changing sign. Others such as Geyi et al. (2000) claim that Foster's theorem does apply. Yaghjian and Best (2005) invoke a far-field "dispersion energy," which is the integral involving the complex far-field pattern and the antenna conductivity. If σ is dispersive, this energy integral adds to the reactance calculation, and voids the Foster theorem for antennas, as expected.

Bandwidth and impedance are the crucial parameters in all antenna considerations. As usual, bandwidth is between upper frequency f_2 and lower frequency f_1, at a center frequency of f_0:

$$\mathrm{BW} = \frac{f_2 - f_1}{f_0} \tag{2.3}$$

From a practical standpoint, the applicability of the Foster theorem to antennas is irrelevant. The proper question to ask is: Does the impedance locus of an antenna always travel clockwise with increasing frequency around the Smith chart? To demonstrate that this is true for dipoles, a computer program was written to calculate dipole impedance versus frequency, using the Carter (1932) impedance formulation involving sine and cosine integrals. Figures 2.1 and 2.2 show the resistance and reactance versus dipole length in wavelengths over the range 0.1–1.5, for length/radius = 500. Similar data for length/radius = 20 are given in Figures 2.3 and 2.4. As asserted by Best, it is immediately obvious that there are impedance regions where the reactance slope does not obey Foster's theorem. However, Figure 2.5 is a Smith chart showing the dipole impedance for length/radius = 500, and it is clear that the rotation is always clockwise. Figure 2.6 is for length/radius = 20, and again the rotation is clockwise. When the impedance locus has resonances with small loops on the Smith (1949) chart, the rotation is still clockwise. We believe that this result holds true for all types of antennas.

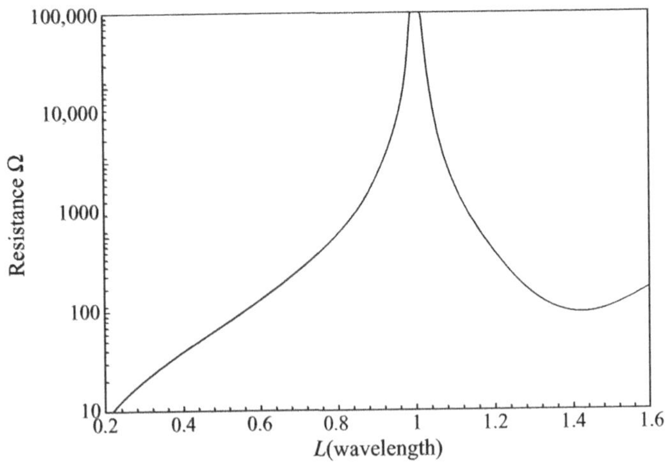

FIGURE 2.1 Resistance versus dipole length for length/radius = 500.

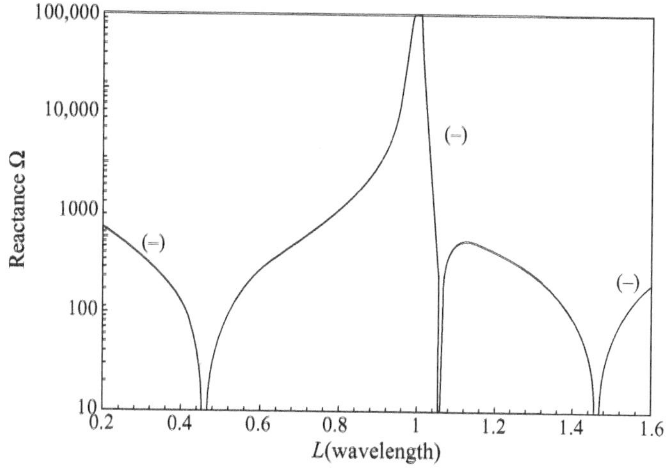

FIGURE 2.2 Reactance versus dipole length for length/radius = 500.

2.3 FANO'S MATCHING LIMITATIONS

The fundamental limits on matching were derived more than 50 years ago by R. M. Fano (1950a, 1950b). Matthaei et al. (1964) in their treatise gave solutions to the simultaneous equations developed by Fano:

$$\frac{\tanh na}{\cosh a} = \frac{\tanh nb}{\cosh b} \tag{2.4}$$

$$\cosh nb = \Gamma \cosh na \tag{2.5}$$

$$\sinh b = \sinh a - 2\delta \sin(\pi/2n) \tag{2.6}$$

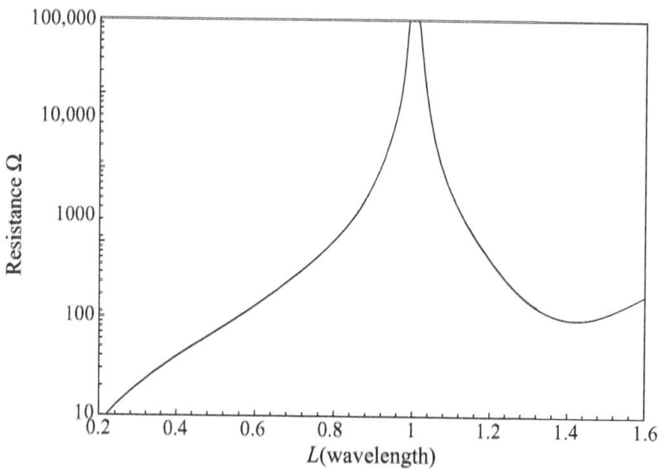

FIGURE 2.3 Resistance versus dipole length for length/radius = 20.

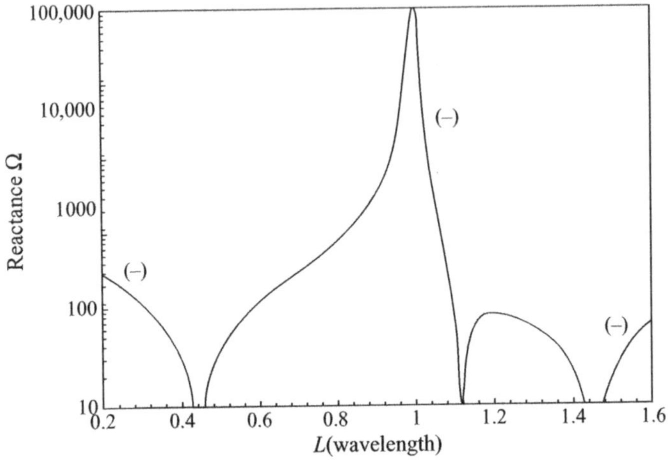

FIGURE 2.4 Reactance versus dipole length for length/radius = 20.

Here, there are n matching sections (including the antenna as one), δ is decrement ($1/Q$ at band edges), and a and b are parameters to be determined. Γ is the reflection coefficient. These simultaneous equations were solved by Matthaei et al. with results presented in graphical form. Subsequently, the Newton–Raphson method

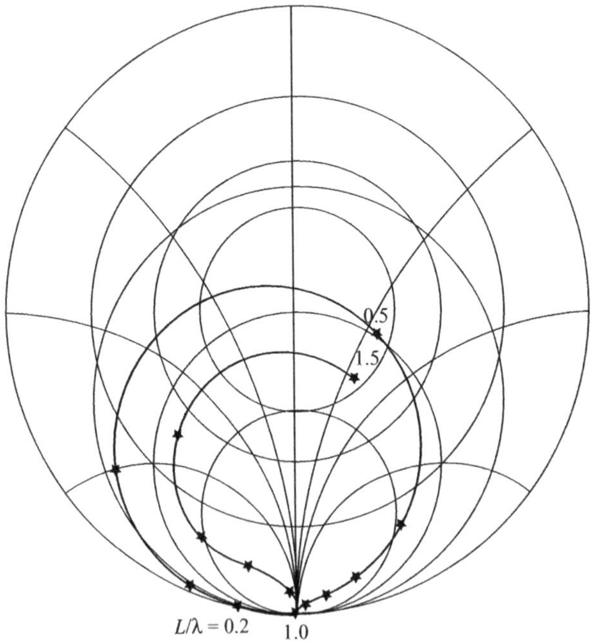

FIGURE 2.5 Smith chart showing dipole impedance for length/radius = 500.

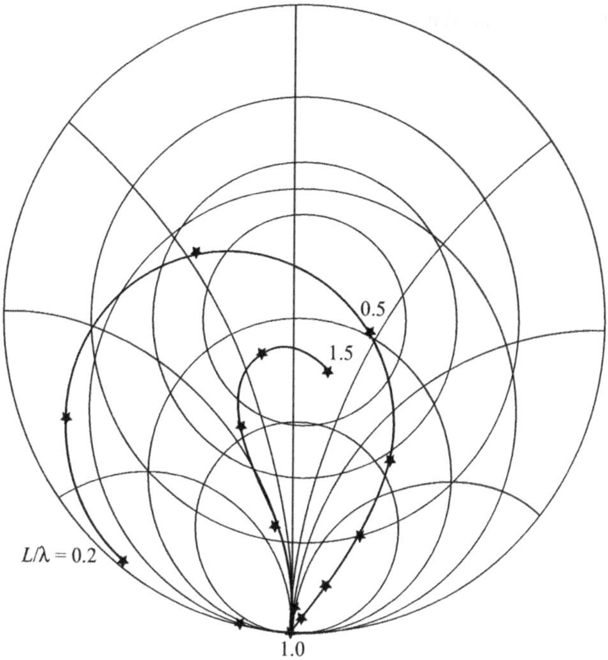

FIGURE 2.6 Smith chart showing dipole impedance for length/radius = 20.

(Stark, 1970) was used by Hansen (2006) to obtain precise values of a and b. Tables 2.1 and 2.2 give values of a, b, and δ for VSWR values of 2 and 5.828 (half-power). Note that $n = 1$ is for the antenna alone. The bandwidth improvement factor (BWIF) is just δ for $n = 1$ divided by δ for $n > 1$: $\text{BWIF} = \text{BW}_{\text{matched}}/\text{BW}_{\text{antenna}}$. Table 2.3 gives these BWIF for several values of VSWR. Actual matching devices will have loss, and this will increase the bandwidth at the expense of efficiency. Note that for the commonly used VSWR = 2, an infinite number of matching circuits will increase bandwidth 3.8 times.

TABLE 2.1 Fano Data: VSWR ≤ 2

N	a	b	δ
1	1.81845	0.32745	1.3333
2	1.03172	0.39768	0.57735
3	0.76474	0.36693	0.46627
4	0.62112	0.33112	0.42416
5	0.52868	0.30027	0.40264
∞	0	0	0.34970

TABLE 2.2 Fano Data: VSWR ≤ 5.828

n	a	b	δ
1	1.14622	0.65848	0.35355
2	0.76429	0.56419	0.17416
3	0.59982	0.47449	0.14394
4	0.50164	0.41026	0.13207
5	0.43483	0.36284	0.12589
∞	0	0	0.11032

A word of caution is needed on the use of the Bode (1945) criterion for BWIF.

$$\text{BWIF} = \frac{\pi}{\ell n(1/\Gamma)} \tag{2.7}$$

Bode assumed constant Γ over the band of interest; this clearly invalidates his results. An exact correction was provided by Hansen (2008), in terms of Γ and VSWR:

$$\text{BWIF} = \frac{1-\Gamma^2}{2\Gamma}\frac{\pi}{\ell n(1/\Gamma)} \tag{2.8}$$

$$\text{BWIF} = \frac{2\,\text{VSWR}}{\text{VSWR}^2-1}\frac{\pi}{\ell n[(\text{VSWR}+1)/(\text{VSWR}-1)]} \tag{2.9}$$

These equations give results that are exactly the same as the Fano results (Table 2.3 for an infinite number of sections). Figures 2.7 and 2.8 give BWIF versus Γ and VSWR.

TABLE 2.3 Maximum Bandwidth Improvement Factors

Number of Additional Matching Circuits	VSWR 1.5	2	3	5.828
1	2.6833	2.3094	2.1213	2.0301
2	3.4202	2.8596	2.5864	2.4563
3	3.8001	3.1435	2.8271	2.6772
4	4.0242	3.3115	2.9699	2.8083
∞	4.6848	3.8128	3.3993	3.2049

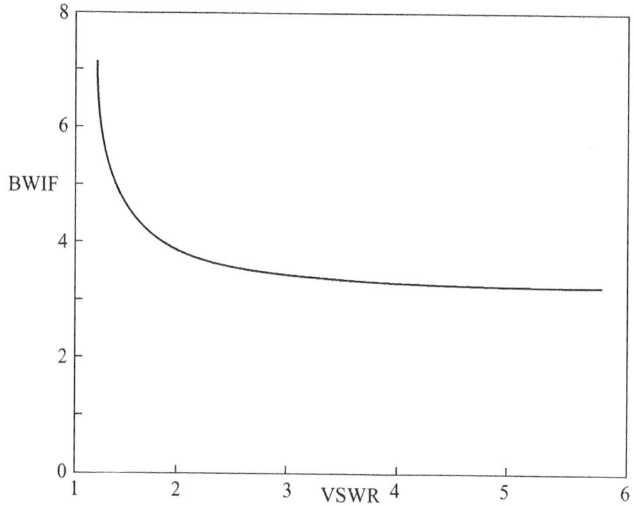

FIGURE 2.7 Bandwidth improvement versus VSWR.

Lopez (2004, 2005) developed a formula for BWIF that closely matches the exact Fano BWIF of Table 2.3. His coefficients were labeled a_n and b_n, but to avoid confusion they are relabeled c_n and d_n, and n has been reduced by 1, to match previous notation. The coefficients are given in Table 2.4. His formula is

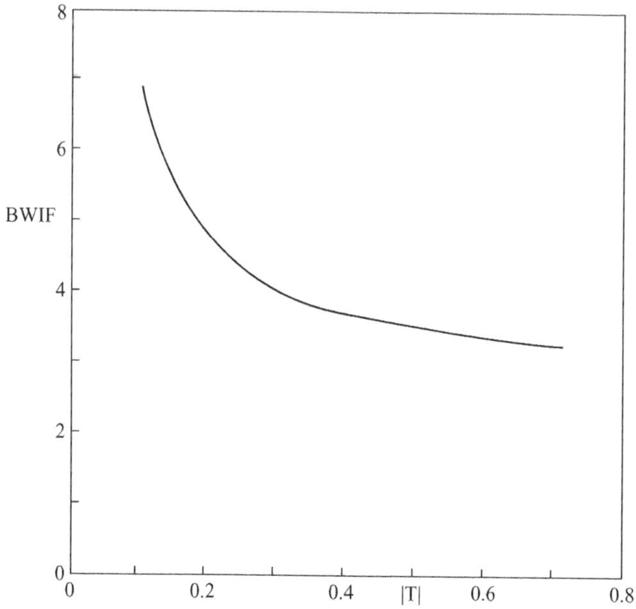

FIGURE 2.8 Bandwidth improvement versus [T].

TABLE 2.4 Lopez Coefficients

η	c_n	d_n
1	2	1
2	2.413	0.678
3	2.628	0.474
4	2.755	0.347
5	2.838	0.264
6	2.896	0.209
7	2.937	0.160
∞	π	0

TABLE 2.5 Bandwidth Improvement Factors

	Fano		Lopez	
Number of Circuits	VSWR = 2	VSWR = 5.828	VSWR = 2	VSWR = 5.828
1	2.3094	2.0301	2.3094	2.0301
2	2.8596	2.4563	2.8608	2.4559
3	3.1435	2.6772	3.1457	2.6772
4	3.3115	2.8083	3.3129	2.8079
∞	3.8128	3.2049	3.8128	3.2049

$$\mathrm{BWIF} = \frac{\sinh[\ell n(1/\Gamma)]}{d_n \sinh[(1/c_n)\ell n(1/\Gamma)] + [(1-d_n)/c_n]\ell n(1/\Gamma)} \tag{2.10}$$

Table 2.5 gives the exact Fano results and those of Equation 2.10. The approximate formula is excellent.

2.4 MATCHING CIRCUIT LOSS MAGNIFICATION

A matching network for an electrically small antenna or for a superdirective array must cancel the large reactance and must transform the small radiation plus loss resistance to the nominal value, usually 50 Ω. Generally, each element of an array will require a different matching network, although symmetric (broadside) arrays need fewer. If the matching network is composed of discrete L and C components, the low-R and high-X requirements will produce very large circulating currents in the overall circuit, consisting of the antenna, the matching network, and the generator. These circulating currents will magnify the intrinsic loss to a much larger realized loss. Similarly, a distributed matching network under low-R and high-X conditions will have very large standing waves along the stubs and transformer sections. Again the intrinsic loss is magnified. In both cases, the larger loss is due to power being proportional to voltage (or current) squared; the circulating currents or standing waves are large.

A transmission line transformer is typical. Let the matched loss be L and the antenna voltage standing wave ratio be VSWR. Then the actual loss L_a is (Moreno, 1948)

$$L_a = \frac{(\text{VSWR}+1)^2L^2 - (\text{VSWR}-1)^2}{4L \cdot \text{VSWR}} \tag{2.11}$$

For VSWR $\gg$ 1, the actual loss is

$$L_a \simeq \frac{\text{VSWR}(L^2 - 1)}{4L} \tag{2.12}$$

Figure 2.9 shows the actual loss versus VSWR for intrinsic (matched) losses of 0.1, 0.2, 0.3, 0.5, 1, 2, 3, and 5 dB (Hansen, 1990). For $R \ll R_0$, VSWR $\simeq (R_0^2 + X^2)/RR_0$, whereas for $X \gg R_0$, VSWR $\simeq X^2/RR_0$. The latter holds also for $R \ll R_0$ and $X \gg R_0$. Some typical values are given in Table 2.6. For example, a case of $R/R_0 = 0.1$ and $X/R_0 = 10$ produces a VSWR = 1010. A matched loss of 0.01 dB becomes an actual loss of 3.35 dB and a matched loss of 0.1 dB becomes an actual loss of 11.01 dB. A more likely case, because the reactance

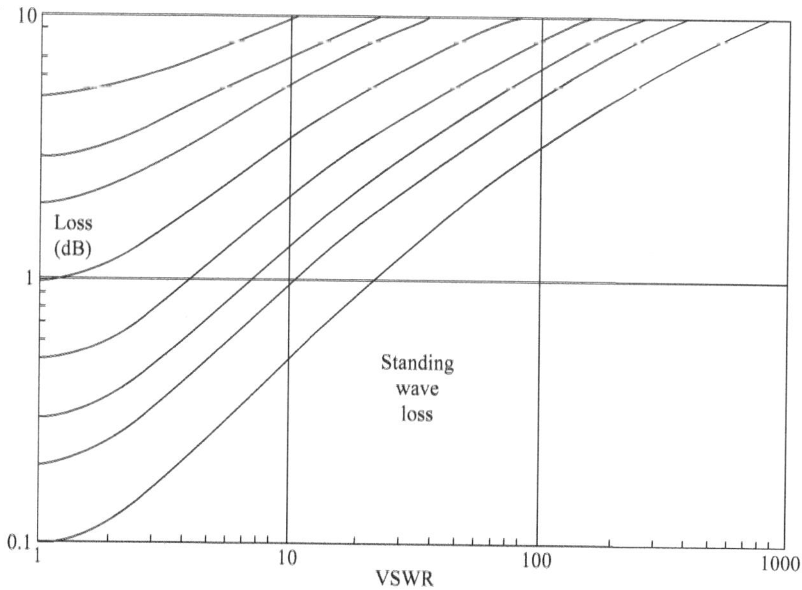

FIGURE 2.9 Line loss with mismatch.

TABLE 2.6 VSWR from Impedance Mismatches

	X/R_0				
R/R_0	1	3	10	30	100
1	1.000	10.91	102.0	902.0	1.00×10^4
0.3	6.820	33.60	337.0	3004.0	3.334×10^4
0.1	20.05	100.0	1010.0	9010.0	1.000×10^5
0.03	66.68	333.4	3367.0	3.003×10^4	3.334×10^5
0.01	200.0	1000.0	1.010×10^4	9.010×10^4	1.000×10^6

ratio is usually higher, is $R/R_0 = 0.1$ and $X/R_0 = 30$, giving VSWR = 9010. A matched loss of 0.01 dB now yields an actual loss of 10.56 dB and a matched loss of 0.1 dB gives an actual loss of 20.2 dB.

2.5 NETWORK AND Z_0 MATCHING

The simplest kind of impedance matching adjusts the load impedance Z_0 such that the mismatch at center frequency equals that at band edges. The load resistance Z_0 is related to the antenna resistance R_0 by a factor T (Pues and Van de Capelle, 1989):

$$T = Z_0/R_0 \tag{2.13}$$

Normally, the Q–bandwidth product QB equals

$$QB = \frac{\text{VSWR} - 1}{2\sqrt{\text{VSWR}}} \tag{2.14}$$

This is maximized for

$$T = \frac{\text{VSWR}^2 + 1}{2\,\text{VSWR}} \tag{2.15}$$

Table 2.7 shows the optimum value of T and the improved QB. It is apparent that the improvement is minor for low values of VSWR.

TABLE 2.7 Bandwidth Improvement Due to Mismatched L

VSWR	T_{opt}	$QB_{\text{opt } T}$	$QB_{T=1}$
1	1	0	0
1.5	1.0833	0.417	0.408
2	1.2500	0.750	0.707
3	1.6667	1.333	1.155
4	2.1250	1.875	1.500
5.828	3.0000	2.828	2.000

A thorough discussion of impedance matching techniques is provided in Chapter 5 of Collin (2001). A general treatment of dissipation matching networks was given by Carlin and LaRosa (1952). Use of the multifunction Darlington equivalent load impedance has been made unnecessary by Youla (1964). Butterworth and Chebyshev bandpass matching networks are discussed by Chen and Chaisrakeo (1980). See also Dedieu et al. (1994). A novel technique for transmit antennas uses a circulator, a directional coupler, and a phaser to adjust the phase of the reflected power and return it to the antenna (Brennan, 1992).

Often the antenna is connected to the receiver or transmitter by a transmission line. When the line has loss, there is a demagnification of VSWR, somewhat the inverse of the matching circuit loss magnification. This problem was worked out by Smith (1949) of Bell Labs. Let the antenna VSWR be $V_{\rm in}$ and the load VSWR be $V_{\rm out}$. The line loss is L; as always, $L =$ power in/power out, so that $L \geq 1$. Loss is given by $\exp(2\alpha l)$, where α is the attenuation coefficient and l is the line length. These are now related by

$$V_{\rm out} = \frac{(L+1)V_{\rm in} + L - 1}{(L-1)V_{\rm in} + L + 1} \tag{2.16}$$

Note: There are typos in Smith's formula. Figure 2.10 shows $V_{\rm out}$ and $V_{\rm in}$ for a lossless line and for line losses from 0.1 to 3 dB. These losses are for the entire line,

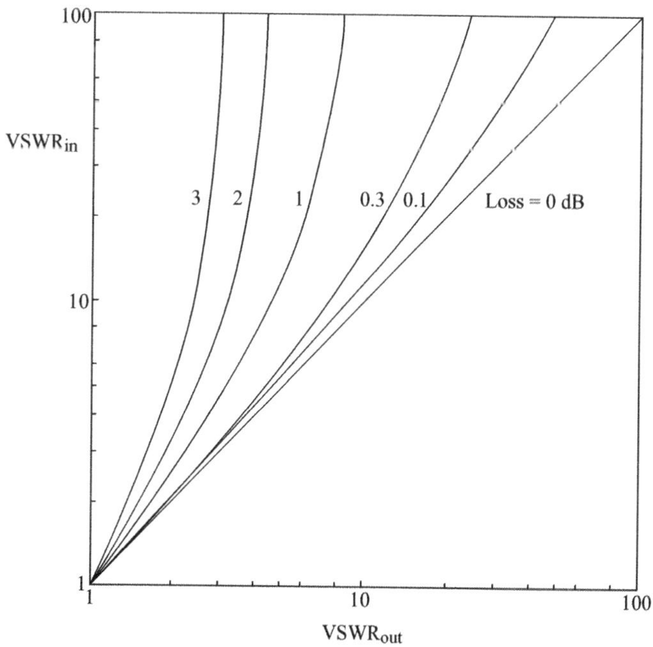

FIGURE 2.10 VSWR reduction due to line loss.

from input to output. Even modest loss (1 dB or more) will substantially reduce half-power V_{in} to low values. The inverse equation is

$$V_{in} = \frac{(L+1)V_{out} - (L-1)}{(-L+1)V_{out} + (L+1)} \tag{2.17}$$

2.6 NON-FOSTER MATCHING CIRCUITS

A new dimension in the impedance matching and impedance transformation of ESA and of superdirective antennas is now offered by non-Foster circuits. These circuits produce negative resistance, inductance, or capacitance. They are non-Foster in that their impedance circles the Smith chart with frequency the wrong way. Thus, they violate Foster's reactance theorem for passive lossless circuits. Such circuits are also called negative impedance converters (NIC). It is also possible to transform impedances with such circuits.

NIC were developed at Bell Labs in the early 1930s and were used for many years in long-line telephony repeaters, where they provided both amplification and modest negative resistance, the latter to offset long-line losses. These repeaters made transcontinental telephone service possible and were used until microwave repeaters came into wide use after World War II. The BTL repeaters, of course, used vacuum tubes. To quote Linvill (1953), "the ideal NIC is an active four-pole with input current equal to output current, and input voltage equal to the negative of the output voltage." Positive feedback is used to obtain these properties. As shown by Linvill, the transistor quickly replaced the vacuum tube in NIC circuits. Applications include raising the Q of filters using negative resistance, and negative and positive inductors at microwave frequencies, where conventional inductors are less attractive (Hara et al., 1988; El Khoury, 1995). The *MTT Transactions* contain many papers on filter applications. See Sussman-Fort (1998) for an extensive reference list. Papers on NIC topology include Tek and Anday (1989) and Sinsky and Westgate (1996). A novel technique was used by Pedinoff (1961) in which a tunnel diode biased into the negative conductance region was used with a slot antenna to provide amplification. Non-Foster circuits have been applied to loudspeakers, where a partial cancellation of voice coil inductance and resistance enhances the lower octaves (Werner and Carrell, 1958).

These concepts were applied to antennas by Mayes and Poggio in 1970. Negative impedance units were placed in series at several locations in the arms of a dipole (Mayes and Poggio, 1973; Poggio and Mayes, 1971). This was extended in an MS thesis under Mayes (Quirin, 1971) using Op Amps. Subsequently, transistors with much improved characteristics became available, allowing operation at microwave frequencies.

An early paper on non-Foster matching used two transistors to provide negative inductance for a loop antenna (Albee, 1976). Whether this was reduced to practice is not known. This was followed by Bahr (1976), who built a current inverting NIC to match a monopole antenna. A related approach uses an Op Amp with feedback to provide a charge amplifier, as opposed to a voltage amplifier (Ryan, 1983), for a short monopole (plate) antenna. Another patent concerns an Op Amp NIC applied to a loop antenna (Sutton, 1994). Op Amps are readily available but are not good choices, because of parasitics, limited unity gain bandwidth, poor efficiency, and noise. A significant advance was made by Skahill et al. (1998), who matched a short dipole with a Khoury current conveyor using three transistors. A patent (Skahill et al., 2000) by these authors provides, in addition to the reactance match, a most important frequency-squared transformer NIC that converts the short dipole radiation resistance to a constant value.

Because NIC employ positive feedback, stability is always a concern; parasitic and implementation defects are critical. Llewellyn (1952) reported on Bell Labs work on non-Foster circuits, and in particular that any two-port circuit needs to be stable with respect to the terminating impedances. Rollett (1962) developed necessary and sufficient conditions for external stability; these were three inequalities in impedance (or admittance) that must be satisfied. However, open- and short-circuit tests are inadequate. More important is that the internal stability is not addressed by Rollett conditions. A non-Foster circuit can have hidden loops and modes that are unstable, yet have external stability. This is a well-known problem for control theory practitioners. The Nyquist (1932) test, again from Bell Labs, looks at the loops in the complex frequency plane. Stearns (2010) has studied the stability of many non-Foster circuits, with special attention to internal loops. His work is the most thorough of any, with additional publications in process.

Sussman-Fort (1994, 1998) terminates an NIC in a one-port network designed to compensate for both active and passive parasitics. A negative inductance circuit was applied to a patch antenna (Kaya et al., 2004; Kaya and Yüksel, 2007). Because patch antennas are easily matched, the extended bandwidth observed was probably due to the NIC losses. Non-Foster matching of a short dipole using a negative capacitance has produced typically a 10 dB improvement over more than an octave of frequency (Sussman-Fort and Rudish, 2005; Rudish and Sussman-Fort, 2005). Figure 2.11 shows the four-transistor NIC used. A VHF monopole with non-Foster match developed for CECOM gave up to 6 dB improvement over a conventionally matched antenna (Sussman-Fort, 2006). Clearly, a judicious use of non-Foster matching can make a major improvement in receiving ESA. For transmitting ESA, the transistor circuit efficiency and the ability to handle the circuit high voltage are critical. Research is ongoing; the use of Class E (switching) circuits in NIC holds promise.

Although not the subject of this book, non-Foster circuits have advantages for phased arrays, either for matching or for connected arrays (see Hansen, 2003, 2004).

2.7 MATCHED AND HIGH-*Z* PREAMP MONOPOLES

Whip antennas for receiving HF and VHF are sometimes provided with a high-impedance preamp at the base. It is useful to compare the S/N performance of these

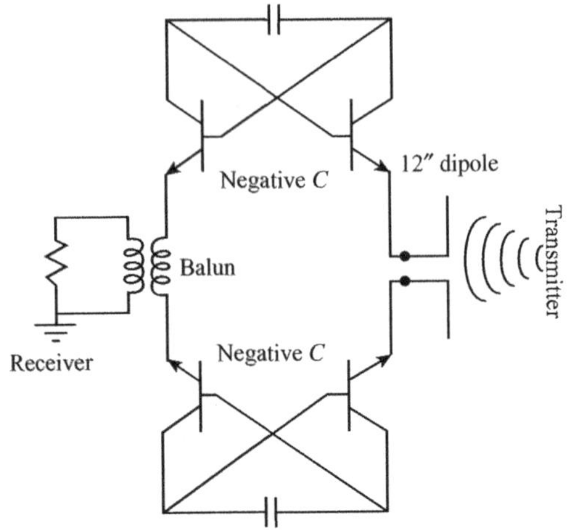

FIGURE 2.11 Measurement of short dipole non-Foster match. Courtesy of Rudish and Sussman-Fort (2005).

with that of an impedance matched monopole. The analysis is in terms of measurable parameters: loss, noise figure, reflection coefficient, and so on. Once these parameters are calculated or measured, the results should be useful regardless of the details of the transistors and accompanying circuits. This work is an extension of Radjy and Hansen (1979) and Sainati (1982).

The frequency at which the atmospheric noise equals city business noise is roughly 20 MHz (Middleton, 2002). Thus, a useful frequency range of interest is 10–30 MHz: 10 MHz is comfortably where atmospheric noise temperature T_a controls and 30 MHz is where standard temperature $T_0 = 290\,\text{K}$ prevails. K is Boltzmann's constant: $-198.60\,\text{dBm}$. Preamplifier noise temperature is as usual, $T = (\text{NF} - 1)T_0$, where NF is noise figure.

2.7.1 A Short Monopole Matched at One Frequency

The monopole is 1 m long and has a radius of 5 mm. Monopole impedances at 10 and 30 MHz are $Z_1 = 0.4386 - j1213.3\,\Omega$ and $Z_2 = 3.948 - j354.97\,\Omega$. Monopole, cable, and connector losses are subsumed in R_{loss}; the corresponding loss is $L_a = (R_{\text{rad}} + R_{\text{loss}})/R_{\text{rad}}$. These parameters are common to the cases considered. Figure 2.12 shows the block diagram, where L_a covers antenna and cable losses. The preamplifier has gain G and noise figure NF. The incident wave has power density P_a and the external noise temperature is T_a. The calculation would be straightforward except for the behavior of the matching network. All matching networks have an intrinsic loss, here called L, which is the loss incurred when both input and output are matched, usually 50 Ω in this case. As shown by Moreno (1948), when a matching network matches an impedance with a large VSWR, the

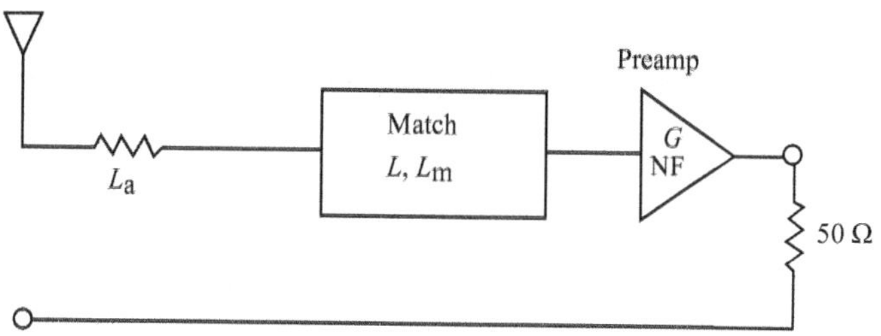

FIGURE 2.12 Narrowband match.

actual loss can be greatly magnified. See Section 2.4. The formula for the actual loss, L_m, is

$$L_m = \frac{(\text{VSWR}+1)^2 L^2 - (\text{VSWR}-1)^2}{4L \cdot \text{VSWR}} \tag{2.18}$$

Total loss is now $L_t = L_a L_m$. Curves and tabular data are given in Section 2.4. The VSWR at the input to the matching section is

$$\text{VSWR} = \frac{|Z_a + 50| + |Z_a - 50|}{|Z_a + 50| - |Z_a - 50|} \tag{2.19}$$

At the preamp output, with preamp gain G, the signal is

$$S = \frac{P_a G}{L_t} \tag{2.20}$$

As usual, T_0 is nominal ambient temperature of 290 K and B is the bandwidth. The noise at the preamp output (Mumford and Scheibe, 1968) is

$$N = \left[\frac{(L_t - 1)T_0 + T_a}{L_t} + (\text{NF} - 1)T_0\right] GKB \tag{2.21}$$

The signal-to-noise ratio is now given by

$$S/N = \frac{P_a}{[(L_t - 1)T_0 + T_a + L_t(\text{NF} - 1)T_0]KB} \tag{2.22}$$

At the low-frequency end of the band, T_a is much greater than T_0 so that

$$S/N \simeq \frac{P_a}{L_t K T_a B} \tag{2.23}$$

At the high-frequency end of the band, typically $T_a < T_0$ and the matching network loss is modest as the impedance mismatch is much smaller. The result is

$$S/N \simeq \frac{P_a}{(L_t\mathrm{NF} - 1)KT_0B} \tag{2.24}$$

At 30 MHz, the antenna VSWR = 651. Assuming an intrinsic matching circuit loss of $L = 0.1$ dB gives an actual matching circuit loss of $L_m = 8.496 = 9.29$ dB. With an antenna loss of 0.1 dB, the total loss is $L_t = 8.896$. For this frequency range, a noise figure of 2 dB is available (Spectrum Microwave); this makes the S/N degradation at 30 MHz equal to 7.2 dB.

2.7.2 Short Monopole with High-Impedance Amplifier

Figure 2.13 sketches this configuration. Parameters are the same as used before except that R_p is the input impedance of the preamp. At the preamp input, the reflection coefficient, where Z_a is the antenna impedance, is

$$\Gamma = \frac{Z_a - R_p}{Z_a + R_p} \tag{2.25}$$

The signal at the preamp input is $S = P_a/L_a$ and the signal at the preamp output is

$$S = \frac{P_aG}{L_a}[1 - |\Gamma|^2] \tag{2.26}$$

The noise at the preamp input is

$$N = \frac{(L_a - 1)T_0 + T_a}{L_a}KB \tag{2.27}$$

The noise at the preamp output is

$$N = \left[\frac{(L_a - 1)T_0 + T_a}{L_a}(1 - |\Gamma|^2) + (\mathrm{NF} - 1)T_0\right]GKB \tag{2.28}$$

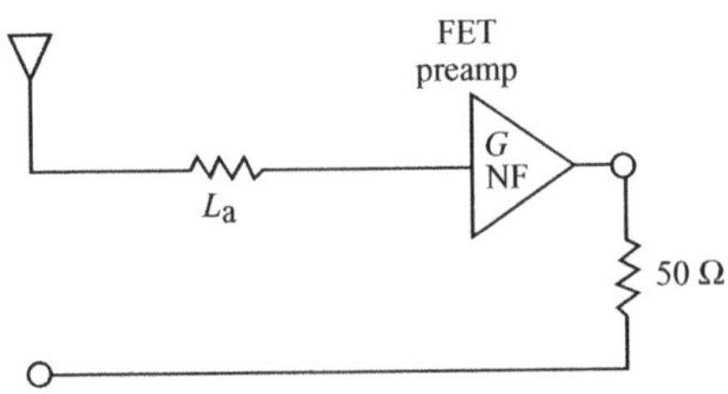

FIGURE 2.13 High-impedance FET.

Finally, the S/N at the preamp output is

$$S/N = \frac{P_a(1-|\Gamma|^2)}{\{[(L_a-1)T_0+T_a][1-|\Gamma|^2]+L_a(\mathrm{NF}-1)T_0\}KB} \tag{2.29}$$

At the low-frequency end of the band, $T_a \gg T_0$ and the reflection coefficient is modest. Thus, S/N is given by

$$S/N \simeq \frac{P_a}{KT_aB} \tag{2.30}$$

At the high-frequency end of the band, typically $R_2 \ll R_p$, so

$$1-|\Gamma|^2 \simeq \frac{4R_2}{R_p} \simeq 0 \tag{2.31}$$

This reduces S/N to

$$S/N \simeq \frac{P_a}{L_a(\mathrm{NF}-1)KT_0B} \tag{2.32}$$

The S/N at the low-frequency end of the band is controlled by the external noise temperature; at the high-frequency end of the band S/N is controlled by To, and is reduced by the inverse of Noise Figure minus one.

Note that the high-Z preamp S/N degradation is much less than that for the matched whip as the latter incurs a large matching section loss. Linearity is a major concern for octave or multioctave preamps; the transistor should be selected to have good IP2 and IP3 (intercept points).

Noise figure can be measured using a signal generator (Montgomery, 1947) or by using hot and cold noise sources (Bryant, 1988).

REFERENCES

Albee, T. K. *Broadband VLF Loop Antenna System*, U.S. Patent No. 3,953,799 April 1976.

Bahr, A. J. Application of Active Impedance Matching to Electrically Small Receiving Antenna. *Proc. ECOM-ARO Workshop on Electrically Small Antennas*, Ft. Monmouth, NJ, October 1976, pp. 199–204.

Best, S. R. The Foster Reactance Theorem and Quality Factor for Antennas. *IEEE Antenn. Wireless Propag. Lett.*, Vol. 3, 2004, pp. 306–309.

Bode, H. W. *Network Analysis and Feedback Amplifier Design*, Van Nostrand, 1945.

Brennan, P. V. Impedance Matching Method Using Passive Lossless Feedback. *Proc. IEE*, Vol. 139H, October 1992, pp. 397–400.

Bryant, G. H. *Principles of Microwave Measurements*, Peregrinus/IEE, 1988, Section 9.1.3.

Carlin, J. H. and LaRosa, R. Broadband Reflectionless Matching with Minimum Insertion Loss. *Proc. Symp. on Modern Network Synthesis*, Vol. 1, Polytechnic Institute, Brooklyn, NY, April 16–18, 1952, pp. 161–178.

Carter, P. S. Circuit Relations in Radiating Systems and Applications to Antenna Problems. *Proc. IRE*, Vol. 20, June 1932, pp. 1004–1041.

Chen, W. K. and Chaisrakeo, T. Explicit Formulas for the Synthesis of Optimum Bandpass Butterworth and Chebyshev Impedance-Matching Networks. *Trans. IEEE*, Vol. CS-27, October 1980, pp. 928–942.

Collin, R. E. *Foundations for Microwave Engineering*, 2nd ed, Wiley, 2001.

Dedieu, H. et al. A New Method for Solving Broadband Matching Problems. *Trans. IEEE*, Vol. CS-41, September 1994, pp. 561–571.

El Khoury, S. The Design of Active Floating Positive and Negative Inductors in MMIC Technology. *IEEE Microwave Guided Wave Lett.*, Vol. 5, October 1995, pp. 321–323.

Fano, R. M. Theoretical Limitations of the Broadband Matching of Arbitrary Impedances. *J. Franklin Inst.*, Vol. 249, February 1950a, pp. 139–154.

Fano, R. M. A Note on the Solution of Certain Approximation Problems in Network Synthesis. *J. Franklin Inst.*, Vol. 249, March 1950b, pp. 189–205.

Foster, R. M. A Reactance Theorem. *Bell Syst. Tech. J.*, Vol. 3, April 1924, pp. 259–267.

Geyi, W., Jarmuszewski, P., and Qi, Y. The Foster Reactance Theorem for Antennas and Radiation *Q*. *Trans. IEEE*, Vol. AP-48, March 2000, pp. 401–408.

Guillemin, E. A. *Communications Networks*, Vol.2, Wiley, 1935, Chapter V.

Hansen, R. C. Superconducting Antennas. *Trans. IEEE*, Vol. AES-26, March 1990, pp. 345–355.

Hansen, R. C. Wideband Dipole Arrays Using Non-Foster Coupling. *Microwave Opt. Technol. Lett.*, Vol. 38, September 2003, pp. 453–455.

Hansen, R. C. Non-Foster and Connected Planar Arrays. *Radio Sci.*, Vol. 39, 2004, RS4004, pp. 1–14.

Hansen, R. C. *Electrically Small, Superdirective, and Superconducting Antennas*, Wiley, 2006.

Hansen, R. C. Bode and Fano Impedance Matching. *Microwave Opt. Technol. Lett.*, Vol. 50, April 2008, pp. 875–877; erratum, Vol. 50, October 2008, p. 2747.

Hara, S. et al. Broad-Band Monolithic Microwave Active Inductor and Its Applications to Miniaturized Wide-Band Amplifiers. *Trans. IEEE*, Vol. MTT-36, December 1988, pp. 1920–1924.

Kaya, A. and Yüksel, E. Y. Investigation of a Compensated Rectangular Microstrip Antenna with Negative Capacitor and Negative Inductor for Bandwidth Enhancement. *Trans. IEEE*, Vol. AP-55, May 2007, pp. 1275–1282.

Kaya, A. et al. Bandwidth Enhancement of a Microstrip Antenna Using Negative Inductance as Impedance Matching Device. *Microwave Opt. Technol. Lett.*, Vol. 42, September 2004, pp. 476–478.

Linvill, J. G. Transistor Negative-Impedance Converters. *Proc. IRE*, Vol. 41, June 1953, pp. 725–729.

Llewellyn, F. B. Some Fundamental Properties of Transmission Systems. *Proc. IRE*, Vol. 40, March 1952, pp. 271–283.

Lopez, A. R. Review of Narrowband Impedance-Matching Limitations. *IEEE Antenn. Propag. Mag.*, Vol. 46, August 2004, pp. 88–90.

Lopez, A. R. Rebuttal to "Fano Limits on Matching Bandwidth". *IEEE Antenn. Propag. Mag.*, Vol. 47, October 2005, pp. 128–129.

Matthaei, G. L., Young, L., and Jones, E. M. T. *Microwave Filters, Impedance-Matching Networks, and Coupling Structures*, McGraw-Hill, 1964, Sections 4.09 and 4.10.

Mayes, P. E. and Poggio, A. J. *Wire Antenna Multiply-Loaded with Active Element Impedances,* U.S. Patent No. 3,716,867, February 1973.

Middleton, W. M., Ed. *Reference Data for Engineers: Radio, Electronics, and Communications*, 9th ed., Newness, 2002, Chapter 32, Figure 10.

Montgomery, C. G. *Technique of Microwave Measurement*, Vol. 11, Radiation Laboratory Series, McGraw-Hill, 1947, Chapter 4.

Moreno, T. *Microwave Transmission Design Data*, Sperry Gyroscope, 1948; Dover, 1958, Chapter 2.

Mumford, W. W. and Scheibe, E. H. *Noise Performance Factors in Communications Systems*, Horizon House, 1968.

Nyquist, H. Regeneration Theory. *Bell Syst. Tech. J.*, Vol. 11, January 1932, pp. 126–147.

Pedinoff, M. E. The Negative-Conductance Slot Amplifier. *Trans. IRE*, Vol. MTT-9, November 1961, pp. 557–566.

Poggio, A. J. and Mayes, P. E. Bandwidth Extension for Dipole Antennas by Conjugate Reactance Loading. *Trans. IEEE*, Vol. AP-19, July 1971, pp. 544–547.

Pues, H. F. and Van de Capelle, A. R. An Impedance-Matching Technique for Increasing the Bandwidth of Microstrip Antennas. *Trans. IEEE*, Vol. AP-37, November 1989, pp. 1345–1354.

Quirin, J. D. *A Study of High-Frequency Solid-State Negative-Impedance Converters for Impedance Loading of Dipole Antennas,* MS Thesis, University of Illinois, 1971.

Radjy, A. H. and Hansen, R. C. S/N Performance of Aperiodic Monopoles. *Trans. IEEE*, Vol. AP-27, March 1979, pp. 259–261.

Rollett, J. M. Stability and Power-Gain Invariants of Linear Twoports. *IRE Trans. Circuit Theory*, Vol. 9, March 1962, pp. 29–32; correction, Vol. 10, March 1963, p. 107.

Rudish, R. M. and Sussman-Fort, S. E. Non-Foster Impedance Matching Improves S/N of Wideband Electrically-Small VHF Antennas and Arrays. *IASTED Conf. on Antennas, Radar, and Wave Propagation*, Banff, Canada, 2005.

Ryan, P. A. *Low Profile Electric Field Sensor,* U.S. Patent No. 4,383,260, May 1983.

Sainati, R. A. Active Antenna Performance Limitation. *Trans. IEEE*, Vol. AP-30, November 1982, pp. 1265–1267.

Sinsky, J. H. and Westgate, C. R. A New Approach to Designing Active MMIC Tuning Elements Using Second-Generation Current Conveyors. *IEEE Microwave Guided Wave Lett.*, Vol. 6, September 1996, pp. 326–328.

Skahill, G., Rudish, R. M., and Piero, J. A. Electrically Small, Efficient, Wide-Band Low-Noise Antenna Elements. *Proc. Antenna Applications Symp.*, Allerton, IL, September 1998, pp. 214–231.

Skahill, G., Rudish, R. M., and Piero, J. A. *Apparatus and Method for Broadband Matching of Electrically Small Antenna,* U.S. Patent No. 6,121,940, September 2000.

Smith, P. H. R-F Transmission-Line Nomographs. *Electronics*, February 1949, pp. 540–546.

Stark, P. A. *Introduction to Numerical Methods*, Macmillan, 1970, p. 130.

STEARNS, S. D. Counterintuitive Aspects of Non-Foster Networks. *ARO Adelphi Antenna Workshop,* July 8, 2010.

SUSSMAN-FORT, S. E. An NIC-Based Negative Resistance Circuit for Microwave Active Filters. *Int. J. Microwave Millimeter-Wave Comput.-Aided Eng.*, Vol. 2, 1994, pp. 130–139.

SUSSMAN-FORT, S. E. Gyrator Based Biquad Filters and Negative Impedance Converters for Microwaves. *Int. J. Microwave Millimeter-Wave Comput.-Aided Eng.*, Vol. 6, 1998, pp. 86–101.

SUSSMAN-FORT, S. E. Matching Network Design Using Non-Foster Impedances. *Int. J. RF Microwave Comput.-Aided Eng.*, March 2006, pp. 135–142.

SUSSMAN-FORT, S. E., AND RUDISH, R. M. Progress in Use of Non-Foster Impedances to Match Electrically-Small Antennas and Arrays. *Proc. Antenna Applications Symp.*, Allerton, IL, 2005, pp. 89–108.

SUTTON, J. F. *Active Antenna,* U.S. Patent No. 5,311,198, May 1994.

TEK, H. AND ANDAY, F. Voltage Transfer Function Synthesis Using Current Conveyors. *Electron. Lett.*, Vol. 25, November 1989, pp. 1552–1553.

WERNER, R. E. AND CARRELL, R. M. Application of Negative Impedance Amplifiers to Loudspeaker Systems. *J. Audio Eng. Soc.*, Vol. 6, October 1958, pp. 240–243.

YAGHJIAN, A. D. AND BEST, S. R. Impedance, Bandwidth, and Q of Antennas. *Trans. IEEE*, Vol. AP-53, April 2005, pp. 1298–1324.

YOULA, D. C. A New Theory of Broad-Band Matching. *Trans. IEEE*, Vol. CT-11, March 1964, pp. 30–50.

CHAPTER 3

ELECTRICALLY SMALL ANTENNAS: CANONICAL TYPES

3.1 INTRODUCTION

The term electrically small antennas (ESA) applies to antennas whose dimensions are small compared to the wavelength. As will be shown, these are mostly dipoles or loops, or minor modifications of these. Various techniques for improving performance are discussed below.

Both short dipoles and small loops are superdirective; the directivity remains at 1.5 as the dimensions decrease, but the radiation resistance decreases and the reactance increases. Thus, the bandwidth is narrow, the efficiency may be low, and the necessary tolerances will be small. See Chapter 6 for more discussion of superdirectivity.

3.2 DIPOLE BASIC CHARACTERISTICS

3.2.1 Dipole Impedance and Bandwidth

First take the case of dipole (or monopole) that is transmitting. A typical equivalent circuit is shown in Figure 3.1. The dipole radiation resistance, loss resistance, and reactance are R_r, R_a, and X_a. The source and matching circuit has loss resistance and reactance R_ℓ and X_ℓ and load resistance R. The important parameters are the radiated

Small Antenna Handbook, Robert C. Hansen and Robert E. Collin.

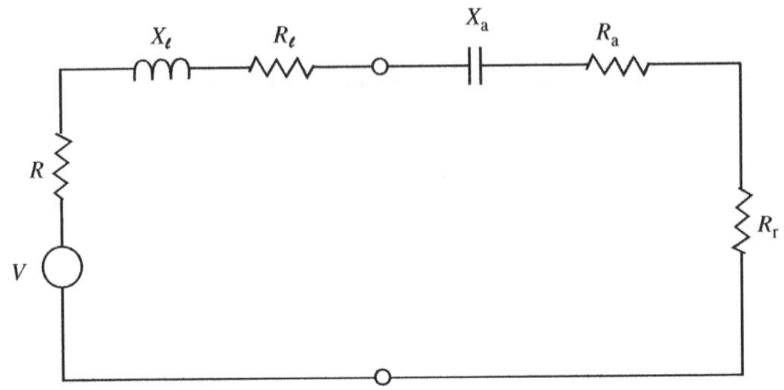

FIGURE 3.1 Transmitting dipole equivalent circuit.

power efficiency and the bandwidth. Assume that at the center frequency the antenna is matched, $X_\ell = -X_a$. Then the efficiency η at center frequency is

$$\eta = \frac{R_r}{R_r + R_\ell + R_a} \tag{3.1}$$

Equation 3.1, of course, states the well-known fact that efficiency is the ratio of radiation resistance to total circuit resistance; clearly, the R_r should be as large as feasible and equal to R_ℓ. For a short dipole of half-length h and a simple monopole also of length h, the radiation resistance is

$$\begin{aligned} R_r &= 20k^2h^2 \qquad \text{dipole} \\ R_r &= 10k^2h^2 \qquad \text{monopole} \end{aligned} \tag{3.2}$$

The loss resistance is a simple function of the dipole dimensions; the reactance also involves the dimensions but is subject to modification through structural changes. The reactance and loss resistance formulas for simple dipoles/monopoles are

$$\begin{aligned} X_a &= 120\left(1 - \ell n\frac{h}{a}\right)\cot kh \qquad \text{dipole} \\ X_a &= 60\left(1 - \ell n\frac{h}{a}\right)\cot kh \qquad \text{monopole} \\ R_a &= \frac{1}{2\delta}R_{dc}\frac{1}{3} = \frac{hR_s}{3\pi a} \qquad \text{dipole} \end{aligned} \tag{3.3}$$

Here δ is the skin depth and R_s is the resistivity in $\Omega/\square$. This formula for loss resistance is valid for short antennas (triangular current distributions) where the loss

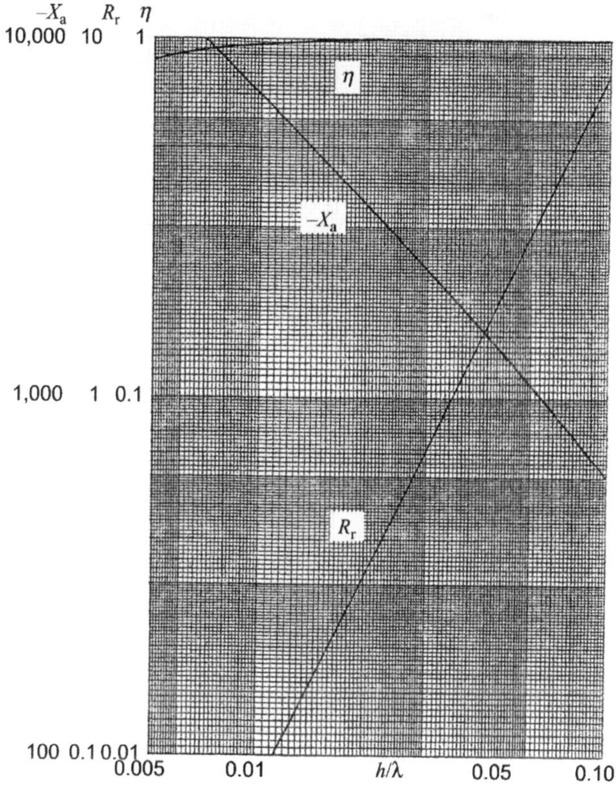

FIGURE 3.2 Dipole parameters.

is one-third that for uniform current and for dipole radius large in skin depths. For a fixed h/a, the resistance is independent of length. Surface resistivity of copper at room temperature is $1.5E{-}7\sqrt{f}$, with f in Hz. A typical value of R_s for copper at 200 MHz is 2.6 mΩ. Figure 3.2 shows the dipole parameters, using a fixed $h/a = 100$. Reactance for other thicknesses can be found by multiplying the reactance from Figure 3.2 by a normalized thickness factor

$$\frac{\ell n(h/a)-1}{\ell n\,100-1} \tag{3.4}$$

Figure 3.3 gives this factor. Note that a flat strip dipole of width w is equivalent to a cylindrical dipole of radius $w/4$. The radiation resistance, over the range of h/λ from 0.005 to 0.01, varies from 0.02 to 8 Ω; the reactance varies from −14,000 to −600 Ω. Not surprisingly, the efficiency is close to 100%, dipping only to 88% at $h/\lambda = 0.005$.

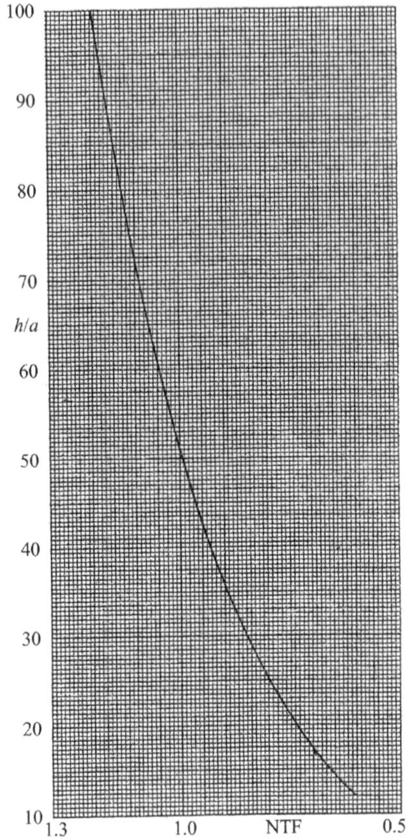

FIGURE 3.3 Normalized thickness factor.

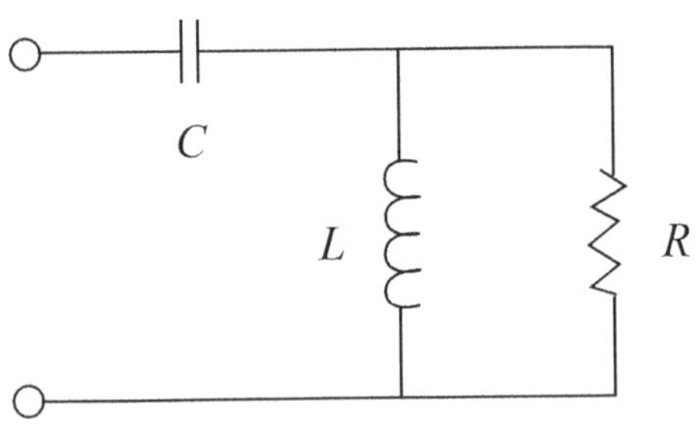

FIGURE 3.4 Chu equivalent circuit for short dipole.

In the simple equivalent circuit of Figure 3.1, the resistance and reactance vary with frequency. It is sometimes useful to have an equivalent circuit whose parameters are fixed. Such a circuit was developed by Chu (1948), valid for electrically short dipoles (below resonance). It is shown in Figure 3.4. The component values are

$$\begin{aligned} C &= |X_{\text{ant}}|/(2\pi f) \\ L &= CR_{\text{rad}}+1/(\omega^2 C) \\ R &= L/(R_{\text{rad}}C) \end{aligned} \tag{3.5}$$

A simple bandwidth calculation assumes that the radiation resistance is constant with frequency, the loss resistance is negligible, and $Q = 1/\omega_0 RC$, where $\omega = 2\pi f$ and the antenna reactance is $X = 1/\omega C$. Using a series equivalent circuit, the impedance divided by R is

$$\begin{aligned} \frac{Z}{R} &= \frac{\omega RC+j(\omega^2 LC-1)}{\omega RC} \\ \frac{Z}{R} &= 1+jQ\left(\frac{\omega}{\omega_0}-\frac{\omega_0}{\omega}\right) \end{aligned} \tag{3.6}$$

The VSWR, abbreviated as V, becomes

$$V = \frac{\sqrt{1+Q^2 v^2}+Qv}{\sqrt{1+Q^2 v^2}-Qv} \tag{3.7}$$

and bandwidth v becomes

$$v = \frac{\omega}{\omega_0}-\frac{\omega_0}{\omega} = \text{BW} \tag{3.8}$$

For a given V, the Q–bandwidth product

$$Q\cdot\text{BW} = \frac{V-1}{2\sqrt{V}} \qquad \text{(antenna only)} \tag{3.9}$$

A calculation useful for short dipoles starts with the reactance derivative formula of Harrington (1968), valid for ESA:

$$Q = \frac{\omega_0\, dX/d\omega}{2\, R_0} \tag{3.10}$$

When the dipole is in the resonance region, Equation 80 of Yaghjian and Best (2005) should be used. For the reactance and resistance, the Carter formula codified by Hansen (1972) is simplified for self-impedance; $x_0 = 0$ in the formula. Note that the series coefficients should be $A(-2) = A(2) = 1$, $A(-1) = A(1) = -4\cos kd$, and $A(3) = 2(1+2\cos^2 kd)$. The half-power bandwidth is given approximately by

$$\mathrm{BW} = 1/Q \tag{3.11}$$

To determine the utility of this result for low-Q dipoles, exact bandwidth was calculated as the difference between two frequencies that satisfy

$$V = \frac{[Z+R_0]+[Z-R_0]}{[Z+R_0]-[Z-R_0]} \tag{3.12}$$

A conjugate impedance match at the center frequency ω_0 is used. The load resistance R_0 equals the antenna radiation resistance at ω_0, and the matching reactance equals the negative of the antenna reactance X_0 at ω_0. No losses are included; matching coil Q could be easily included.

Figure 3.5 shows $\mathrm{BW} = 2/Q$ and exact bandwidth for half-power: $\mathrm{VSWR} = 3+\sqrt{8} = 5.828$; the $2/Q$ results are surprisingly accurate for Q as small as 2.

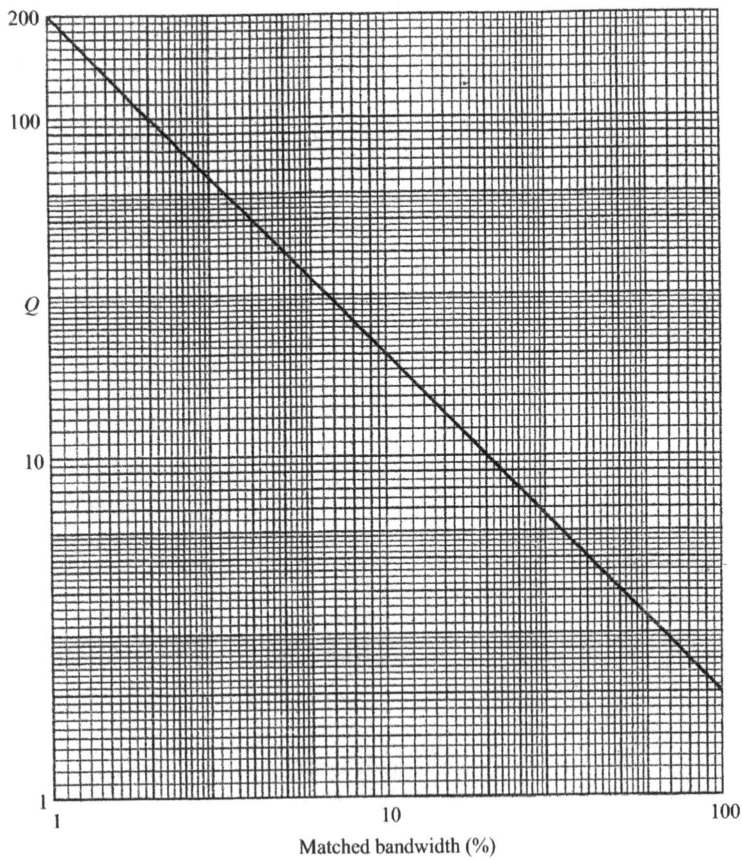

FIGURE 3.5 Q versus bandwidth, half-power.

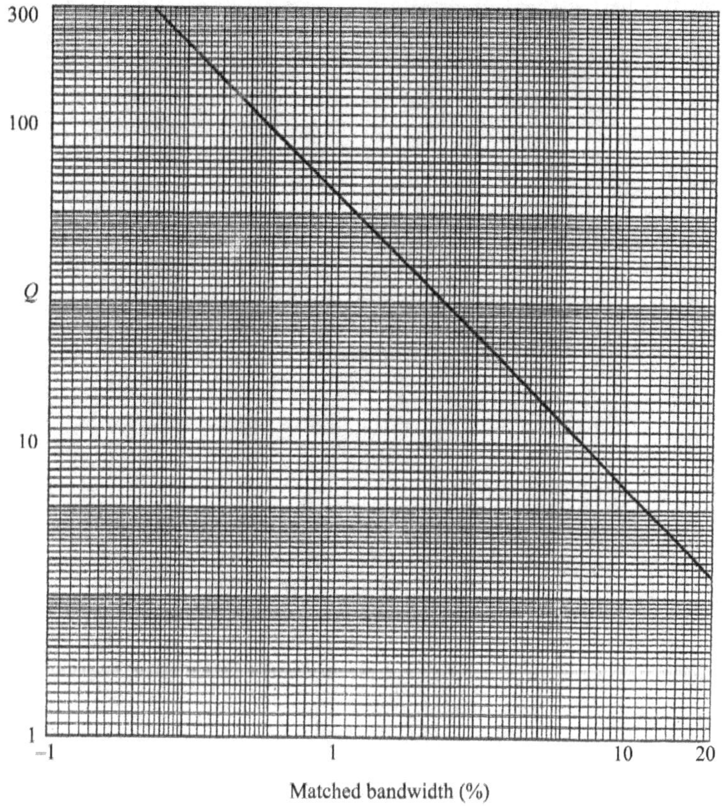

FIGURE 3.6 Q versus bandwidth, VSWR = 2.

Similar calculations for VSWR = 2 are given in Figure 3.6. Again $1/Q$ is accurate for $Q \geq 3$.

Next is a look at bandwidth versus dipole length. Given the dipole half-length h/λ, h/a, and VSWR, the $\omega_2 > \omega$ and $\omega_1 < \omega_0$ that satisfy Equation 3.12 can be found using a root finder, in this case Wegstein (1960) rooter. This subroutine is advantageous in that it does not require explicit derivatives. Figure 3.7 gives % bandwidth for VSWR ≤ 2 for dipole lengths from 0.1λ to 0.5λ. Two curves are shown for $L/a = 50$ and $L/a = 200$. $1/Q$ is also shown in the figure (Hansen, 2007); $1/Q$ and BW are a close match for ka up to 0.5.

These calculations assume that the matching inductance is lossless. In practice, it will have losses, and this loss will widen the bandwidth and lower the efficiency. Also, a transformer is often used to transform the low radiation resistance to a convenient value such as 50 Ω. All transformers have loss and again the bandwidth is improved at the cost of efficiency. To reduce the losses, both the matching inductor and transformer can be made of HTS material; see Chapters 2 and 7. To convert bandwidth for any VSWR, given bandwidth for VSWR = 2,

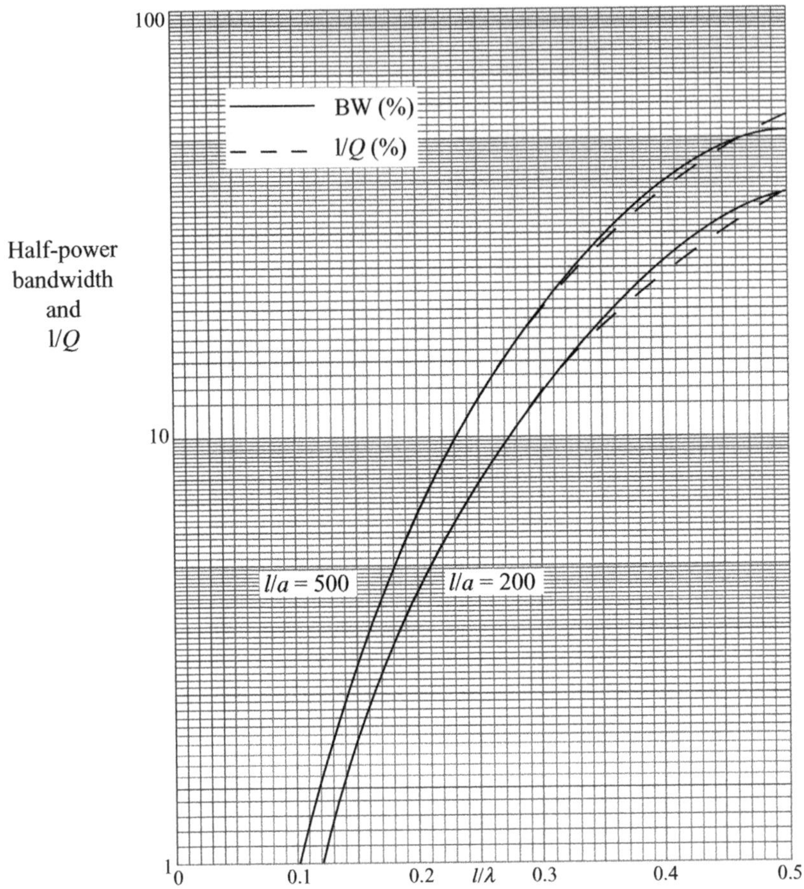

FIGURE 3.7 Dipole bandwidth.

$$\mathrm{BW} = \mathrm{BW}_2 \frac{\sqrt{2}(\mathrm{VSWR}-1)}{\sqrt{\mathrm{VSWR}}} \tag{3.13}$$

The quantity most useful for evaluating short receiving antennas is effective length l_e, which gives the open-circuit voltage produced by a unit electric field strength: $V_{oc} = l_e E$. For a short dipole, $l_e = h$ is independent of frequency. The equivalent circuit is the same as Figure 3.1 for transmitting. The low R_r and large X lead, of course, to a narrow bandwidth. For a receiving antenna, the goal is to deliver as much of the incident power as possible to the load. The receiving case is thus the same as the transmitting case: minimize losses and utilize a good matching network.

Bandwidth is easily measured by connecting the antenna to a network analyzer, such as the HP 8510. This gives return loss (RL), which is

$$\mathrm{RL} = \frac{(\mathrm{VSWR}+1)^2}{(\mathrm{VSWR}-1)^2} \tag{3.14}$$

$$\mathrm{VSWR} = \frac{\sqrt{\mathrm{RL}}+1}{\sqrt{\mathrm{RL}}-1} \tag{3.15}$$

Q can be measured with a time-domain procedure (Liu et al., 1999).

3.2.2 Resistive and Reactive Loading

Distributed resistive loading can be used to make a broadband dipole (Wu and King, 1965; Shen and Wu, 1967; Shen, 1967; Taylor, 1968; Lally and Rouch, 1970). Unfortunately, this technique is not effective for short dipoles; results are good for dipole lengths of the order of a quarter-wavelength. Distributed loading also helps short pulse response (Kanda, 1978, 1994; Esselle and Stuchly, 1990), but again only for longer dipoles. To avoid the drop in efficiency produced by resistive loading, distributed capacitive loading has been used (Rao et al., 1969). Again short dipoles do not benefit. Other loadings include impedance loads at two locations (Lin et al., 1970), multiple loads (Fanson and Chen, 1973), and loading with a resonant circuit at one location (Smith, 1975). Again these are not useful for a short dipole. S/N performance has been analyzed by Maclean and Saini (1981).

It has been accepted practice for many years to improve the performance of a short monopole (whip) by inserting an inductor into the whip. A pioneering work was done by Bulgerin and Walters (1954). They performed a series of experiments on short, fat monopoles of various lengths at 100 MHz. For each length, a number of loading points were used. Because of the relatively large gaps where the loading coil was located, these data have not been widely circulated. Also, modern ferrite toroid coils offer much higher Q values. Harrison (1963) analyzed the loaded monopole (dipole) with superposition of asymmetrically excited dipoles. However, because the coupled integral equations for the current distribution on an asymmetric dipole have not been solved except in principle, the current distribution used for each asymmetrically excited dipole was obtained from a zero-order solution. With this approach, Harrison calculated input impedance, current, and loading inductor currents for several lengths and for a number of loading coil Q values. The results showed a gradual increase in efficiency as the load point moved closer to the dipole ends; the data essentially stopped at a 2/3 load point, that is, 2/3 from the feed to the end. These results have been superceded by the more accurate moment method results discussed below. Czerwinski (1965, 1966) measured monopoles with distributed inductance where the monopole is a helix of small diameter with tapered pitch. For narrowband operation, a discrete coil offers better performance than a helix (see Figure 3.8). Moment method calculations, described below, have indicated that the maximum efficiency point occurs closer to the feed than the 2/3 value predicted by the approximate theories.

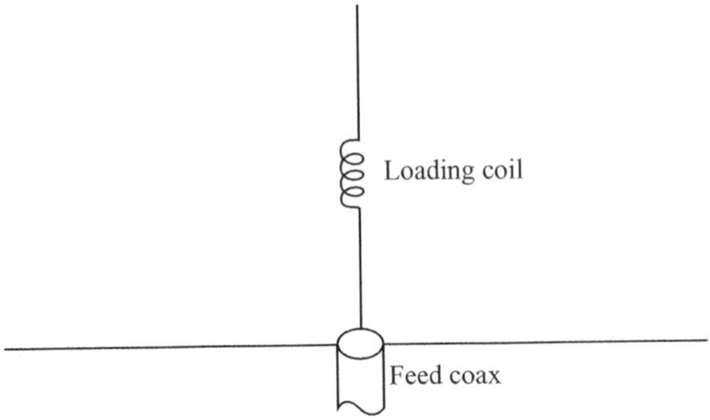

FIGURE 3.8 Inductively loaded monopole.

Moment method calculations (Hansen, 1975a, 1975b) used piecewise sinusoidal expansion and test functions, a Galerkin stationary form. Carter mutual impedances were used, in a convenient subroutine (Hansen, 1972). A simultaneous equation solver provided the currents, from which input impedance was found. Discrete loading was included by adding the loading $R+jX$ to the matrix of mutual impedances at the proper segment junction. The load was added to the self-impedance on the diagonal for that segment. A restriction on the use of the moment method occurs for fat monopoles; each segment must have a length-to-diameter ratio sufficiently large that the filamentary current approximation can be employed. At the same time, there must be sufficient segments to accurately represent the current distribution. When the wire is too fat for the number of segments used, the susceptance exhibits a shift. Also, the number of segments must be chosen such that a segment junction occurs at each desired loading point. If N is the number of symmetrical segments on the dipole wire and I is the segment number at the loading point to dipole half-length (or monopole length),

$$\frac{N}{2I} = \frac{1}{1-\gamma} \tag{3.16}$$

The ratio of distance from feed to load to dipole half-length is γ. All data presented here were obtained by using $N = 12$, which allowed loading points of 1/6, 1/3, 1/2, 2/3, and 5/6 to be considered. Results were spot checked with $N = 6$ and 48 to validate the runs.

In principle, the loading inductor functions by keeping the current distribution nearly constant from the feed to the load point, thereby increasing the current moment. Because the radiation resistance varies as current moment squared and the effective length varies with current moment, it is clear that inductive loading will improve short monopole performance (Fournier and Pomerleau, 1978). There is a value of loading reactance that allows the current to approximate a constant value out to γh with a linear drop-off beyond. This value of inductance is, however, insufficient to produce input impedance resonance. The resonant value of load produces a modest

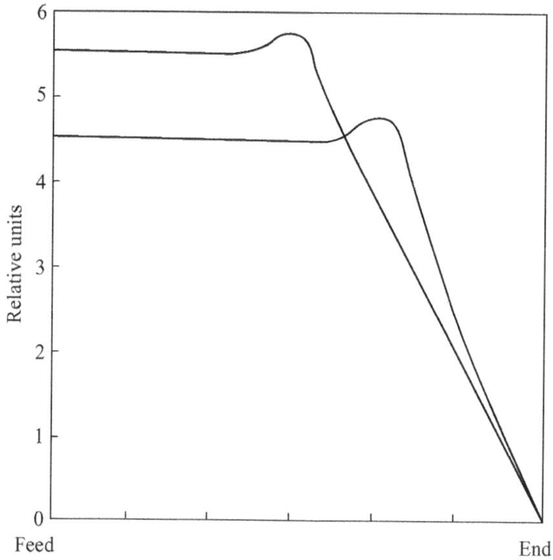

FIGURE 3.9 Current distributions for loading at 1/2 and 2/3 points. Courtesy of Hansen (1972).

current peak just beyond γh so that the current moment is increased by an additional amount over that expected from the constant current model. Figure 3.9 shows two typical current distributions calculated for a thin dipole using $N = 40$. Figures 3.10–3.12 show the resonant input resistance as a function of the load point γ for lengths of monopole $h/\lambda = 0.05$, 0.1, and 0.15, for loading coil $Q = \infty$, 300,

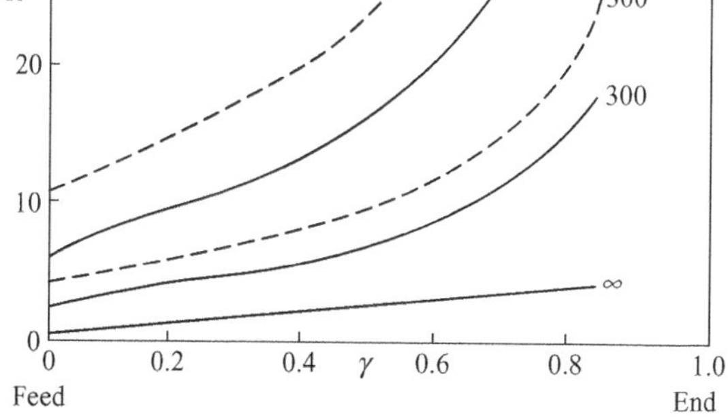

FIGURE 3.10 Input resistance versus loading point. Courtesy of Hansen (1972).

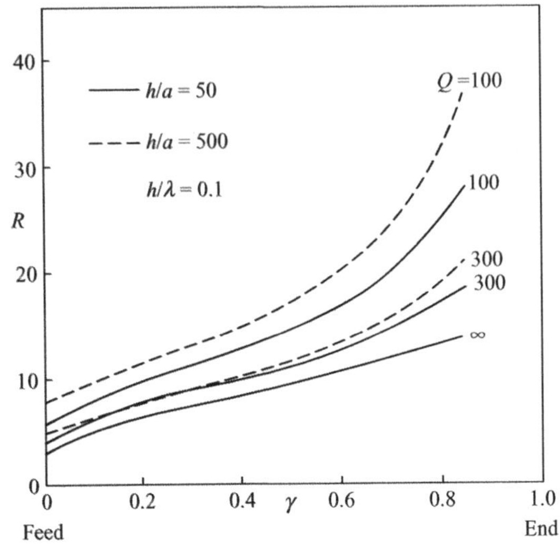

FIGURE 3.11 Input resistance versus loading point. Courtesy of Hansen (1972).

and 100, and for two monopole length-to-radius ratios $h/a = 50$ and 500. Radiation resistance improvement factor β is given in Figures 3.13 and 3.14 for the same cases. It should be noted in Figures 3.10–3.12 that a significant part of the input resistance is due to coil losses, and this is reflected in the efficiency values. An empirical fit has been made to the radiation resistance enhancement factors based on the model of a

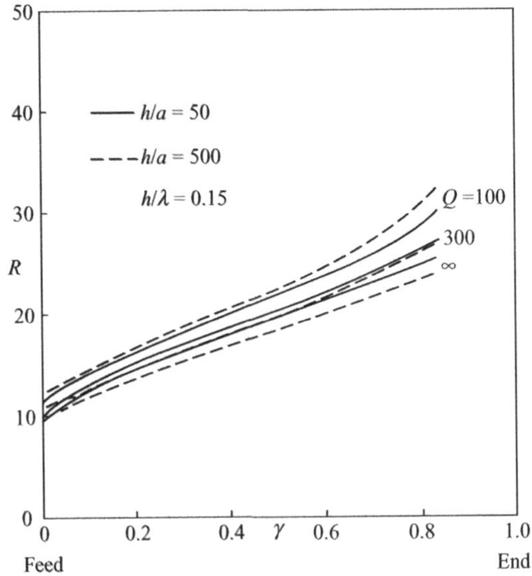

FIGURE 3.12 Input resistance versus loading point. Courtesy of Hansen (1972).

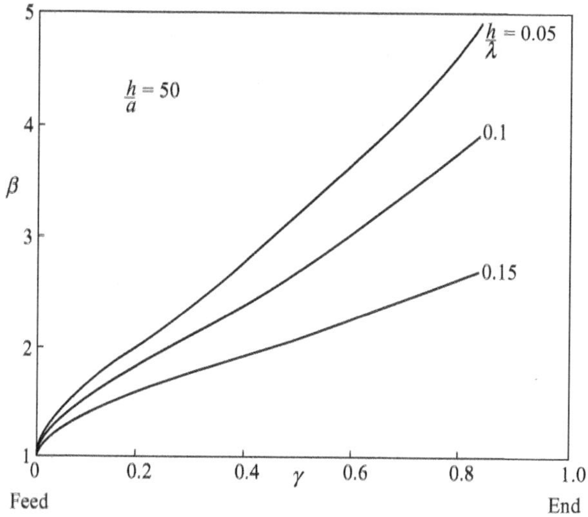

FIGURE 3.13 Radiation resistance improvement factor. Courtesy of Hansen (1975a).

current rising to a peak and then linearly dropping to zero at the wire end. The improvement factor β is the ratio of loaded to unloaded radiation resistance, and is

$$\beta = \left[1+\gamma+\frac{\gamma^{1/2}}{2}\left(1-\frac{h/\lambda}{0.25}\right)\right]^2 \tag{3.17}$$

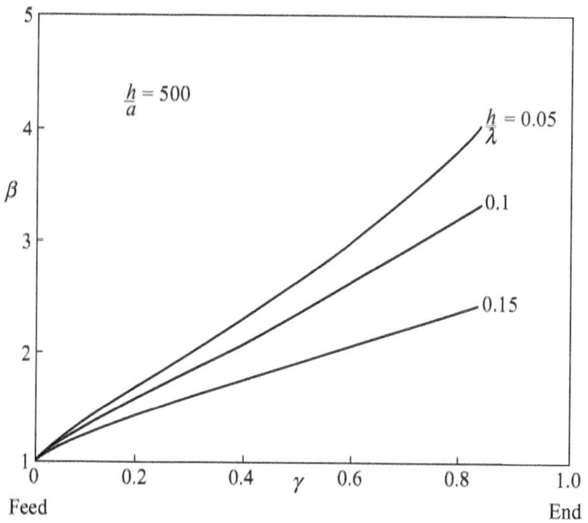

FIGURE 3.14 Radiation resistance improvement factor. Courtesy of Hansen (1975a).

This will be used later in finding efficiency transition values. It is interesting to note that the radiation resistance improvement can for short antennas be larger than 4, the value that results with uniform antenna current. As mentioned, these results are due to the current peaking at or just beyond the load point.

As the load point moves toward the end, the loading reactance must increase to maintain resonance, finally becoming infinite at the end. For monopole lengths below roughly 0.1λ, the antenna reactance increases linearly (inversely) with length; the loading reactance does also, as seen in Figure 3.15. Here values for three loading points are shown. Figure 3.16 gives the factor $\alpha = X_{\text{LOAD}}/X_{\text{ANT}}$ for the same three load points. An empirical fit to be used later is just $\alpha = 1/(1-\gamma)$.

Efficiency versus load point γ is of interest. Figures 3.17 and 3.18 give efficiency for $h/a = 50$ and 500, $h/\lambda = 0.05$, 0.1, and 0.15, and loading coil $Q = \infty$, 300, and 100. With ferrite toroid coils, Q values approaching 300 are practical over the HF–UHF range; the higher permeability cores are typified by the $Q = 100$ values. Lengths below 0.05λ are usually not practical as the efficiency is well below 50%. At the other extreme, monopoles longer than 0.15λ are seldom loaded as their radiation resistance and bandwidth are more tractable. Of course, the fatter monopoles have higher efficiency as there is less antenna reactance to offset. Maximum efficiency occurs for a loading point between 0.3 and 0.4 from the feed, although the efficiency

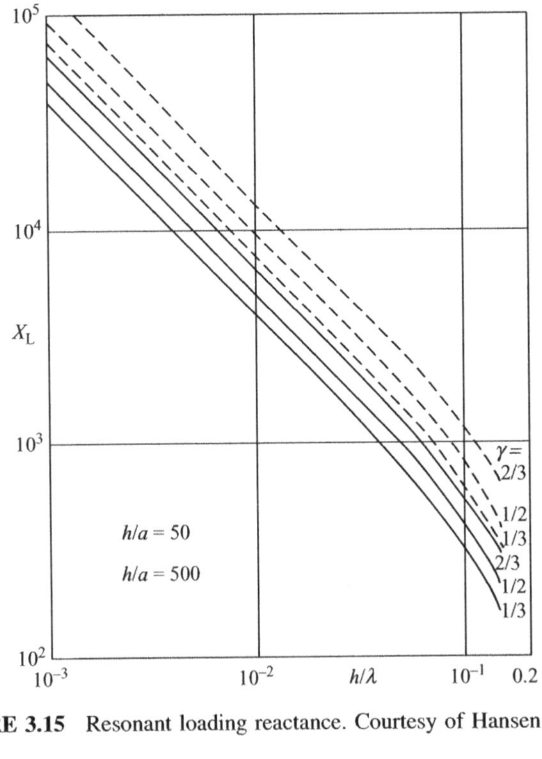

FIGURE 3.15 Resonant loading reactance. Courtesy of Hansen (1975a).

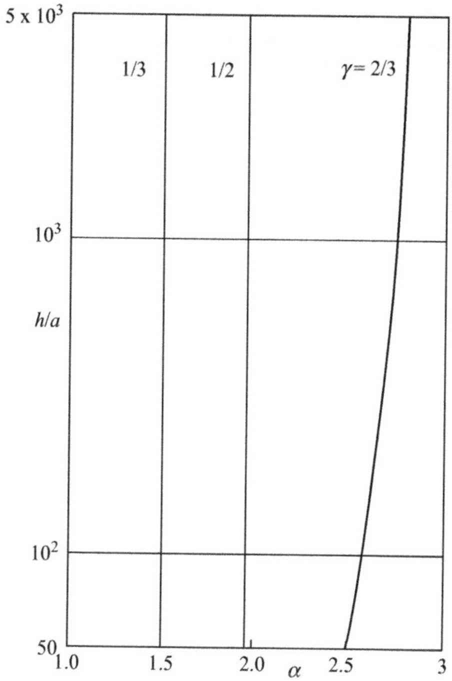

FIGURE 3.16 Ratio of loading reactance lo antenna reactance. Courtesy of Hansen (1975a).

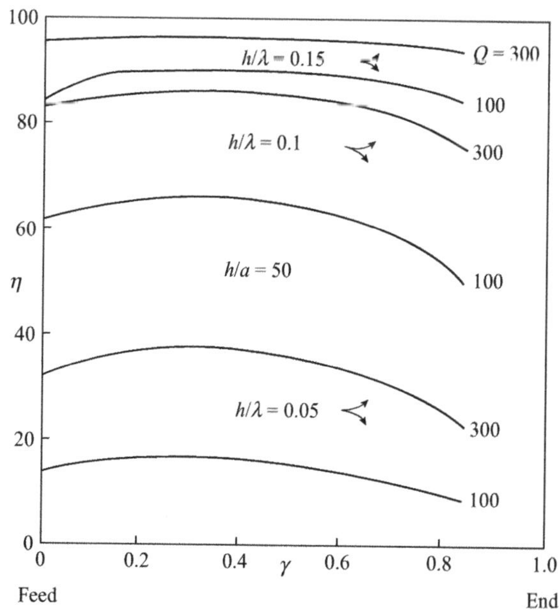

FIGURE 3.17 Efficiency versus loading paint. Courtesy of Hansen (1975a).

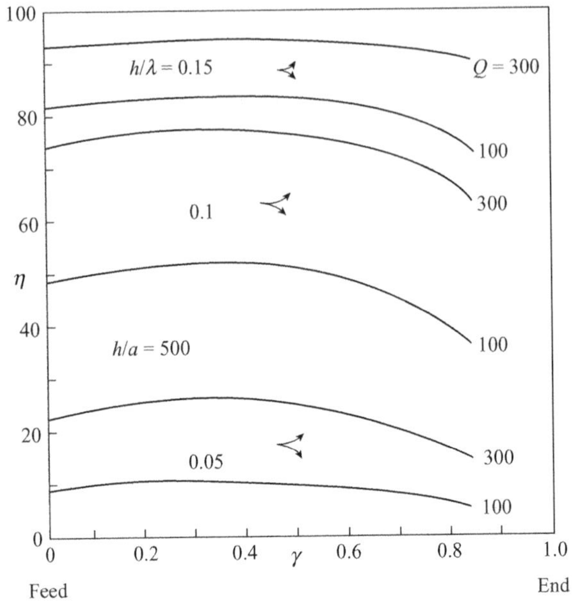

FIGURE 3.18 Efficiency versus loading point.

varies slowly with load point. There is no apparent variation of the maximum point with Q or with monopole length. It should be noted that when the loading coil is located at the feed, the efficiency is always lower. Thus, for maximum efficiency, the inductive load should be located at roughly $0.4h$. However, in most cases, a small sacrifice in efficiency should be made to obtain a higher input resistance. Figures 3.10–3.12 show the input (radiation) resistance versus γ, and the trade-off may be made between these figures and Figures 3.17 and 3.18. Antenna copper loss is not included as it is always much smaller than the loading coil loss.

An efficiency transition formula has been derived. Efficiency η is

$$\eta = \frac{Q\beta R_r}{Q\beta R_r + \alpha X} \tag{3.18}$$

where for short monopoles $R_r = 10k^2h^2$ and $X = 60(\ell n\, h/a - 1)/kh$. Putting $\eta = 0.5$ and using the empirical formulas for α and β gives the h/λ for 50% efficiency:

$$\frac{h}{\lambda} = \frac{1}{2\pi}\left(\frac{6(\ell n\, h/a - 1)}{Q(1-\gamma)[1+\gamma+\gamma^{1/2}(1-4h/\lambda)/2]^2}\right)^{1/3} \tag{3.19}$$

As the $(1-4h/\lambda)$ factor has a weak influence on the result, the equation is readily solved by iteration. Note that the transition length varies as the cube root of Q and even more slowly with h/a. For lengths below the transition length, the efficiency drops rapidly, in the limit as f^3, or as length/wavelength cubed. Use of HTS materials can improve η, but it is the loading coil that needs to be HTS. Even with the coil at the base, this conclusion is still true. Thus, the important overall conclusion: With ordinary inductances, loading monopoles shorter than 0.06λ is not practical because of the low efficiency. There are, of course, special cases where the application is sufficiently important to warrant acceptance of low efficiencies, for example, VLF transmitting antennas.

A remark on bandwidth is in order. With an ideal loading coil ($Q = \infty$), the bandwidth is essentially unchanged by loading. Although the radiation resistance is substantially increased by loading, the slope of reactance is correspondingly increased, so the overall Q, which is the inverse of fractional bandwidth, is essentially unchanged. Real loading coils will have losses, with Q in the 100–300 range for ferrite toroids and in the 50–100 range for air core coils. These coil losses will, of course, decrease antenna Q and increase bandwidth.

To further quantify inductive loading, a dipole of length 0.2λ and radius 0.001λ was studied, with lumped inductive loads at either 1/2 or 2/3 point; the latter is with loads near the ends. The inductive loads were adjusted to give resonance, while the load Q was adjusted to give an input resistance of $50\,\Omega$. Table 3.1 gives loading reactance with its Q, input resistance, efficiency due to loads, and VSWR $= 2$ bandwidth, all with inductive load at the 1/2 point of each dipole arm. Table 3.2 gives similar data for the loads nearer to dipole ends, at the 2/3 points. It can be seen that 1/2 point loading gives higher efficiency, while 2/3 point loading gives larger bandwidth. In both cases, the inductor Q needed for resonance is easily within the ferrite core range: 40–66.

Inductive loading of a closely spaced parasitic dipole is also effective. A parasite of equal length to the dipole and parallel to it is evaluated. Two cases are computed:

TABLE 3.1 Dipole with Inductive Loading at 2/3 Points

Q	Loading X (Ω)	Input R (Ω)	η	BW (%)
∞	753.80	22.56	1	1.363
200	753.95	31.63	0.7773	1.910
100	754.20	41.72	0.6917	2.457
50	754.96	59.02	0.6184	3.556
66.304	754.54	50.00	0.6459	3.015

TABLE 3.2 Dipole with Inductive Loading at 1/2 Points

Q	Loading X (Ω)	Input R (Ω)	η	BW (%)
∞	538.25	18.76	1	1.546
200	538.33	25.04	0.7996	2.063
100	538.45	31.32	0.7139	2.581
50	538.77	43.92	0.6359	3.618
40.306	538.97	50.00	0.6156	4.112

TABLE 3.3 Loaded Parasitic Dipole

Resonant Resistance (Ω)	Loading Reactance (Ω)	Q	Efficiency	Bandwidth (%)
28.5	871.1	∞	1.0000	–
50.0	871.4	81.28	0.6993	3.44
44.3	324.0	∞	1.0000	–
50.0	324.0	362.00	0.9653	2.66

a single inductance at the parasite center, and nine equal loads, spaced equally, with one at the center and four on each side except at parasite ends, of course. Table 3.3 gives resonant resistance, load reactance, Q, efficiency, and bandwidth. A summary of these data is as follows: center-loaded parasite gives best bandwidth ($\mathrm{BW} \cdot \eta = 2.41$) and distributed loads in the parasite give higher efficiency ($\mathrm{BW} \cdot \eta = 2.57$). These bandwidth × efficiency numbers are slightly better than those of Tables 3.1 and 3.2. Where the physical arrangement favors use of a parasitic dipole, the results are roughly equivalent to loading the dipole alone.

3.2.3 Other Loading Configurations

A half-bowtie monopole is an excellent antenna; a wire frame monopole can be used (Wong and King, 1986), as sketched in Figure 3.19. This is a zigzag meander construction and adjacent parasitic monopole is of similar shape. Performance is roughly equal to a $\lambda/4$ monopole, with half the length.

Capacitive loading is also effective. If space allows, a flat disk or plate at the monopole end and a perpendicular top hat can be used. An extreme example is the top-loaded VLF antenna described in Appendix A. A novel, but less effective, capacitive load is the trefoil knot monopole of Figure 3.20 (Hong et al., 2010). The arms are connected by vias to sublayer straps. Capacitive loading, where a wire

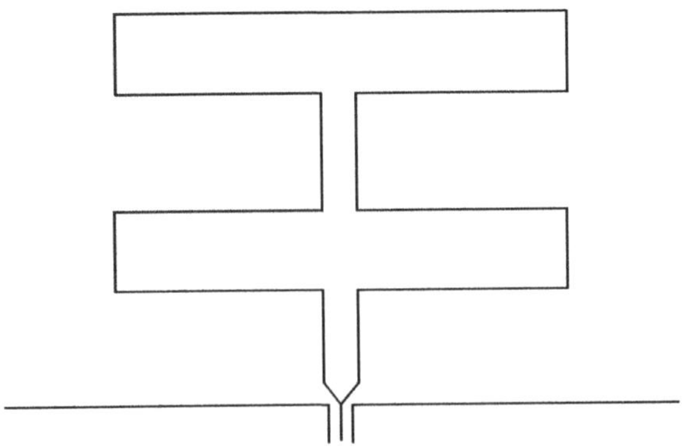

FIGURE 3.19 Wire frame monopole. Courtesy of Wong (1986).

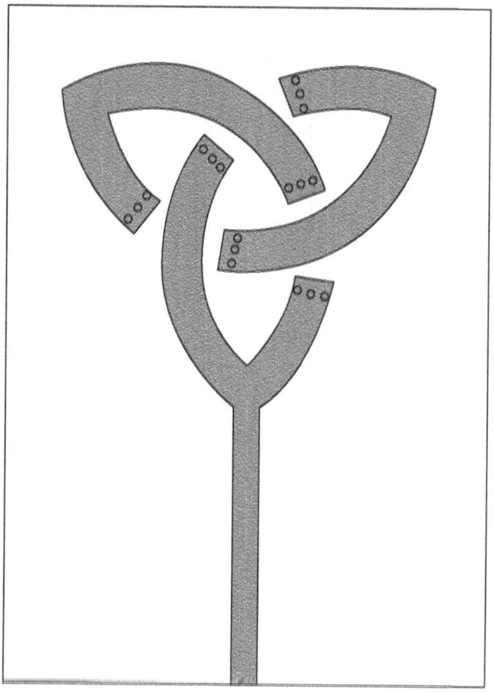

FIGURE 3.20 Monopole with trefoil load. Courtesy of Hong et al. (2010).

antenna is enclosed in a dielectric, is not effective. Resonant length is reduced but efficiency and bandwidth are severely reduced. See Section 4.4 for details.

A folded dipole may also be loaded to reduce its size. The folded dipole is often used because it offers a 4:1 impedance step up over a dipole, and this ratio can be changed over wide limits by making the two arms of different diameters (Guertler, 1950; Hansen, 1982). When the folded dipole is short, the transmission line currents see a low impedance, thus reducing the effectiveness of the element. One way of overcoming this shorting out effect, and making the folded dipole usable at lower frequencies, is to use a series impedance at the ends where the two arms join, in the center of the folded arm, or both. Figure 3.21 shows the general configuration. A short folded dipole has radiating currents in a triangular distribution, like a dipole, with maximum radiating current at the feed (and shorted) end and zero current at the other end where the arms join. Insertion of an inductor reduces the transmission line currents but probably has no effect on the radiating currents if the dipole is short. Analyses have been made of the end-loaded folded dipole (Harrison and King, 1961) and of the base-loaded folded monopole (Leonhard et al., 1955), but calculations of the radiation resistance as a function of loading are not available. Basically, the short loaded folded dipole will have a more useful input reactance—not very large as that of the same length dipole nor very small as that of the unloaded folded dipole. Radiation properties are probably not much affected. The ability to produce

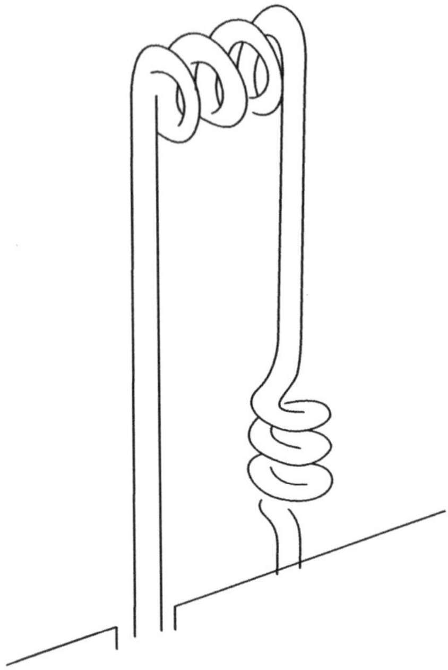

FIGURE 3.21 Loaded folded monopole. Courtesy of Harrison (1961).

broadband behavior through judicious choice of loading is as yet a very poorly understood situation.

To sum up, inductively loading a short dipole gives the best trade of performance versus space. Top hat loading is effective, if space allows (Francavilla et al., 1999). Loops are inefficient, but ferrite-cored loops are good for receiving. Fat dipoles such as bowtie dipoles offer good bandwidth. If space allows, the Goubau antenna offers the best performance for a given small volume; see Section 3.2.6. There are no ultrawideband electrically small antennas.

An antenna on a ground plane that fits within a hemispherical volume uses eight arc arms, fed in parallel at the ground plane and ending at the apex of the sphere. See Figure 3.22. Each arm is inductively loaded to increase the radiation resistance, but the parallel connection reduces it by a factor of 8 (Adams and Bernhard, 2008; Adams et al., 2011).

3.2.4 Short Flat Resonant Dipoles

Some applications require a flat dipole-type antenna. These can be made resonant by adding extra wire length. Several geometries for adding length are compared here. All dipoles are 0.2λ long and 0.05λ in width. Conducting strips are replaced by equivalent round wires of radius a, where $a = w/4$ (Lo, 1953). The flat dipoles are analyzed using a 64-bit piecewise sinusoidal Galerkin moment method code (MBC) (Tilston and Balmain, 1990). The code is also available in a 32-bit version. The candidate flat dipoles are evaluated on resonant resistance and VSWR = 2 bandwidth. Some lossy versions are also evaluated on efficiency η.

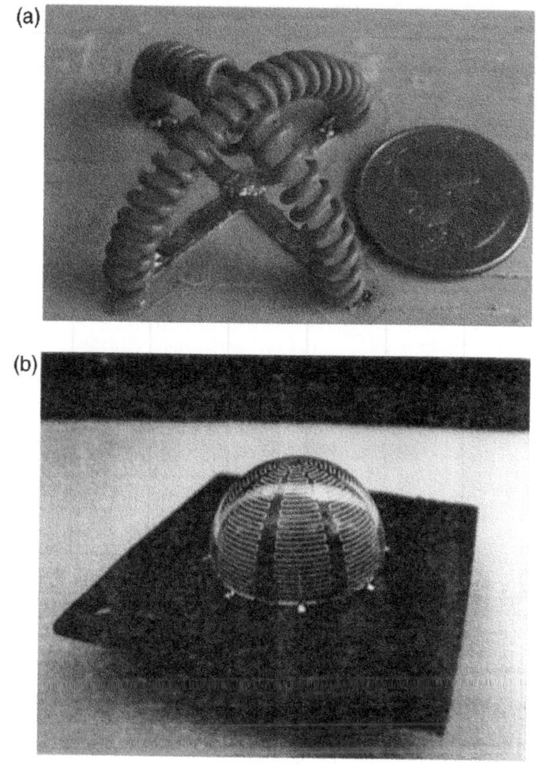

FIGURE 3.22 Helical Antenna. (a) 4 arms. Courtesy of Adams and Bernhard (2008); (b) 8 arms. Courtesy of Adams et al. (2011).

The flat dipole candidates are four- and eight-arm meander dipoles, 10- and 20-section zigzag dipoles, three- and four-folded-arm (Tai) dipoles (Rashed and Tai, 1991), and a simple fat dipole. Figure 3.23 shows a four-arm meander dipole, with a small gap between folds and dipole ends. Figure 3.24 shows a corresponding eight-arm meander dipole. Figure 3.25 shows a 10-section zigzag antenna; note that the dipole size is now $0.2\lambda \times 0.05\lambda$. Figure 3.26 shows a 20-section zigzag antenna and now the width is 0.038λ. The zigzag widths were chosen to give close to a 50 Ω resonant impedance. Figure 3.27 shows a Tai three-folded-arm antenna; this is not a folded dipole, and for want of a more descriptive term, it is called here "folded-arm"

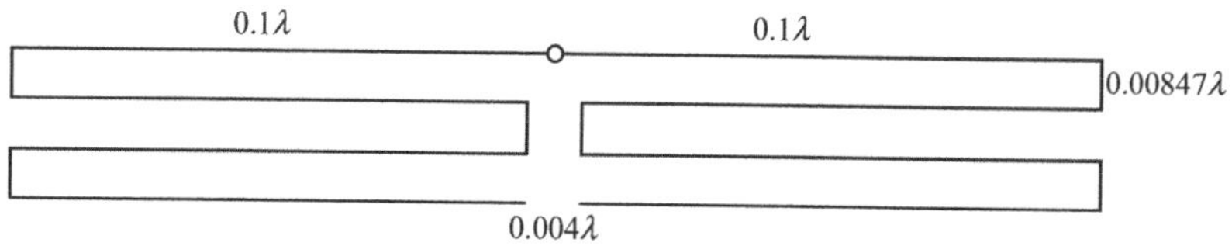

FIGURE 3.23 Four-arm meander antenna, $a = 0.2016$, $ka = 1.267$

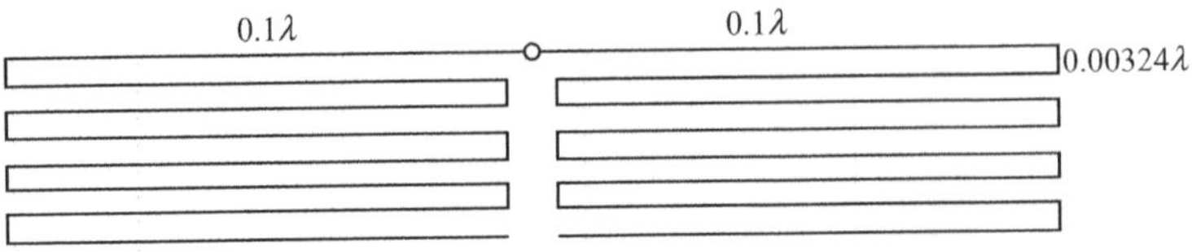

FIGURE 3.24 Eight-arm meander antenna, $a = 0.20128$ $ka = 1.265$.

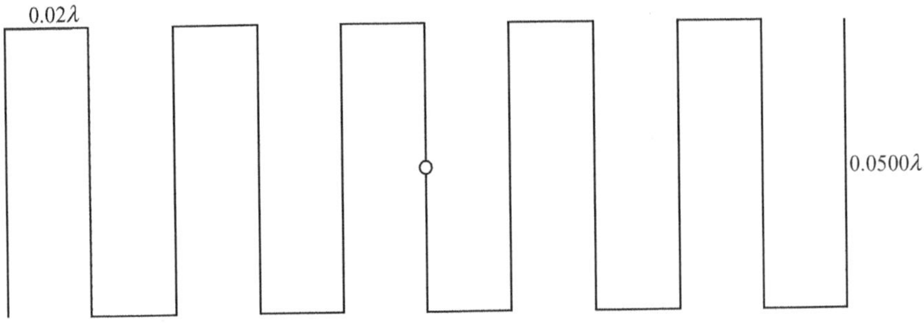

FIGURE 3.25 Ten-section zigzag antenna, $a = 0.10308\lambda$, $ka = 0.6477$.

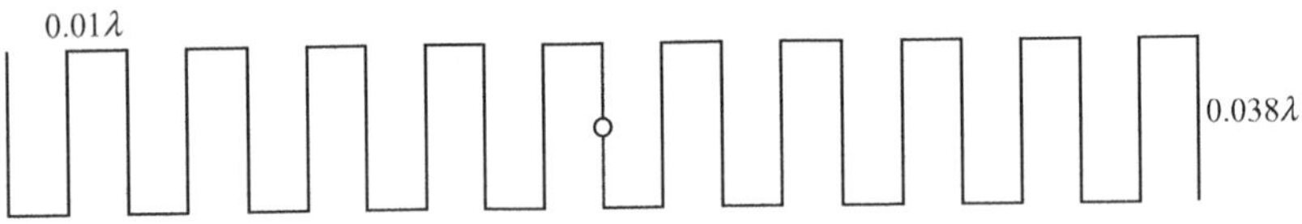

FIGURE 3.26 Twenty-section zigzag antenna, $a = 0.10179\lambda$, $ka = 0.6396$.

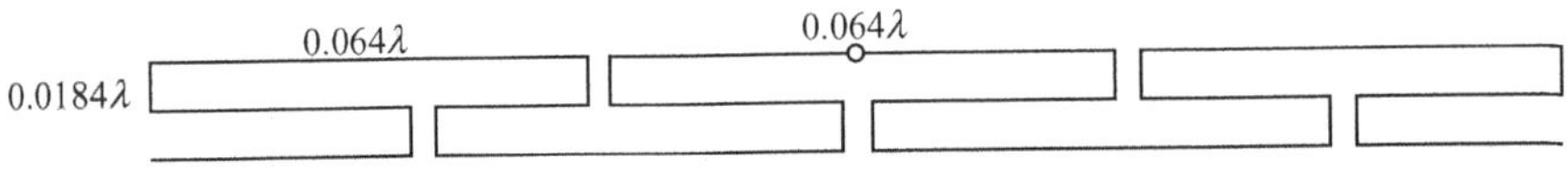

FIGURE 3.27 Three-folded-arn antenna, $a = 0.19549\lambda$, $ka = 1.228$.

antenna. Figure 3.28 shows a four-folded-arm antenna; both are approximately 0.2λ long. Again the width is adjusted to give a resonant $50\,\Omega$ impedance. These folded-arm antennas have three sections along the length of the dipole; one could construct dipoles with five, seven, or more segments but these have not been investigated. Table 3.4 gives the calculated results for these flat dipoles: bandwidth, resonant resistance, width in wavelengths, and wire length in wavelengths. It can be seen that the zigzags are superior, probably because they have no parallel wires with opposing currents, and the extra wires are very short and are cross-polarized. They also have a modest wire length compared to meander dipoles. Current alignment effects are discussed by Best and Morrow (2003).

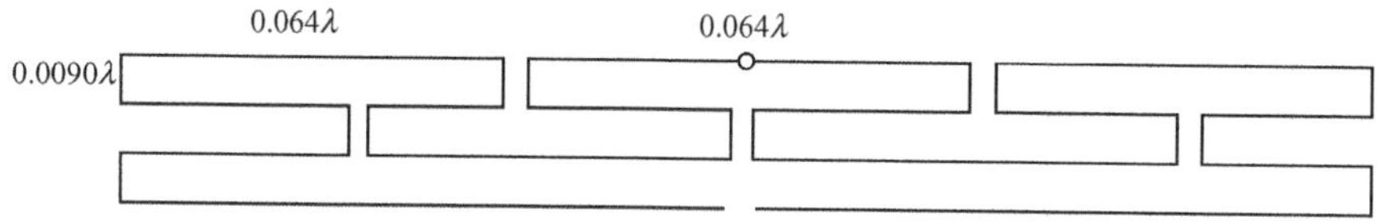

FIGURE 3.28 Four-folded-arm antenna, $a = 0.19389\ \lambda$, $ka = 1.218$.

TABLE 3.4 Short Flat Dipole Results

Dipole	BW (%)	R (Ω)	Width (λ)	Wire length (λ)
Four-arm meander	2.00	8.6	0.025	0.84
Eight-arm meander	2.00	8.7	0.023	1.62
Ten-section zigzag	2.81	16.4	0.050	0.75
Twenty-section zigzag	2.69	15.9	0.038	0.99
Three-folded-arm	2.18	12.8	0.037	0.79
Four-folded-arm	1.82	13.0	0.027	0.89
Simple dipole	1.41	$7.9 - j586$	0.050	0.20

Of the short flat dipoles evaluated, the 10-section zigzag gave the best performance. An option to increase bandwidth is to use lossy wire or lossy strips. Cases were run with the total resistance from 1 to 10 Ω. Table 3.5 gives the bandwidth in percent, the efficiency, and the input resistance in ohms. Also shown is the bandwidth ratio and efficiency ratio, all with respect to the perfectly conducting case. As the loading resistance increased, the bandwidth increased and almost doubled at 10 Ω resistance. The efficiency almost decreased to half. The net result was that the bandwidth–efficiency product was roughly constant at 2.75. As expected, the input resistance increased but did not achieve a 50 Ω value. When efficiency can be traded

TABLE 3.5 Zigzag with Lossy Conductor

R (Ω)	BW (%)	η	BW Ratio	η Ratio	η (dB)	Input (Ω)
0	2.184	0.9747	1		0	16.4
1	2.964	0.9249	1.053	1.054	−0.23	17.3
2	3.194	0.8605	1.135	1.133	−0.54	18.6
3	3.403	0.8046	1.209	1.211	−0.83	19.9
4	3.621	0.7558	1.287	1.290	−1.10	21.2
5	3.838	0.7127	1.364	1.368	−1.36	22.5
6	4.054	0.6743	1.441	1.445	−1.71	23.7
7	4.339	0.6261	1.542	1.557	−2.03	25.6
8	4.608	0.5927	1.638	1.645	−2.27	27.0
9	4.855	0.5722	1.725	1.734	−2.50	28.5
10	5.605	0.5344	1.814	1.824	−2.61	30.0

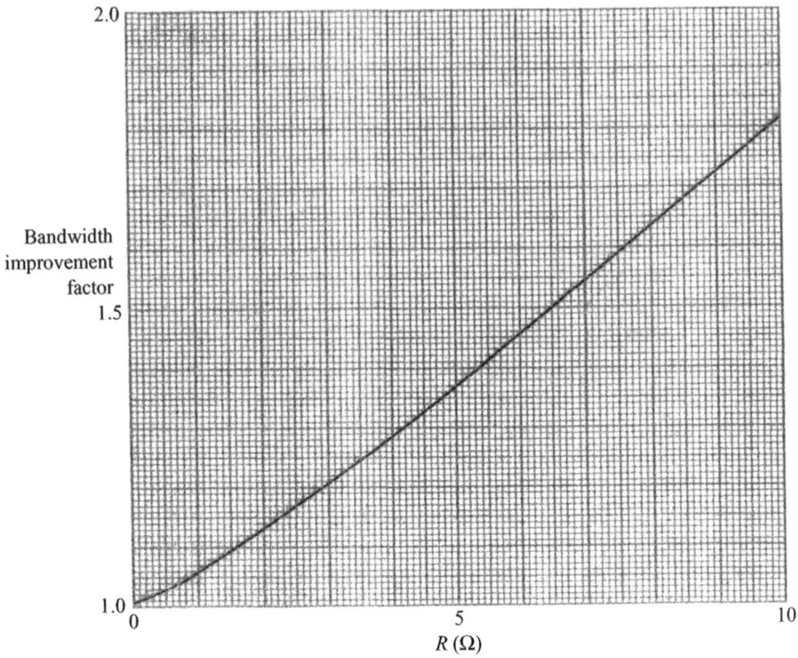

FIGURE 3.29 Zigzag dipole BWIF versus wire resistance.

for bandwidth, use of lossy film conductors is a useful option. Figure 3.29 shows the bandwidth improvement factor versus loading resistance, and for all but very small loading values the improvement is almost linear with resistance.

3.2.5 Spherical Helix Antennas

Any attempt to approach the Chu limit should make full use of the radian sphere. One clever way of doing this is the spherical helix antenna. As developed by Best (2004–2007), the spherical helix can be nonfolded or folded. In the nonfolded one, all of the turns are in the same direction. Figure 3.30 shows a one-turn nonfolded antenna. Best

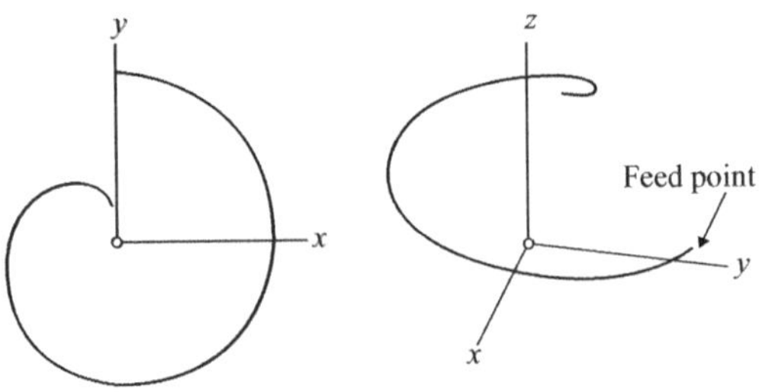

FIGURE 3.30 One-turn, nonfolded spherical helix antenna.

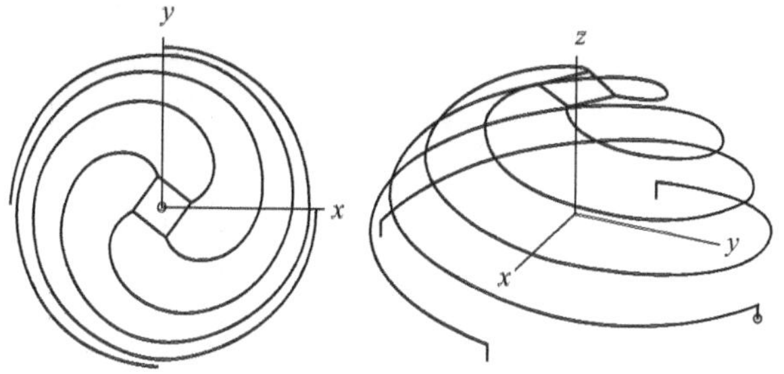

FIGURE 3.31 Depiction of the electrically small one-turn, four-arm, folded spherical helix monopole. The antenna height is 5.77 cm and the wire diameter is 2.6 mm. The antenna exhibits a first series resonance at 300.3 MHz with a resonant resistance of 43.1 Ω.

evaluated these antennas with 1/2, 1, and 3/2 turns. In contrast, the folded spherical helices have one or more arms circularizing in one direction, and then reversing. Figure 3.31 shows a four-arm spherical helix antenna, where each arm makes one turn. It will be seen below that the four-arm antennas perform exceedingly well. These folded antennas exhibit a single resonance and can be readily designed to match 50 Ω. Cross-polarization is small, and the folded spherical helix operates as a dipole. Table 3.6 gives data for both folded and nonfolded spherical helix antennas (Best, 2004). Of primary interest here are Q, ka, and input resistance. All the antennas in the table have $a = 5.89$ cm.

A spherical antenna with one turn in each of two planes provides a more isotropic pattern (Mehdipour et al., 2008). With $ka = 0.624$, it provides $Q = 12.14$, but this is significantly higher than the Chu–Thal limit value, probably due to using only one turn in each plane.

A novel spherical antenna is a magnetic dipole (TE mode) excited by a curved dipole (Kim, 2010). Figure 3.32 sketches this antenna. An eight-turn version has $ka = 0.254$ and $Q = 207$. The Chu–Thal $Q = 195$, so this antenna performance

TABLE 3.6 Spherical Helix Properties

Turns	f (MHz)	Q	ka
Nonfolded			
1/2	468.9	28	0.578
1	274.25	92	0.338
3/2	194.35	184	0.240
Folded: two arms			
1/2	469.3	15.6	0.579
1	284.95	50.0	0.352
3/2	203.80	114.0	0.251
Folded: four arms			
1/2	515.8	5.6	0.636
1	300.3	32.0	0.370
3/2	210.0	88.0	0.259

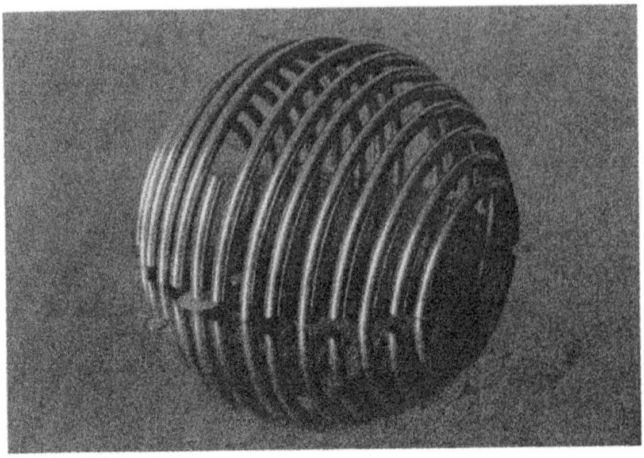

FIGURE 3.32 Fabricated prototype of the SSR antenna. Courtesy of Kim (2010).

FIGURE 3.33 Depiction of the four-arm folded slot spherical helix magnetic dipole. Courtesy of Best (2009).

is within 6% of the fundamental limit. A folded spherical slot antenna with eight arms has $ka = 0.28$ and $Q = 193.8$ (Best, 2009). As the Chu–Thal $Q = 147$, the Q ratio is 1.18. A four-arm version is shown in Figure 3.33.

3.2.6 Multiple Resonance Antennas

3.2.6.1 Spherical Dipole; Arc Antennas A significant advance in spherical antennas was made by Stuart and Tran (2005, 2007). The basic structure is a dipole with spherical arcs connected to the dipole ends; the arcs define the radian sphere. One or two coplanar dipoles are fed; other dipoles plus arcs are parasitically coupled. The four-arm antenna has two dipole arcs in one plane, and two more in a plane at 90°. The six-arm antenna uses three planes, and so on. Figure 3.34 shows the dipole arcs. Figure 3.35 shows four- and six-arm antennas. Stuart (2008) has analyzed these multielement spherical antennas in terms of eigenmodes and eigenvalues. Table 3.7

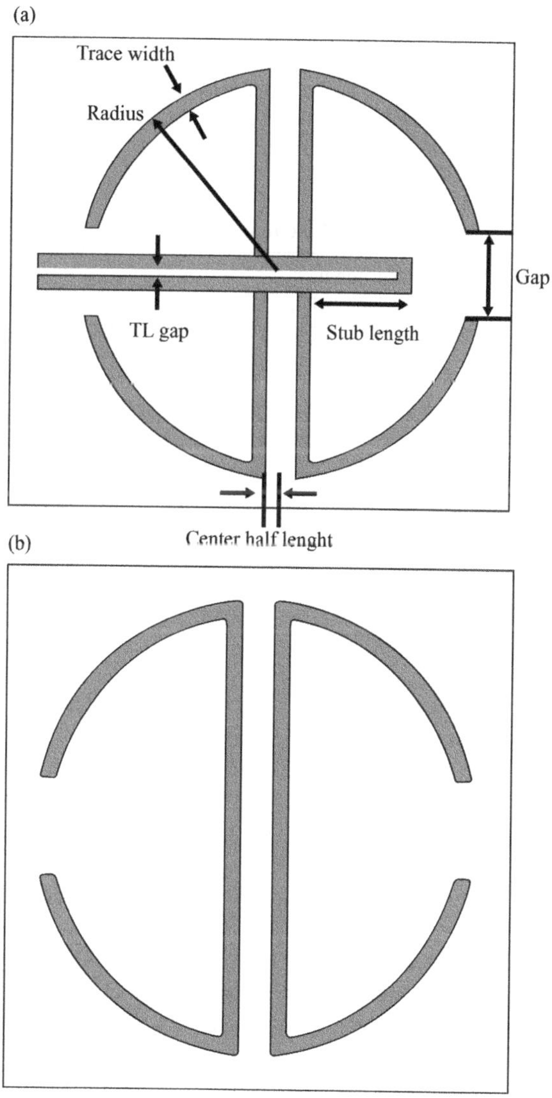

FIGURE 3.34 Dipole arc Antenna.

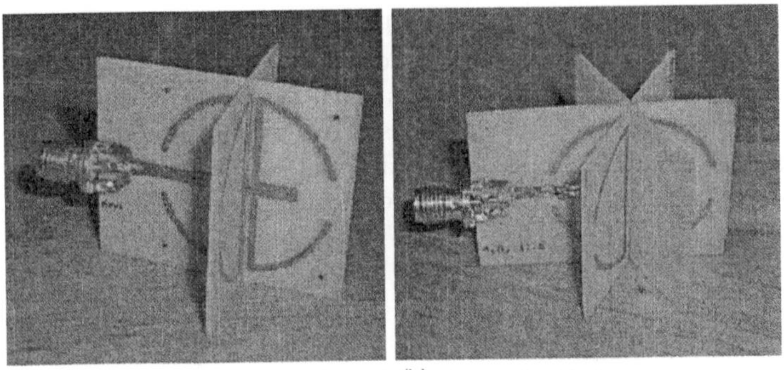

(a) (b)

FIGURE 3.35 Examples of (a) the four-arm antenna and (b) the six-arm antenna. A chip balun is used to interface the balanced feed to the unbalanced coaxial cable. A pair of series chip capacitors are visible along the feed line in the six-arm antenna.

TABLE 3.7 Spherical Antenna Eigenmodes

	Resonant Frequency (MHz)	*ka*	*Q*
One-arm	38.5	0.60	19.6
Two-arm	47.7	0.75	10.6
	54.9		90.2
Four-arm	43.0	0.67	9.61
	55.0		89.60
	57.4		3864.00
Six-arm	40.2	0.63	9.91
	52.7		72.20
	57.6		2422.00
	59.0		13,938.00

gives the resonant frequencies, ka, and Q for one-, two-, four-, and six-arm antennas. The number of resonant frequencies appears to equal the number of arms; the higher resonances appear to have very high Q values. Stuart et al. (2007) have shown that in the vicinity of resonance Q and bandwidth are no longer simply related. For a six-arm antenna, with $ka = 0.54$, the $Q = 13.94$. For this case, the Chu–Thal $Q = 10.84$, so this antenna is 1.29 times the fundamental limit, a very good performance.

3.2.6.2 *Multiple Mode Antennas*

The simple formulas for Q given in Chapter 1 are for an antenna with a single TM or TE mode. Exciting both modes in an ESA increases the bandwidth. As shown by McLean (1996), the Chu Q is now

$$Q = \frac{1}{ka} + \frac{1}{2k^3a^3} \tag{3.20}$$

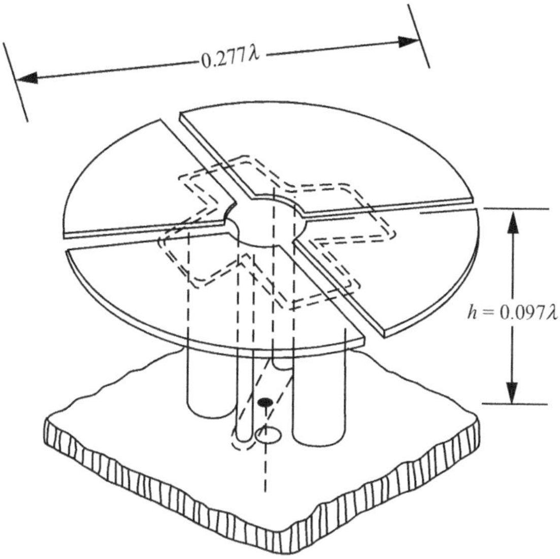

FIGURE 3.36 Goubau antenna.

Attempts to excite both modes using a single-turn loop and a coplanar short dipole have not been successful due to the low radiation resistance of the small loop and the pattern incompatibility. See Section 5.4. An improved configuration uses a capacitively loaded bowtie dipole interconnected with a dual loop (McLean and Crook, 2001).

Grimes and Grimes (1995, 1996) claimed that an antenna exciting both TM and TE modes would provide a Q value below that prescribed by Chu. Alas, there is no free lunch! Collin (1998) integrated the time-dependent Poynting vector to get reactive power, and then used an incorrect energy conservation law. Further, the constant of integration for the integral of the time derivative of standing energy density was incorrectly chosen.

An effective dual-mode design using three dimensions is a cloverleaf dipole with coupling loops over a ground plane (or double cloverleaf dipole without ground plane) developed by Goubau (1976). This antenna, as sketched in Figure 3.36, in symmetric form requires $ka = 1.04$ and gives an octave bandwidth, or $Q = \sqrt{2}$. An improved version has been developed by Friedman (1985). Another version, but with less bandwidth, is developed by Jung and Park (2003).

3.2.6.3 Q Comparisons It is useful to compare the values of Q achieved with various ESA previously described. The most useful comparison is Q versus the Chu–Thal limit, the latter of course being a function of ka (see Equation 1.27). Antennas compared are

- flat meander dipoles;
- flat zigzag dipoles;

- nonfolded spherical helix;
- two-arm folded spherical helix;
- four-arm folded spherical helix;
- two-plane folded spherical antenna;
- multiple resonance spherical antenna.

Figure 3.37 shows the results; the curve is the Chu–Thal limit of Q versus ka. The remarkable accomplishment is that the four-arm spherical helix almost exactly fits the curve. The two-arm folded spherical helix antennas are moderately higher than the curve; the nonfolded spherical helix antennas are moderately higher than the two-arm. For ka>0.5, the multiple resonance antennas have Q close to the limit, as does the two-plane folded antenna. Zigzag models have Q moderately above the Q limit; however, meander antennas have Q way above the limit.

The Goubau antenna is not shown in the figure as its ka is too large (1.04). Reasonable recommendations would be that four- or six-arm (not shown) spherical helix antennas are excellent; the multiple resonance spherical antennas and the two-plane spherical antennas are very good. This, of course, assumes that the spherical volume is available. For flat (planar) antennas, the zigzag dipole is best.

3.2.7 Evaluation of Moment Method Codes for Electrically Small Antennas

The most useful simulation codes for ESA are moment method codes. The emphasis is on piecewise sinusoidal codes as these are superior. The choice of expansion (basis) and test functions is critical. The Galerkin solutions are important, due to reduction of first-order errors. Subroutines for piecewise sinusoidal codes are critical, and are discussed. Finally, seven codes are compared for thin half-wave and short dipoles. These codes include 32- and 64-bit RCH codes, 64-bit MBC code, and 32- and 64-bit NEC-2 and NEC-4 codes. Finally, cautions and recommendations are made. Most ESA involve linear, as opposed to planar, conductors. The moment method for electromagnetics was apparently invented by Mei and Van Bladel (1963). They represented the current on a conductor by a sequence of short, contiguous, constant values (pulses). Then delta functions were used as test functions (point matching). At roughly the same time, Andreasen (1964) represented the current as a sum of entire trig functions: $\sin Ax + \cos Ax +$constant. This was a carryover from his graduate work with R. W. P. King at Harvard. Andreason, at MB Associates, developed the precursor of the NEC code. The NEC-2 and NEC-4 versions use delta test functions. In NEC, the current flows on the conductor surface, while the testing is performed on the conductor axis. The choice of NEC expansion function requires an odd number of functions. Meanwhile, Richmond (1965) at OSU was developing piecewise sinusoidal expansion and test functions, including three-dimensional wire geometries. He used current on the conductor axis, with testing on the conductor surface. His papers date from 1965, but detailed OSU unclassified reports were published circa 1963. Harrington (1967), who erroneously claimed to

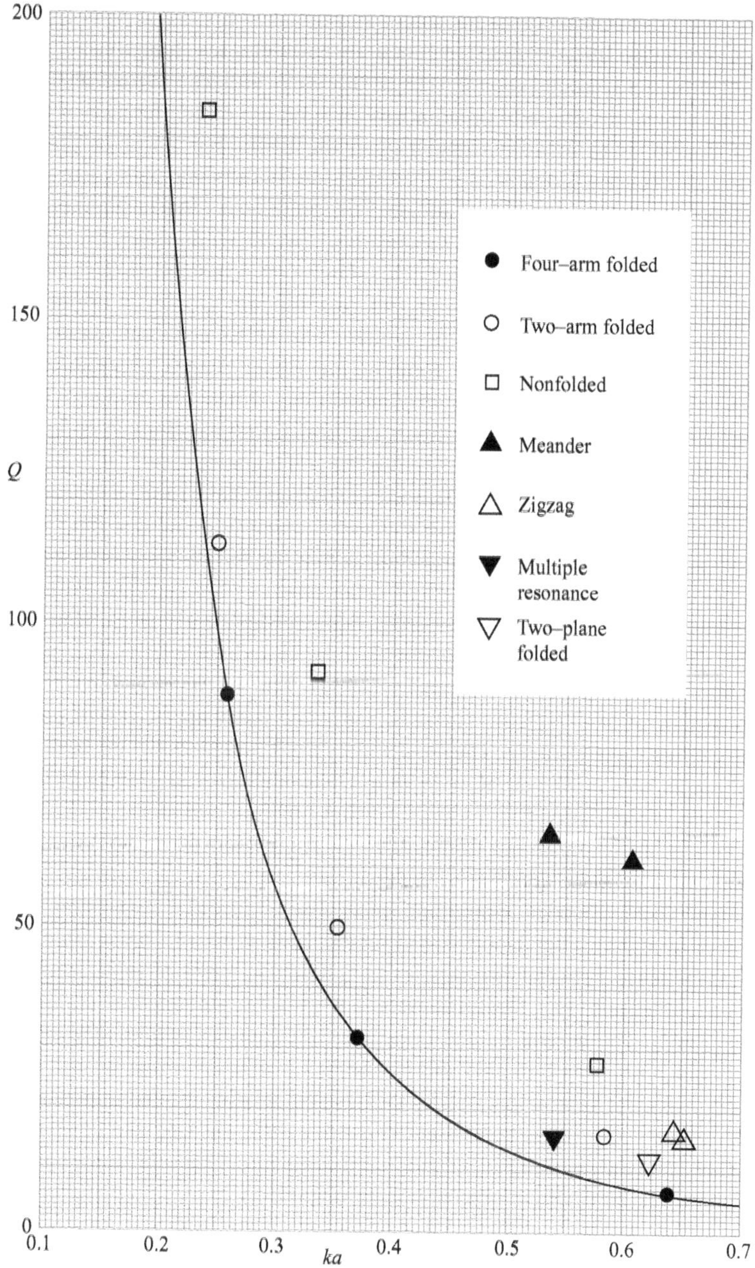

FIGURE 3.37 Q comparisons.

have invented moment methods for antennas, published his first reports circa 1966. He used piecewise linear expansion functions and delta test functions.

The electric field integral equation (EFIE) for wires utilizes a current on the axis of a wire, but calculates the field on a narrow strip on the circumference of the wire. This, of course, is to avoid a singularity in the EFIE. For linear or planar wire structures, this is not a problem as all the offsets are in the same plane. But for three-dimensional structures, the radius offsets will produce a small asymmetry in the resulting impedance matrix. This was resolved by the multiradius bridge current (MBC) code of Tilston and Balmain (1990). This code is basically a Richmond code with piecewise sinusoidal expansion and test functions, with a bridge current modification to produce a symmetric matrix. The MBC code calculates the four monopole–monopole impedances, rotating the offset planes to produce matrix symmetry. The rotation produces discontinuities, which are bridges. For wire antennas and scattering problems, this is the best frequency-domain code.

There are several procedures for solving integral equations, such as least squares, but the most useful is the Galerkin. This uses expansion (basis) and test functions that are the same, for example, piecewise sinusoidal expansion and test functions. The major advantage of the Galerkin procedure is that the results are stationary; that is, results have second-order errors for first-order errors in the currents. Formulations that are not stationary, hence less accurate, include Mei's early work, Harrington's, and NEC. The Galerkin procedure is closely related to the Rayleigh–Ritz procedure and to the reaction concept of Rumsey (1954). For antennas and scatterers that employ flat or curved metallic surfaces, the same rules apply, but of course the calculations are more intricate. The most widely used, and probably the best, moment method codes here utilize the Rao–Wilton–Glisson (Rao et al., 1982) formulation. This Galerkin formulation uses as an expansion function two planar triangles with a common side; the triangles can form a dihedral angle and a faceted 3D structure can be represented by these dihedral triangles.

Moment method codes employing piecewise sinusoidal expansion and test functions use mutual impedances based on the 1932 work of Carter. He expressed mutual impedance between two dipoles in echelon in terms of finite sums of sine and cosine integrals. His formulas were put in form suitable for computers by Hansen (1972). The several mutual impedance subroutines require sine and cosine integrals, and the choice of this subroutine has a marked effect on the resulting accuracy. The basic subroutine used by RCHI in calculating mutual impedances is an echelon dipole code. It has been used with three different sine and cosine integral subroutines, and the resulting accuracies were compared. One subroutine has double-precision results and single-precision argument. It is based on Abramowitz and Stegun (1970). Another subroutine was originally in the IBM Scientific Subroutine Library; it was based on calculations published by Luke and Wimp (1961). This subroutine has output and argument all single precision. A third subroutine is based on coefficients in a book by Luke (1975); this is a completely double-precision routine. Finally, the Richmond OSU code, and the derivative MBC code, used a table of coefficients to calculate exponential integrals; the coefficients are based on Abramowitz and Stegun (1970).

TABLE 3.8

No. of Segments	64-Bit RCH	32-Bit RCH	64-Bit MBC
2	$73.13+j42.17$	$72.73+j42.17$	$73.40+j42.96$
4	$81.19+j41.32$	$80.82+j41.39$	$81.54+j42.10$
6	$82.79+j42.03$	$82.18+j42.19$	$83.14+j42.81$
10	$84.05+j43.17$	$88.05+j40.73$	$84.42+j43.95$
14	$84.61+j43.89$	$66.37+j93.58$	$84.99+j44.67$
20	$85.11+j44.59$	$33.99-j7.54$	$85.48+j45.36$
24	$85.32+j44.91$	$44.81-j16.11$	$85.69+j45.69$
30	$85.55+j45.28$	$72.65+j39.97$	$85.93+j46.05$
34	$85.68+j45.47$	$14.77-j130.1$	$86.06+j46.25$
40	$85.84+j42.71$	$56.40-j345.3$	$86.22+j46.49$
Carter	$73.13+j42.17$		
	Dipole: $l/\lambda=0.5$	$a/\lambda=0.001$	

Comparative results for $L=\lambda/2$.

Comparative results are given in Tables 3.8 and 3.9. Table 3.8 gives dipole impedance for dipole lengths of 0.5λ and Table 3.9 for dipole lengths of 0.05λ, and $a=0.001\lambda$, for RCH codes using the best 32- and 64-bit sine and cosine integral subroutine. Results from the 64-bit MBC code are also shown. Subroutines tested include a 32-bit code from the IBM Scientific Subroutine Library, a 32-bit code from Abramowitz and Stegun, and a 64-bit code from Luke (1975). The IBM/Luke code was the best of the 32-bit codes, but as Table 3.8 shows, results are useful only for roughly 10 or less segments, both for resonant and for short dipoles. The 64-bit subroutines, which use Chebyshev expansions via Luke (RCH code) and numerical integration (MBC code), give excellent results. $N=2$ gives the Carter impedances.

Tables 3.10 and 3.11 give results for NEC-2 and NEC-4 codes, for both 32- and 64-bit versions. For the half wave dipole, all four NEC results are good and are close

TABLE 3.9

No. of Segments	64-Bit RCH	32-Bit RCH	64-Bit MBC
2	$0.495-j1724$	$0.520-j1724$	$0.507-j1722$
4	$0.420-j1695$	$0.395-j1695$	$0.430-j1693$
6	$0.396-j1637$	$0.671-j1637$	$0.406-j1634$
10	$0.364-j1561$	$3.21-j1561$	$0.374-j1558$
14	$0.341-j1500$	$2.22-j1500$	$0.349-j1498$
20	$0.310-j1420$	$0.752-j1418$	$0.317-j1418$
24	$0.290-j1369$	$40.1-j1364$	$0.297-j1367$
30	$0.260-j1290$	$173.0-j1319$	$0.266-j1289$
34	$0.239-j1236$	$14.7-j1242$	$0.245-j1234$
40	$0.209-j1150$	$214-j1510$	$0.214-j1148$
Carter	$0.495-j1095$		
	Dipole: $l/\lambda=0.05$	$a/\lambda=0.001$	

Comparative results for $L=\lambda/20$.

TABLE 3.10

No. of Segments	32-Bit NEC-2	64-Bit NEC-2	32-Bit NEC-4	64-Bit NEC-4
3	81.25 43.89	81.25 43.89	81.25 43.89	81.25 + j43.89
5	81.95 45.54	81.95 45.54	81.95 45.54	81.95 + j45.54
7	82.70 46.31	82.70 46.31	82.70 46.31	82.70 + j46.31
11	83.67 47.13	83.67 47.13	83.07 47.12	83.67 + j47.12
15	84.26 47.58	84.26 47.59	84.26 47.59	84.26 + j47.59
21	84.83 48.13	84.82 48.03	84.82 48.02	84.82 + j48.02
25	85.09 48.33	85.08 48.24	85.08 48.22	85.08 + j48.22
31	85.38 48.57	85.38 48 46	85.38 48.45	85.38 + j48.45
35	85.60 49.00	85.54 48.58	85.53 48.57	85.53 + j48.56
41	85.65 47.72	85.73 48.72	85.72 48.70	85.72 + j48.70
	Dipole:	$l/\lambda = 0.5$	$a/\lambda = 0.001$	

Comparative results for $L = \lambda/2$.

TABLE 3.11

No. of Segments	32-Bit NEC-2	64-Bit NEC-2	32-Bit NEC-4	64-Bit NEC-4
3	0.0710 −2188	0.07101 −2188	0.07100 −2189	0.07100 − j2189
5	0.0596 −1963	0.05966 −1965	0.05963 −1966	0.05963 − j1966
7	0.05418 −1848	0.05428 −1859	0.05420 −1861	0.05420 − j1861
11	0.04838 −1741	0.04837 −1743	0.04813 −1746	0.04813 − j1746
15	0.04472 −1664	0.04471 −1670	0.04428 −1676	0.04428 − j1676
21	0.04102 −1579	0.04073 −1589	0.04002 −1599	0.04002 − j1599
25	0.03851 −1544	0.03853 −1543	0.03768 −1554	0.03768 − j1554
31	0.03548 −1472	0.03551 −1478	0.03453 −1489	0.03453 − j1489
35	0.03422 −1446	0.03357 −1435	0.03254 −1447	0.03254 − j1447
41	0.03006 −1409	0.03067 −1369	0.02961 −1380	0.02961 − j1380
	Dipole:	$l/\lambda = 0.05$	$a/\lambda = 0.001$	

Comparative results for $L = \lambda/20$.

to the piecewise sinusoidal expansion function results of Table 3.8. However, data (not shown) indicate that 32-bit NEC-2 for 81 and 101 segments are unstable. NEC-2 64-bit is all right for these larger segments numbers. For short dipoles, things are different. The values of input resistance are way off: roughly 0.03–0.07 instead of the correct values, roughly 0.2–0.5 Ω. Reactance values are also slightly off. Clearly, all NEC codes should be used with caution for electrically short antennas.

These results indicate why the choice of codes is critical. NEC-2, for example, is poor for short dipoles, and poor for many segments. Commercial variations on NEC-2 and NEC-4 must be carefully examined, and probably should not be used. NEC-4 is superb for earth or substrate problems where Sommerfeld integrals are involved, but is not the best choice for other geometries. When many expansion functions are needed, the 32-bit codes must be used carefully. In Table 3.8, results are good for half-wave dipoles with number of segments ≤10; for short dipoles, 64-bit codes are recommended. The NEC reactances are somewhat different from those calculated with piecewise sinusoidal Galerkin codes. This is probably due to the NEC use of delta "function" (point matching) test functions. All moment method codes are less accurate (Cebik, 1998), and results must be examined carefully, when

- wires have large steps in radius;
- wires are joined with a small included angle;
- parallel wires are closely spaced.

3.3 PARTIAL SLEEVE, PIFA, AND PATCH

3.3.1 Partial Sleeve

The partial sleeve antenna, which is a two-dimensional transmission line antenna, was developed by Nash and others at the University of Illinois Antenna Lab in the early 1950s, but data were published only in reports to USAF ASD. The antenna consists of a square or rectangular plate located adjacent to a ground plane and shorted to the ground plane. A feed is connected at a point that gives a convenient impedance level, as shown in Figure 3.38. A capacitor may be used to tune the end of the sleeve. The ground plane and the sleeve may be curved in either dimension.

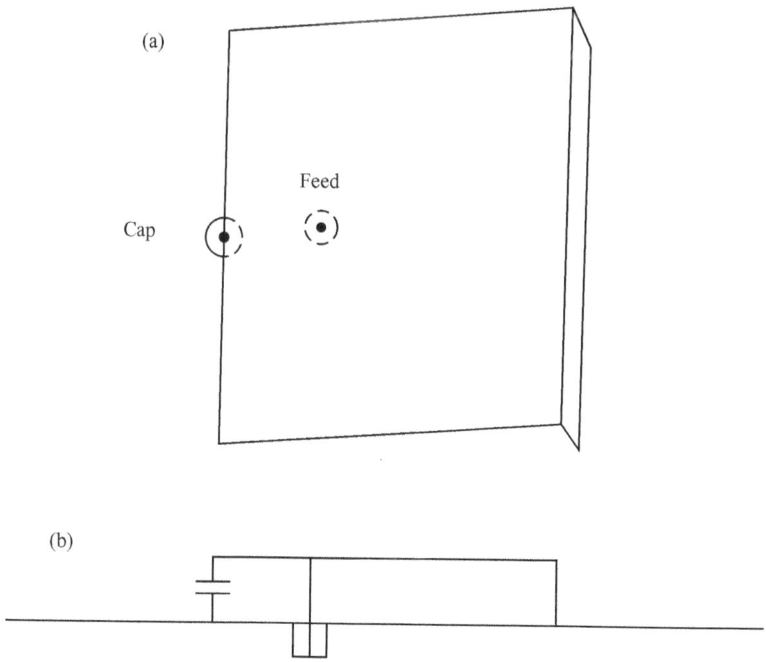

FIGURE 3.38 Partial sleeve antenna: (a) top view; (b) side view.

The term sleeve arose as such an antenna may be used on a curved aircraft surface. Typical sizes are separation 1/10 of the length, and length from 0.01λ to 0.1λ. The radiating portion of the antenna consists of a U-shaped slot, which for small lengths will have a nearly triangular distribution of current. Radiation from the two side slots will mostly cancel out, leaving the end slot as the primary source. However, the side slots act as a form of loading to increase the current moment of the primary radiator. Assuming a square plate or sleeve, the current moment of the end slot is $5L/6$, which is 1.67 times better than that of a short dipole of length L. This is the same improvement factor (2.78 in radiation resistance) produced by the dipole with optimum inductive loading in the dipole. Both the transmission line antenna and the partial sleeve are useful when $\lambda/4$ of length is available but are not attractive when short because of the low impedance.

The important fact about the partial sleeve is that it led to the invention of the ubiquitous patch antenna by Deschamps and Sichak (1953). It is also the PIFA (planar inverted-F antenna): a flat plate shorted at one end and fed near the open end. See Section 3.3.2.

3.3.2 PIFA Designs

As previously mentioned, the PIFA is simply a small flat plate parallel to and closely spaced to a ground plane, and shorted to the ground plane at one end. A feed is located along the plate at a point to provide the desired impedance. Figure 3.39 sketches a PIFA with a shorting tab at one corner of the antenna. Taga (Hirasawa and Haneishi, 1992) has given extensive design data for the basic patch. This includes electric field distributions over the PIFA as a function of aspect ratio. Figure 3.40 gives bandwidth (for VSWR $\leq$ 1.5) versus height of the plate above the ground screen for several aspect ratios. The shorting plate is along one edge. Pinhas and Shtrikman (1988) also give data on bandwidth versus height for several widths. Figure 3.41 gives relative bandwidth for a square plate as a function of the width of the shorting tab and the height of the plate above the ground screen. He analyzed

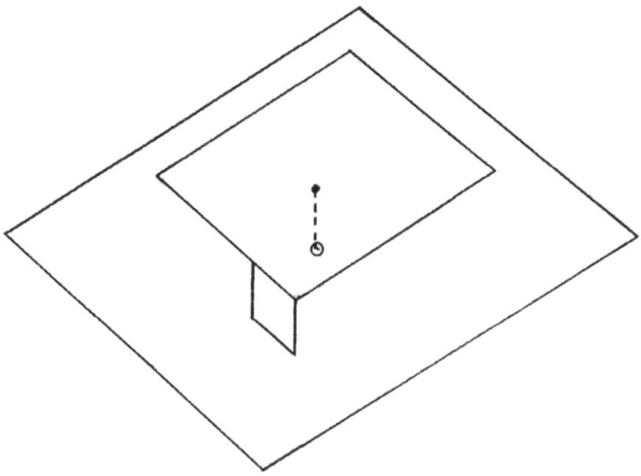

FIGURE 3.39 PIFA with tab short. Courtesy of Hirasawa and Haneishi (1992).

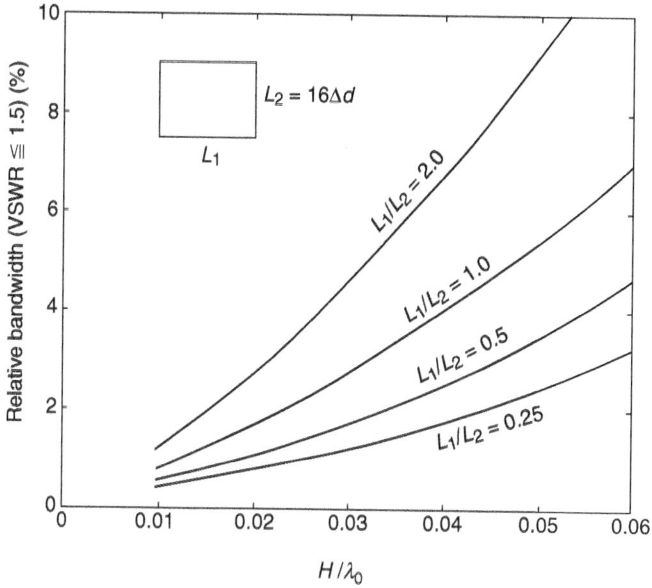

FIGURE 3.40 Bandwidth versus aspect ratio. Courtesy of Hirasawa and Haneishi (1992).

these antennas by representing the antenna as a three-dimensional lattice network, solved in the time domain. Figure 3.42 shows normalized resonant frequency of a square plate PIFA versus width of the shorting strap compared to the width of the plate. It can be seen that the calculations and measurements agree well.

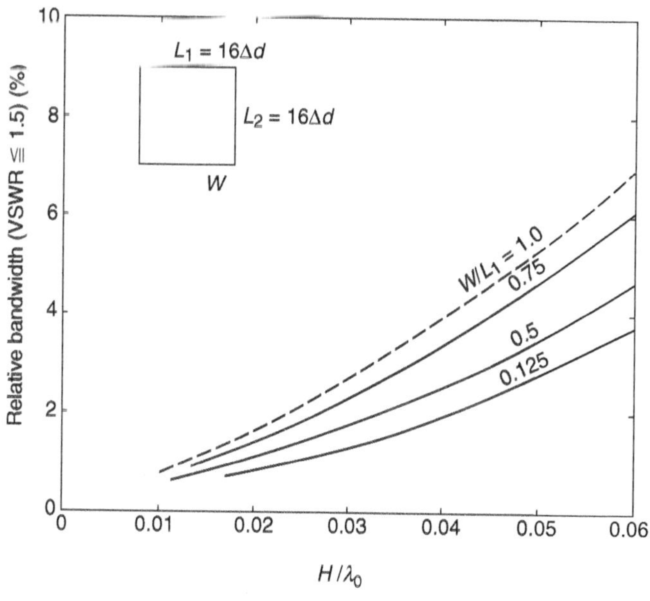

FIGURE 3.41 Bandwidth versus tab width. Courtesy of Hirasawa and Haneishi (1992).

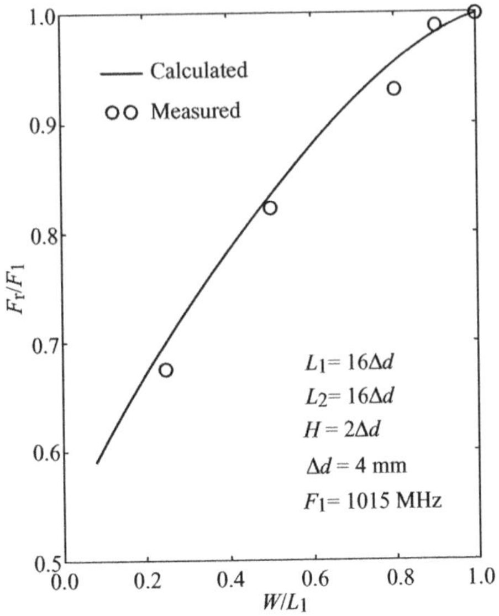

FIGURE 3.42 Resonant frequency versus shorting tab width. Courtesy of Hirasawa and Haneishi (1992).

In general, the best way of numerically analyzing a PIFA is through the use of a patch-type moment method code. This allows the plate and shorting straps to be accurately represented. And of course the probe feed can be located as desired. This applies to antennas on a large ground plane. Of course, most PIFA are employed on handheld or portable electronic devices where the ground plane is not flat and not large in wavelengths. When the feed terminal is accessible, measurements can be made with a conventional network analyzer.

Most applications for PIFA are not single frequency but multiband. There are many two-, three-, and four-band designs. Bands of interest are 824–894 MHz (GSM), 1710–1880 MHz (GSM), 2.4–2.5 GHz (Bluetooth and WLAN), 5.15–5.25 GHz (WLAN), and 8–12 GHz (SATCOM). Dual- or multiple-band PIFA can be designed in either of the two ways. First, individual PIFA of successively smaller size can be nested together to provide multiple band coverage. Figure 3.43 is an

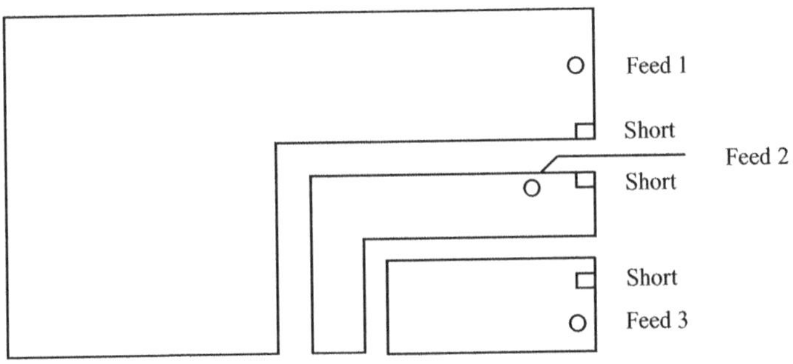

FIGURE 3.43 Three-band PIFA. Courtesy of Wong (2003).

example of a three-band PIFA covering roughly 900, 1800, and 2450 MHz (Wong, 2003). The contrasting approach, which is more commonly used, incorporates one or more slots in the PIFA and perhaps divides part of the PIFA into fingers. Figure 3.44 shows a two-band PIFA employing a single L-shaped slot, for a 1.8 and 2.45 GHz band (Wong, 2002). Another dual-band design is shown in Figure 3.45 where there are now two slots, one L-shaped and one reentrant (Hsiao et al., 2002). This is a 900–1800 MHz design. A PIFA with a U-shaped slot is developed by Salonen et al. (2000). A folded design where the folded section acts as a capacitor with a ground plane provides dual band at 900 and 2300 MHz (Villeger et al., 2003) (Figure 3.46). A quad-band design, shown in Figure 3.47, utilizes a capacitor end load plate, and three U-shaped slots in the top plate. This covers bands at roughly 900, 1800, 2450, and 5200 MHz (Nashaat et al., 2005). A more radical design for quad-band incorporates one finger and a truncation of one side of the PIFA (Figure 3.48). By appropriately adjusting the dimensions, this covers 2, 5, and 8–12 GHz (Sim and Cheng, 2010). For additional information, good references are the Hirasawa–Haneishi book referenced earlier and books by K. L. Wong (2002, 2003).

FIGURE 3.44 PIFA with L slot. Courtesy of Wong (2002).

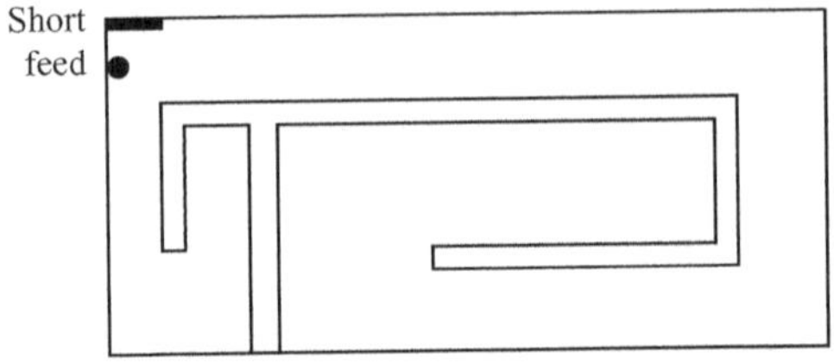

FIGURE 3.45 PIFA with folded and L slots. Courtesy of Hsiao et al. (2002).

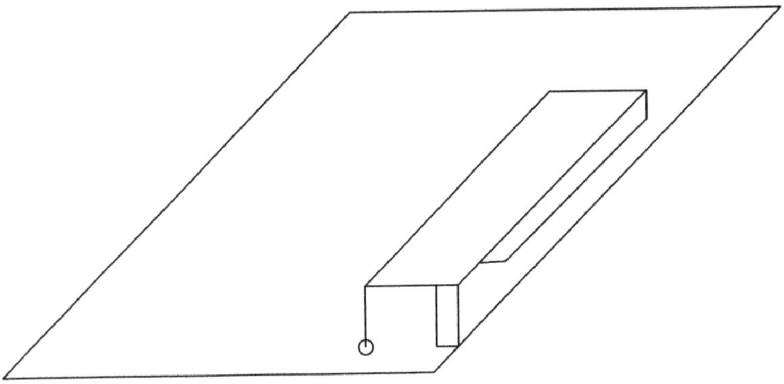

FIGURE 3.46 Folded PIFA. Courtesy of Villeger et al. (2003).

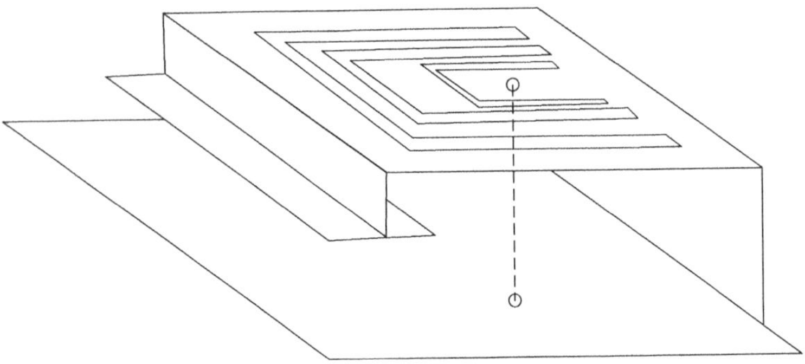

FIGURE 3.47 Quad-band PIFA. Courtesy of Nashaat et al. (2005).

3.3.3 Patch with Permeable Substrate

The conventional patch antenna is not electrically small, but it may be so when constructed with a high-ε substrate. Ceramic substrates embodying barium, strontium, and calcium titanates, among others, may have a large range of ε, perhaps

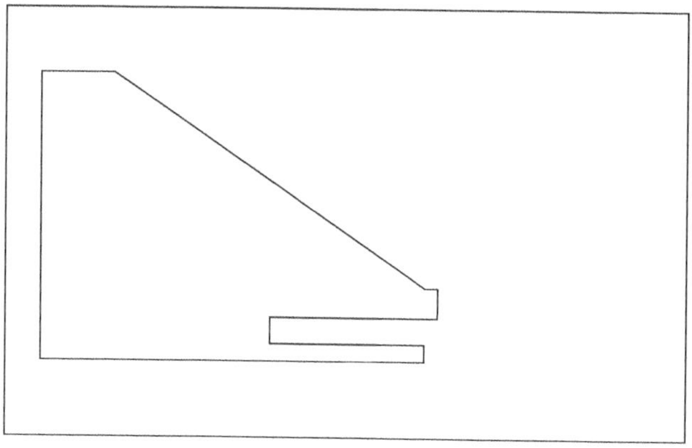

FIGURE 3.48 Truncated quad-band PIFA. Courtesy of Sim and Cheng (2010).

ε>100 (Middleton, 2002). When used as a substrate for a patch, the resonant width is decreased by $\sqrt{\varepsilon}$. Such a patch can be electrically very small, and along with the reduced size goes a narrower bandwidth and tighter tolerances. Bandwidth, for VSWR ≤ 2, is approximately $4t(\sqrt{2\varepsilon_r}\lambda_0)$, where t is the substrate thickness. A thorough treatment of patch antennas is given by Balanis (2005). A closely related antenna, the dielectric resonator antenna (DRA), is discussed in Section 3.5.

What if a low-loss material with both μ and ε is used as a patch substrate; how would the characteristics change? The first answer was given by Hansen and Burke (2000), where a square patch with μ, ε substrate was analyzed via a transmission line model. This zero-order analysis assumed that the patch length was resonant, $a/\lambda = 1/\sqrt{2\mu\varepsilon}$, and the edge susceptance was omitted. The edge (radiation) conductance is given by (Wheeler, 1965)

$$G = \frac{1}{120\pi^2}\int_0^\pi \frac{\sin^2[(\pi/2\sqrt{\mu\varepsilon})\cos\theta]\sin^3\theta\, d\theta}{\cos^2\theta} \tag{3.21}$$

In many works, it is stated that an approximation valid for small a/λ (large $\mu\varepsilon$) is

$$G \simeq \frac{a^2}{90\lambda^2} = \frac{1}{360\mu\varepsilon} \tag{3.22}$$

A similar approximation stated to be valid for large a/λ (small $\mu\varepsilon$) is

$$G \simeq \frac{a}{120\lambda^2} = \frac{1}{240\sqrt{\mu\varepsilon}} \tag{3.23}$$

It will be shown that Equation 3.23 for large $\mu\varepsilon$ is good, whereas Equation 3.22 for small $\mu\varepsilon$ is poor. Figure 3.49 shows the exact integral, along with the two

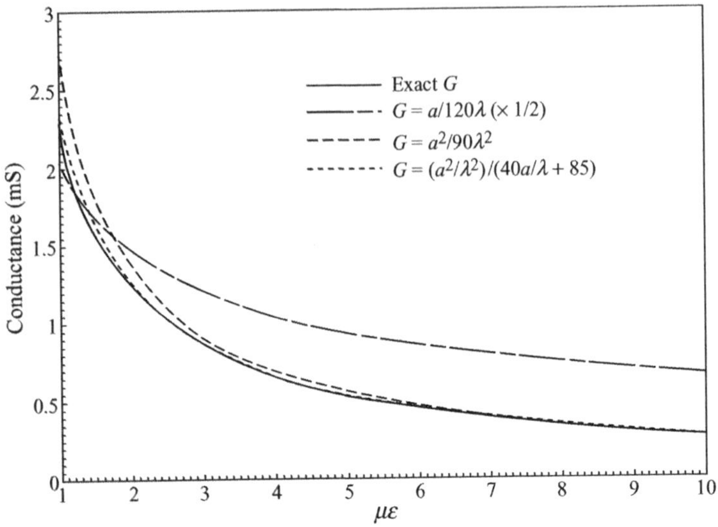

FIGURE 3.49 Exact and approximate square patch radiation conductance. Courtesy of Hansen and Burke (2000).

approximations, for $\mu\varepsilon = 1\text{–}10$; a 128-point Gaussian integrator was used. In this figure, half of the conductance is used as the patch has two radiating slots. Note that the large $\mu\varepsilon$ approximation is useful for $\mu\varepsilon > 3$. The small approximation is good for $\mu\varepsilon < 1$. Even with an added factor of 1/2 as shown in Figure 3.49, the approximation is poor for $\mu\varepsilon = 1$. A good fit is given for the entire range of $\mu\varepsilon = 1\text{–}10$ by

$$G \simeq \frac{1}{40\sqrt{\mu\varepsilon} + 170\mu\varepsilon} \tag{3.24}$$

This conductance will be used with the transmission line analysis. From Wheeler (1965), the characteristic admittance of a wide microstrip line is

$$Y_0 = \frac{a\sqrt{\varepsilon}}{\eta t\sqrt{\mu}} = \frac{\lambda}{2\eta\mu t} \tag{3.25}$$

where $\eta = 120\pi$ and t is the dielectric thickness. It should be noted that, for the resonant patch, the characteristic impedance involves only μ. The radiation quality factor is $Q = \pi Y_0/4G$ (Wheeler, 1947), and the VSWR = 2 bandwidth BW is $1/\sqrt{2}Q$ (with matched load). The result is that the zero-order bandwidth is

$$\mathrm{BW} = \frac{96\sqrt{\mu/\varepsilon}\,t/\lambda_0}{\sqrt{2}[4 + 17\sqrt{\mu\varepsilon}]} \tag{3.26}$$

Unfortunately, this result is not simply a function of μ/ε. The bandwidth result (Balanis, 2005; Hansen, 1998)

$$\mathrm{BW} \simeq \frac{4\sqrt{\mu t/\lambda_0}}{\sqrt{2\varepsilon}} \tag{3.27}$$

is seen to be fair for $\varepsilon = 1{-}2$ and poor for $\varepsilon > 2$. The other approximate result

$$\mathrm{BW} \simeq \frac{16t\lambda_0}{3\sqrt{2\varepsilon}} \tag{3.28}$$

is fair for $\varepsilon > 4$ and poor for $\varepsilon < 4$. However, Equation 3.26 is accurate for the entire range of $\mu\varepsilon$.

The resonant patch length is reduced by $\sqrt{\mu\varepsilon}$. Two cases are of interest.

- *Case 1*: $\mu = \varepsilon$. If μ is put equal to ε and the small $\mu\varepsilon$ result were applicable, then this shortened patch would enjoy the same bandwidth as the air larger patch. A more accurate evaluation is obtained by setting $\mu = \varepsilon$ in Equation 3.26; the result is

$$\mathrm{BW} \simeq \frac{96t/\lambda_0}{\sqrt{2}[4+17\varepsilon]} \tag{3.29}$$

  A comparison of values from Equation 3.26 with Equation 3.29 where $\varepsilon = 10$ and $\mu = 1$ shows that only a small improvement in bandwidth results. Specifically, for $\mu = \varepsilon = 10$, the improvement in bandwidth is only 5%. However, the patch length is shortened by $1/\mu$.
- *Case 2*: $\varepsilon = 1$, $\mu > 1$. A substrate with $\varepsilon = 1$ and $\mu > 1$ would offer a modest increase in bandwidth over the $\varepsilon = 1 = \mu$ patch. For $\mu \gg 1$, the increase in bandwidth would be $21/17 = 1.235$. Compared to a typical patch with $\varepsilon = 3$ and $\mu = 1$, the improvement in bandwidth is 330%.
  Metaferrites is the term given to a lamination of layers, where each thin layer has a pattern of short segments of a mixture of high-μ materials. The objective is to realize a large μ and a modest ε, with low loss (Walser, 2001).

3.4 LOOPS

3.4.1 Air Core Loops, Single and Multiple Turns

It is usually desirable for small loops to have a pattern that is omnidirectional in the plane of the loop. This requires the loop diameter D to be less than roughly 0.1λ (Balanis, 2005). When the loop has multiple turns, the winding length must be $\leq 0.1\lambda$. The equivalent circuit is simply the inductive reactance X in series with the radiation resistance R_r and the loss resistance R_ℓ. For a loop with N turns, the radiation resistance is proportional to N^2 and to the square of effective permeability:

$$R_r = \frac{5\pi^2 N^2 k^4 D^4 \mu_e^2}{4} \tag{3.30}$$

Magnetic core effects are discussed below. The single-turn loop will be treated first. Now a single conductor, single turn, will act as a very thin tube of conductor because of skin effect. Rosa and Grover (1916) give the inductance of an infinitely thin tube bent into a circle, from which the reactance in ohms is

$$X = 60\pi k D\left[\left(1+\frac{a^2}{D^2}\right)\ell n\frac{4D}{a}-2\right] \tag{3.31}$$

Here D is the mean diameter of the loop and a is the tube radius. Most practical single-turn loops will have $D \gg a$, so the reactance is well approximated by

$$X = 60\pi k D\left(\ell n\frac{4D}{a}-2\right) \tag{3.32}$$

If the loop diameter is much larger than the wire diameter, the proximity effect, which tends to concentrate the current on the inside of the loop, can be neglected (Smith, 1972). Then

$$R_\ell = \frac{D}{2a\sigma\delta} = \frac{R_s D}{2a} \tag{3.33}$$

Here R_s is the surface resistivity, in $\Omega/\square$. Figure 3.50 gives $R_s D/a$ versus D/a and frequency. The efficiency formula is the same as for dipoles:

$$\eta = \frac{R_r}{R_r + R_\ell} \tag{3.34}$$

Immediately, the efficiency is found to be

$$\eta = \frac{5\pi^2 k^4 D^4}{5\pi^2 k^4 D^4 + 2R_s D/a} \tag{3.35}$$

Efficiency is shown in Figure 3.51 for $D/a = 50$. The Q is given by

$$Q = \frac{240\pi k D[\ell n(4D/a)-2]}{5\pi^2 k^4 D^4 + 2R_s D/a} \tag{3.36}$$

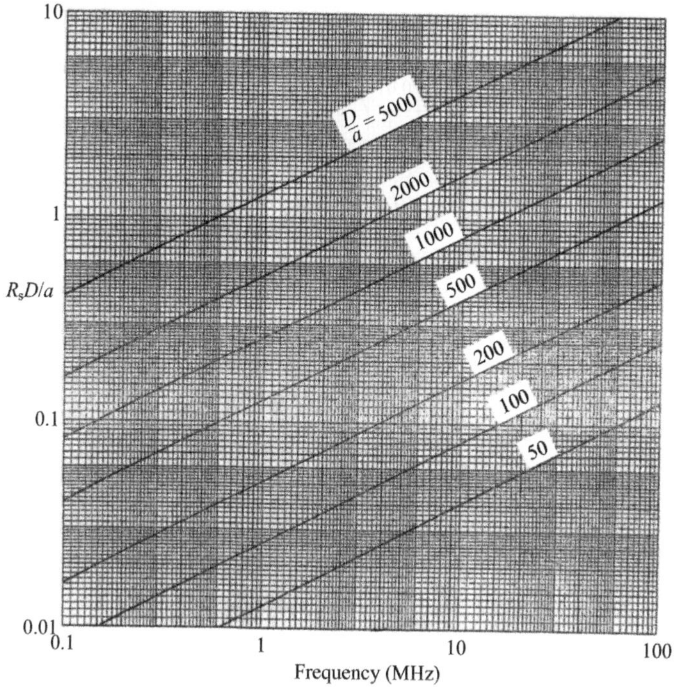

FIGURE 3.50 Loss factor for copper.

Because small loops have high Q, half-power bandwidth becomes

$$\mathrm{BW} \simeq 1/Q \tag{3.37}$$

Half-power bandwidth is given in Figure 3.52. It is a minimum for

$$\frac{D}{\lambda} = \frac{1}{2\pi}\left[\frac{2R_sD/a}{15\pi^2}\right]^{1/4} \tag{3.38}$$

The bandwidth × efficiency merit factor is

$$\mathrm{BW} \cdot \eta = \frac{R_r}{X} = \frac{\pi k^3 D^3}{48[\ell n(4D/a) - 2]} \tag{3.39}$$

For the largest feasible small loop, $D/\lambda = 0.1$, and for $D/a = 50$, the bandwidth × efficiency product is

$$\mathrm{BW} \cdot \eta = 0.004922 \tag{3.40}$$

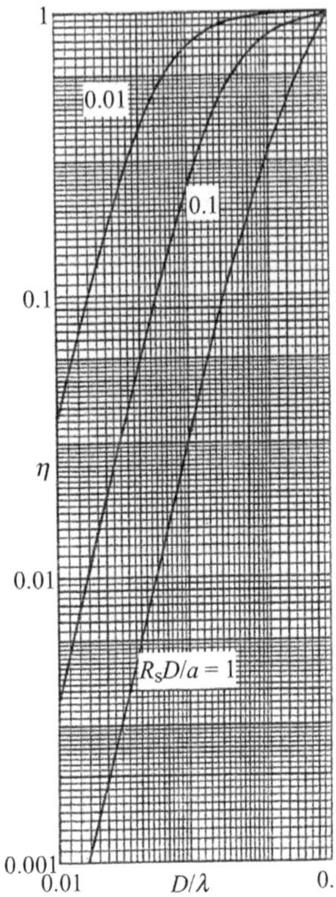

FIGURE 3.51 Loop efficiency, $D/a = 50$.

Figure 3.53 shows R_r, X, and $\mathrm{BW} \cdot \eta$ for $D/a = 50$. The range of D/λ is 0.01–0.1. As expected, the radiation resistance and bandwidth × efficiency product are small. Because the loop resistivity enters into both, data are shown for $R_s D/a = 1, 0.1$, and 0.01. All of these graphs are for $D/a = 50$; the R_s values are 0.02, 0.002, and 0.0002. For copper, these values cover the range of 0.6 MHz to 6 GHz.

The air core loop may be squished into a narrow rectangle to allow convenient mounting on a platform. It must be remembered that loop performance is proportional to area.

Since the total wire length in a loop is limited, it is useful to write the radiation resistance in terms of wire length (WL):

$$R_r = \mathrm{WL}\frac{kD}{4} \tag{3.41}$$

When the maximum WL is used, the kD factor means that D/λ should be as large as possible, that is, a single-turn loop. Nonetheless, physical constraints sometimes require a smaller diameter, multiturn loop.

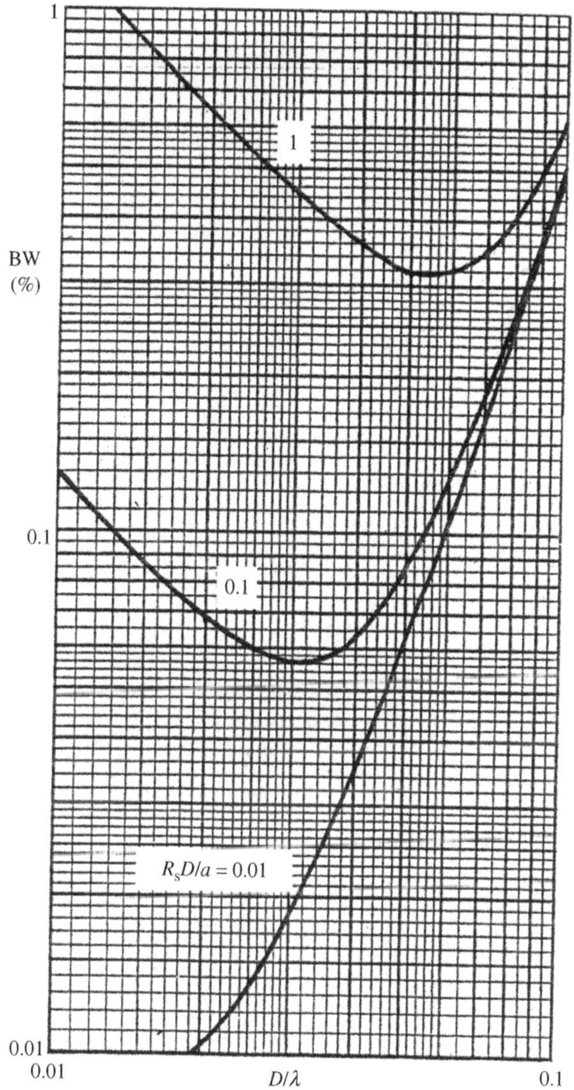

FIGURE 3.52 Loop bandwidth, $D/a = 50$.

A typical multiturn air core loop is either circular or square in shape, with a winding of rectangular cross section, that is, many layers with each layer having many turns. Such a loop is a short multilayer solenoid, with reactance given approximately by

$$X = 60\pi N^2 kD\left(\ell n\frac{9D}{b} - 1\right) \tag{3.42}$$

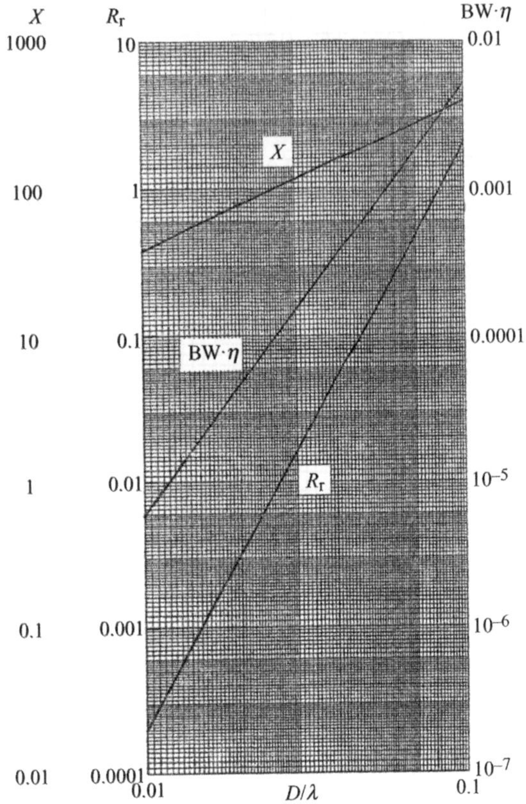

FIGURE 3.53 Loop parameters, $D/a = 50$.

Solenoid diameter and axial length are D and b, respectively. This result, derived by Maxwell (Rosa and Grover, 1916), is adequate for comparative purposes. Such coils are usually designed for optimum Q; the only parameters that significantly affect radiation resistance are diameter and number of turns. Various formulas are available for optimum design of multilayer solenoids. That of Butterworth (Terman, 1943) requires b, D, and winding radial thickness t to obey

$$3t + 2b = D \tag{3.43}$$

Examination of the loss formulas for short multilayer solenoids shows that the loss of the optimum designs varies slowly as the ratio t/b changes around unity. Because a coil with square winding cross section is physically attractive, the loss results will be quoted for $t = b$. Then for the optimum, $t/D = 0.2$ and the Butterworth parameter KbD. Assuming the wires are spaced with one wire diameter air gap, the ratio of AC and DC resistance, using Butterworth's H and G functions, is

$$\frac{R_{\text{ac}}}{R_{\text{dc}}} = H + 0.49GN \simeq H + \frac{GN}{2} \tag{3.44}$$

If now the wire diameter is large in skin depths, which may not be a valid assumption but yields an upper bound, H and G may be approximated by $H \simeq a/2\delta \simeq 2G$, giving

$$\frac{R_{\text{ac}}}{R_{\text{dc}}} = \frac{a}{2\delta}\left(1 + \frac{N}{4}\right) \tag{3.45}$$

The lower limit is, of course, $R_{\text{ac}} = R_{\text{dc}}$. $R_{\text{dc}} = ND/a^2\sigma$, which gives the coil resistance

$$R_\ell = \frac{NR_sD}{2a}\left(1 + \frac{N}{4}\right) \tag{3.46}$$

Although this value was derived under a number of assumptions, it is weak only in the assumption of wire diameter large in skin depths. The corresponding reactance is given by

$$X = 168\pi N^2 kD \tag{3.47}$$

With these reactance and resistance results, the efficiency and bandwidth× efficiency merit factors may be obtained. The efficiency is

$$\eta = \frac{5\pi^2 Nk^4 D^4}{5\pi^2 Nk^4 D^4 + 2R_s(D/a)(1 + N/4)} \tag{3.48}$$

The Q is found to be

$$Q = \frac{672\pi NkD}{5\pi^2 Nk^4 D^4 + 2R_s(D/a)(1 + N/4)} \tag{3.49}$$

The bandwidth–efficiency merit factor is independent of N, and is

$$\text{BW} \cdot \eta = \frac{5\pi k^3 D^3}{672} \tag{3.50}$$

This result is similar to that for a single-turn loop with $D/a = 50$.

3.4.2 Permeable Core Loops

The most common loop with a magnetic core is the "loop stick," a single layer winding on a cylindrical ferrite rod. Equation 3.30 gives radiation resistance in terms

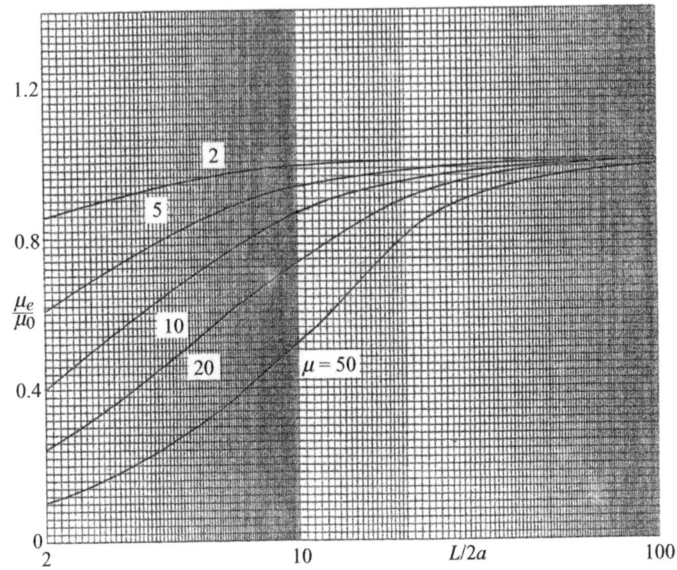

FIGURE 3.54 Effective permeability of solid core.

of effective permeability μ_e and turns N. The effective permeability of a ferrite or other magnetic core depends on the length-to-diameter ratio of the core, with large ratios required to realize μ_e close to the intrinsic permeability of the material. Therefore, ferrite core loops will usually be long but not large in diameter. Figure 3.54 gives effective permeability, normalized to permeability, as a function of core length-to-diameter ratio (Wait, 1953a). Effective permeability, normalized by the shape factor $L/2a$, is given in Figure 3.55. Whether the coil is distributed over most of the length of the core or lumped at the center is not critical and affects mainly distributed capacity and the flux enhancement of the end turns. The core need not be solid, as thin ferrite walls will give a significant increase of flux density.

For a long single-layer solenoid of many turns and with ferrite core, the reactance is given by

$$X_a = \frac{30\pi^2 N^2 k L \kappa \mu_e}{(L/2a)^2} \tag{3.51}$$

where L is now the length of the core and a the radius of the winding rather than the radius of the wire; κ is Nagaoka's constant (Langford-Smith, 1953), which for large L/a is approximately $1-0.08/(L/a)$, and thus for $L/a \gg 1$ is nearly unity. There exist formulas for calculating loss of long air core solenoids, but formulas for computing loss of ferrite core solenoids are very poor. Because the derivation of formulas of sufficient generality to be of use here is too arduous a task, a simpler but adequate scheme will be used. If low-loss ferrites are utilized, and if the coil dimensions are in the neighborhood of optimum, the coil Q will vary slowly with

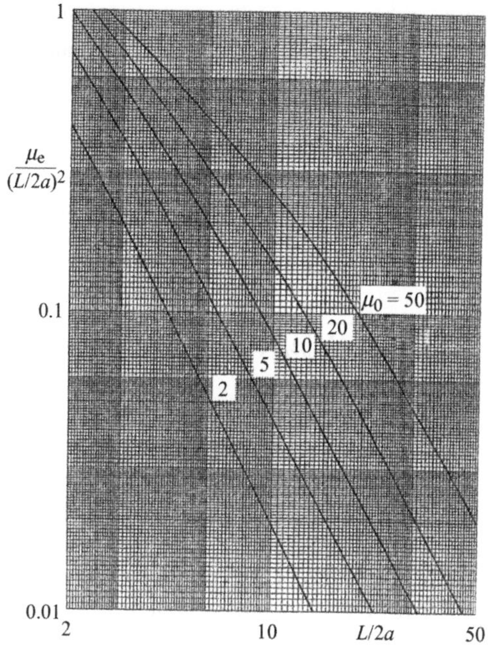

FIGURE 3.55 Shape factor for cylindrical core.

dimensions, number of turns, and so on. That is, the Q_{maximum} is very broad. Table 3.12 gives parameters for TDK materials, and it may be observed that the maximum Q is roughly independent of frequency from 50 to 200 MHz. Of course, at higher frequencies the ferrite permeability is lower so that copper and dielectric losses tend to play a larger role.

Using the previously defined symbols, the principal parameters, starting with the radiation resistance, are

$$\begin{aligned} R_r &= \frac{5\pi^2 N^2 k^4 L^4 \mu_e^2}{4(L/2a)^4} \\ X &= \frac{30\pi^2 N^2 k L \mu_e}{(L/2a)^2} \end{aligned} \tag{3.52}$$

TABLE 3.12

	K5	K6a	K8
Upper frequency (MHz)	8	50	200
μ	290	25	16
tan δ	0.008	0.004	0.004
Curie temperature (°C)	>280	>450	>500

TDK ferrite cores.

The core length is L; the core radius is a and the coil radius is also a, as the coil wire is usually of small diameter. The factor $L/2a$ is separated as it controls μ_e/μ. Loss resistance is

$$R_\ell = \frac{30\pi^2 N^2 k L \mu_e \tan\delta}{(L/2a)^2} \tag{3.53}$$

Efficiency and bandwidth × efficiency merit factors are, again, neglecting the effect of distributed capacity on bandwidth:

$$\eta = \frac{k^3L^3[\mu_e/(L/2a)^2]}{k^3L^3[\mu_e/(L/2a)^2]+24\tan\delta} \tag{3.54}$$

$$\mathrm{BW}\cdot\eta = \frac{k^3L^3\mu_e}{24(L/2a)^2} \tag{3.55}$$

In all these formulas, the shape of the coil and the core effect are contained in the factor $\mu_e/(L/2a)^2$. In the curves of Figure 3.55, this shape factor is plotted as a function of length-to-diameter ratio of the core. It can be seen that most cores will have shape factors between 0.1 and 1, with a few perhaps as low as 0.01. Accordingly, calculations of loop antenna parameters were made with three values of shape factor: 1, 0.1, and 0.01. In Figure 3.56 are plotted R/N^2, X/N^2, and R_r/N^2

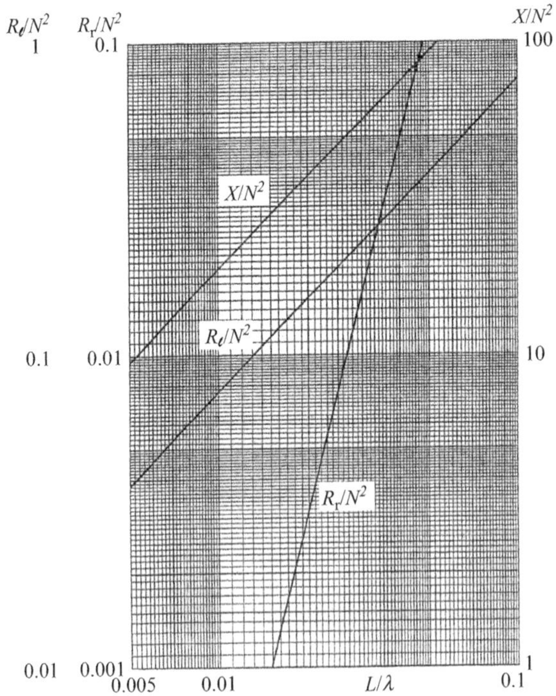

FIGURE 3.56 Magnetic core loop parameters, $L/2a = 1$.

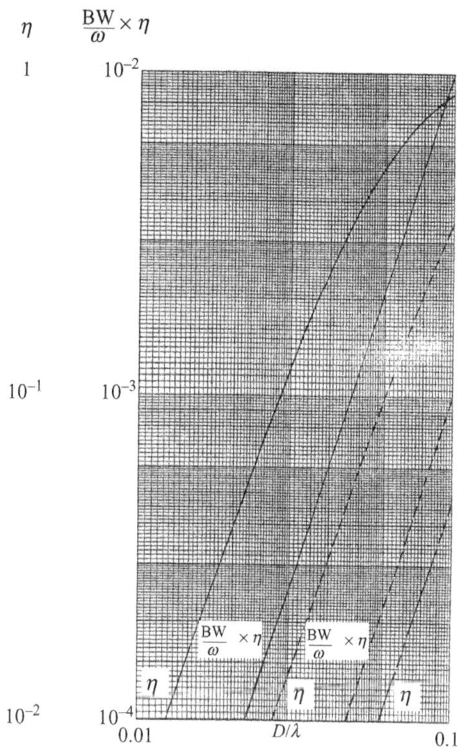

FIGURE 3.57 Ferrite loop parameters.

for a shape factor of 1. Note that the number of turns is included by multiplying all these values by N^2. The value of $\tan\delta$ for R_ℓ is 0.004. Also the vertical scales for R_ℓ and R_r are different, so the crossing of these two in Figure 3.56 does not represent $\eta = 50\%$. Efficiency and bandwidth × efficiency merit factors are given in Figure 3.57. In the figures, values of $\mu_e/(L/2a)^2$ are 1 (solid line), 0.1 (dashed line), and 0.01 (dot-dashed line).

The effective permeability data in Figure 3.54 were obtained by Wait (1953a). He assumed that the core was a prolate spheroid, with a coil around the center of the spheroid. For thin rod cores, this is good approximation. Using the spheroidal functions, the boundary conditions are matched, resulting in

$$\frac{\mu_e}{\mu} = \frac{1}{1+(\mu-1)(\eta_0^2-1)Q_1(\eta_0)} \tag{3.56}$$

Note that the spheroidal coordinates are (η, δ). Q_1 is a spheroidal wave function

$$Q_1 = \frac{\eta}{2}\ell n\frac{\eta+1}{\eta-1} - 1 \tag{3.57}$$

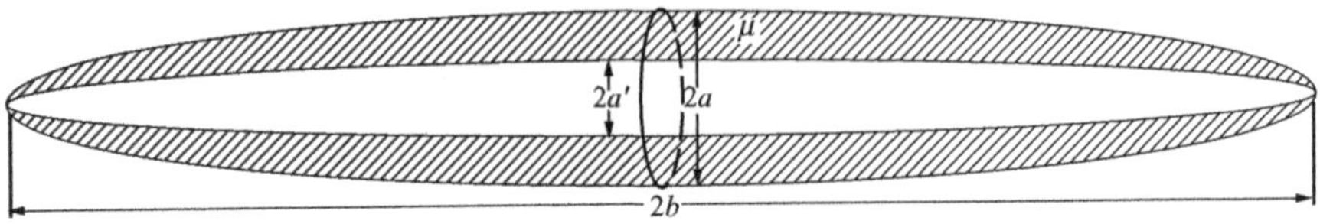

FIGURE 3.58 Loop with hollow prolate spheroidal core.

The coordinate η_0 is

$$\eta_0 = 1/\sqrt{1-4a^2/L^2} \tag{3.58}$$

where L is core length and a is core radius. These result in Figure 3.54, where μ_e/μ is plotted versus $L/2a$. Higher permeability cores need to be longer to reach the intrinsic permeability. It was discovered that most of the flux density is near the surface of the core, so that hollow cores could be used. This problem was also solved by Wait (1953b), using prolate spheroidal geometry. In Figure 3.58, core external and internal radii are a and a'. The results are

$$\frac{\mu_e}{\mu} = 1+BQ_1' \tag{3.59}$$

where

$$B = \frac{B_1[Q_1(\eta_0)-\mu\eta_0 Q'(\eta_0)]-\eta_0(\mu-1)}{B_1(\mu-1)Q_1(\eta_0)Q_1'(\eta_0)+\mu Q_1(\eta_0)-\eta_0 Q_1'(\eta_0)} \tag{3.60}$$

$$B_1 = \frac{(\mu-1)\eta_1}{Q_1(\eta_1)-\mu\eta_1 Q_1'(\eta_1)} \tag{3.61}$$

$$Q_1' = \frac{1}{2}\ell n\frac{\eta+1}{\eta-1} - \frac{\eta}{\eta^2-1} \tag{3.62}$$

Figure 3.59 gives μ_e versus $L/2a$ for $\mu = 50$, and Figure 3.60 is for $\mu = 200$. These design data have proved satisfactory for loop stick antenna design. In summary, a ferrite loop antenna should first use as high a permeability ferrite as possible, and second, the length (of core) should be as large as feasible. When length L is fixed, then the core should be made stubby, that is, with low $L/2a$ as the improvement in coil area increases faster than effective permeability decreases.

It should also be noted that both efficiency and radiation Q are independent of number of turns. Thus, the low efficiencies of Figure 3.57 are inescapable and are a result of the intrinsically poor performance of a loop small in wavelengths. A short folded dipole has similar low efficiency. Parameters for $L > 0.1\lambda$ are not shown, as the loop is resonant just above $L = 0.1\lambda$.

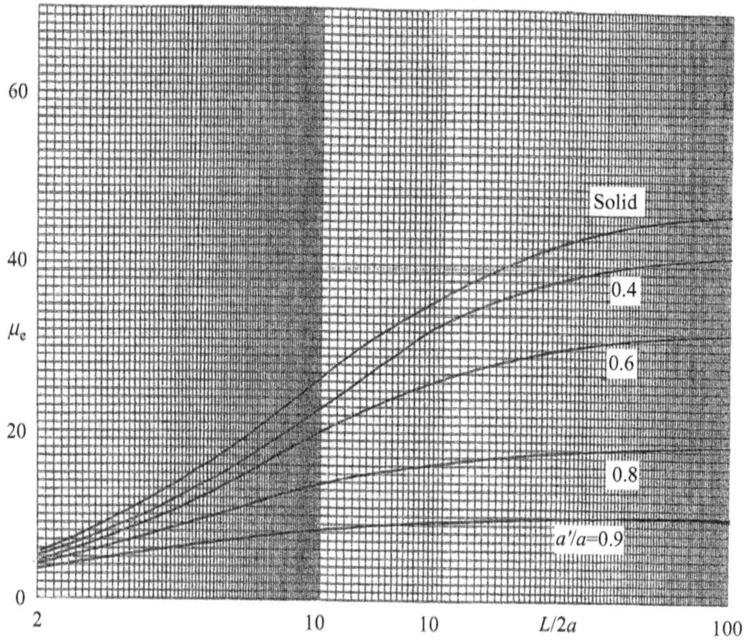

FIGURE 3.59 Effective permeability of hollow core, $\mu_0 = 50$.

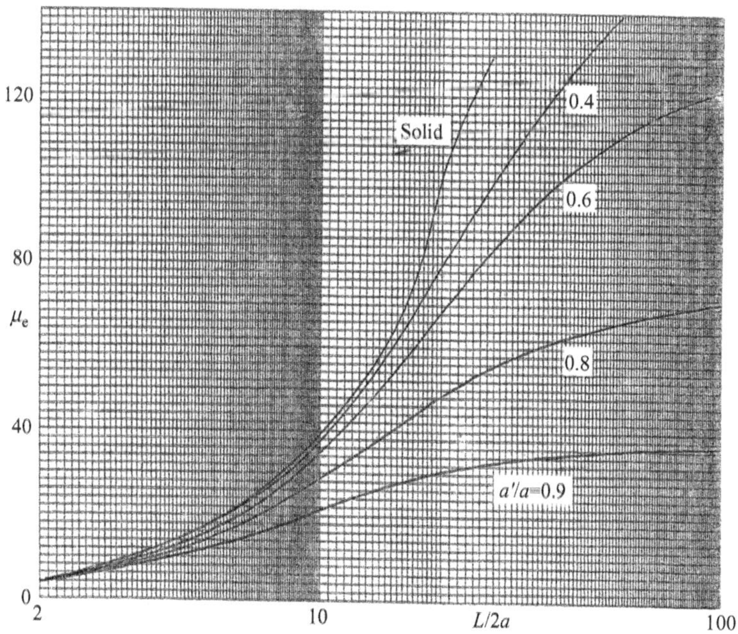

FIGURE 3.60 Effective permeability of hollow $\mu_0 = 200$.

3.4.3 Receiving Loops

The effective length for a single-turn air core receiving loop is

$$l_{\text{eff}} = \frac{\pi k D^2}{4} \tag{3.63}$$

and when this is normalized to open-circuit voltage per unit field strength in volts per wavelength,

$$\frac{l_{\text{eff}}}{\lambda} = \frac{V_{\text{oc}}}{\lambda E} = \frac{k^2 D^2}{8} \tag{3.64}$$

Now for low distributed capacitance, as before, the voltage at the preamp is simply the open-circuit voltage multiplied by the circuit Q:

$$\frac{V}{\lambda E} = \frac{k l_{\text{eff}} X}{2\pi(R_\ell + R_{\text{r}})} \tag{3.65}$$

where the tuning capacitor loss and input circuit loss are assumed to be small with respect to R_ℓ and R_{r}. With current technology, this assumption is quite valid. Using values of X, R_ℓ, and R_{r} previously derived, the output voltage for a single-turn loop is

$$\frac{V}{\lambda E} = \frac{30\pi k^3 D^3 [\ell n(4D/a) - 2]}{5\pi^2 k^4 D^4 + 2R_{\text{s}}(D/a)} \tag{3.66}$$

This merit factor, normalized output voltage, is plotted in Figure 3.61 for three values of the loss factor $R_{\text{s}}D/a$ as before. Effective length depends only on D/λ as shown. The output voltage peaks for

$$\frac{D}{\lambda} = \frac{1}{2}\left[\frac{6R_{\text{s}}D/a}{5\pi^2}\right]^{1/4} \tag{3.67}$$

Just as the lower loss factor gave higher efficiency, the lower loss factor here gives a higher output voltage, with a higher peak.

For a magnetic core loop, the effective length is now

$$l_{\text{eff}} = \frac{\pi}{4} N k D^2 \mu_{\text{e}} \tag{3.68}$$

and this normalized as before, using the shape factor $\mu_{\text{e}}/(L/2a)^2$, is

$$\frac{l_{\text{eff}}}{\lambda} = \frac{V_{\text{oc}}}{\lambda E} = \frac{N k^2 L^2 \mu_{\text{e}}}{8(L/2a)^2} \tag{3.69}$$

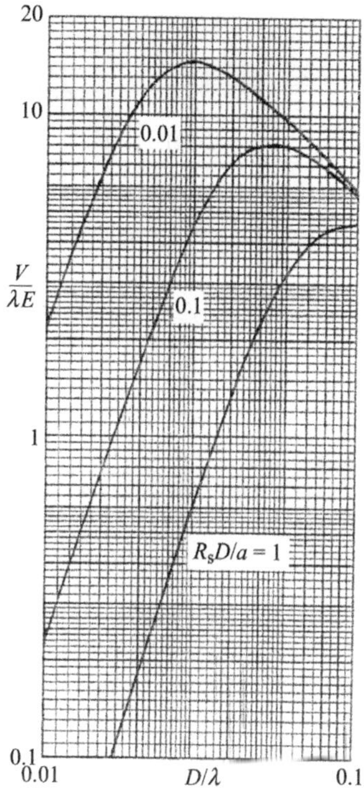

FIGURE 3.61 Receiving loop merit factor, $D/a = 50$.

The normalized output voltage is simply

$$\frac{V}{\lambda E} = \frac{l_e Q}{\lambda} = \frac{Nk^2L^2\mu_e X}{8(L/2a)^2(R_r + R_\ell)} \tag{3.70}$$

Using previously developed results for R_ℓ, X, and R_r, the output voltage merit factor becomes

$$\frac{V}{\lambda E} = \frac{3Nk^2L^2\mu_e}{(L/2a)^2[k^3L^3\mu_e/(L/2a)^2 + 24\tan\delta]} \tag{3.71}$$

This merit factor is shown in Figure 3.62 for three values of the shape factor $\mu_e/(L/2a)^2$, as before. The value of $\tan\delta$ used is 0.01. The output voltage merit factor is larger for larger D/λ.

Noise reception must be discussed, as it is sometimes said that loop antennas pick up less noise. At sufficiently low frequencies (VLF) and close to electrical storms, the antenna may be in the near-field of the source such that the wave impedance is higher

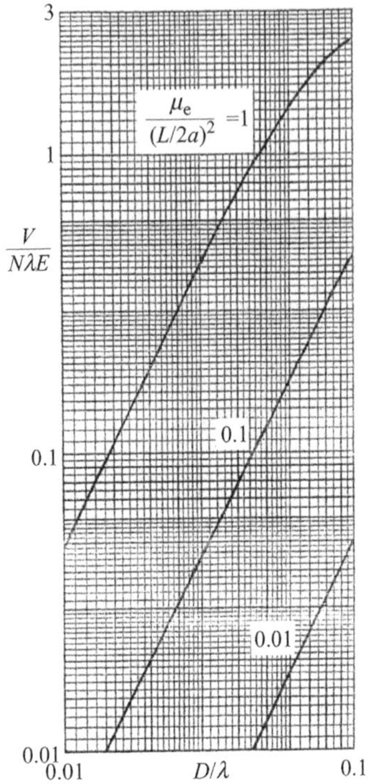

FIGURE 3.62 Receiving loop merit factor, tan $\delta = 0.01$.

than 120π. In these cases, the dipole/monopole will pick up more noise than the loop. But at any usable communications frequency, this does not happen; loops treat signal and noise equally. Precipitation static may also affect a dipole more than a loop. In many cases, the noise of a loop is low because its signal is also low.

3.4.4 Vector Sensor

A vector sensor has six antennas, three of which are electric dipoles (dipole antennas) and the other three magnetic dipoles (loop antennas). Typically, the loops and dipoles are all orthogonal and are co-centered. Figure 3.63 is a sketch of such a vector sensor. This configuration has been called the CART antenna, after Compact Array Radiolocation Technology (Hatke, 1993). Here, the capabilities and limitations of vector sensors, using electrically small antennas, are investigated. Emphasis is on the practical problems of impedance matching, bandwidth, and gain degradation.

For angle-of-arrival (AOA) applications, the vector sensor should be isotropic, that is, be effective for all angles of incidence. Single-turn loops should be no larger than 0.1λ in diameter, as larger loops have patterns that are not omnidirectional and have impedance swings.

Many papers on AOA estimation using vector sensors state that loops respond to magnetic field whereas dipoles respond to electric field, thereby implying that this is

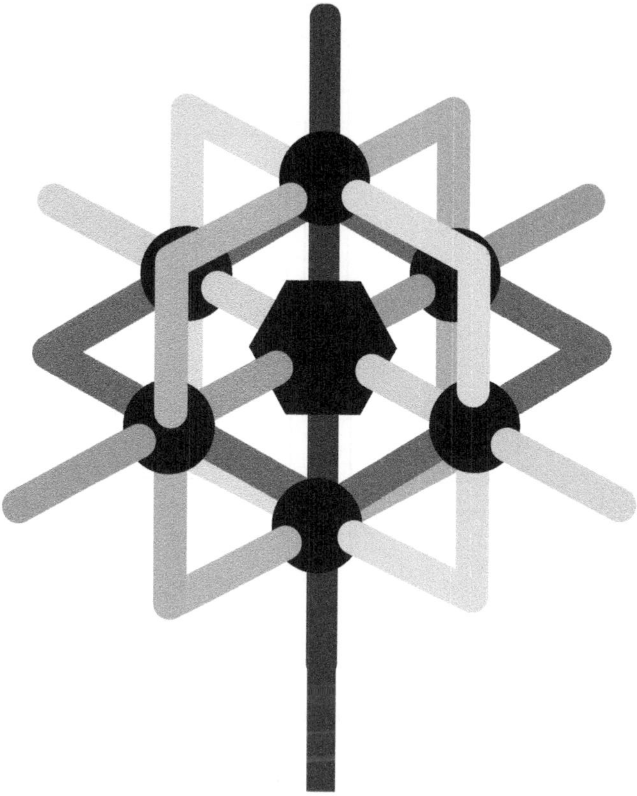

FIGURE 3.63 Notional vector sensor. Courtesy of Hatke (1993)

why performance is better than that obtained by tripoles, for example. This is a misconception: Loops are magnetic field sensors and dipoles are electric sensors only in the near-field region, where distance between source and antenna is less than a wavelength. For all practical applications, vector sensor sources are in the far-field; both loops and dipoles couple to the incident field, which has both E and H, with $E/H = 120\pi$.

The salient advantage of a vector sensor is that it allows Poynting vector (AOA) to be determined. This is possible because the response of a dipole depends simply on the angle between the dipole axis and AOA, and the response of a small loop depends simply on the angle between the loop axis and AOA. By taking ratios of the three dipole outputs, the scale factors are removed, leaving angular functions of AOA. Similarly, ratios of the three loop outputs yield additional angular functions; these together allow AOA to be estimated. Collectively, the six antennas are isotropic, so that polarization nulls are effectively removed. Now some ambiguity suppression and signal discrimination is feasible. As each antenna has its own preamplifier, the gains can be adjusted to provide equal sensitivities, taking into account the dipole and loop efficiencies. The behavior of a coplanar loop–dipole pair has been analyzed by

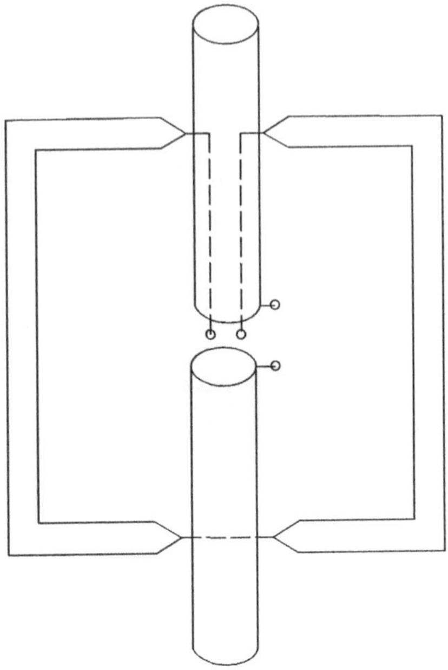

FIGURE 3.64 Coplanar segmented loop and dipole.

Overfelt (1998). For the loop and dipole to act independently, the relative phasing must be 90°.

It is crucial that the loop arms for a coplanar loop and dipole not contact the dipole arms. This means that the loop arms must connect across the dipole arms either by an insulated feed-through or by an insulated bypass outside the dipole conductor. Figure 3.64 sketches a possible configuration in which the dipole arms are tubes, with a twin line feed for the loop contained in one of the tubes. The dipole plane and the loop plane could be offset to avoid intersections, but although this is easy for one loop plus one dipole, it does not alleviate the problem for three loops and three dipoles. An easy fabrication for one loop plus one dipole would be to print a loop and feed line on one side of a board and print a dipole on the other side. But three boards cannot be "egg-crated" together without cutting some conductors, thereby necessitating pigtail or wire bond connections.

Figure 3.65 shows a loop and two dipoles, all coplanar. Each dipole arm would need to allow two loop sides to connect through the arm; maintenance of symmetry will be a challenge. Each loop will connect through four dipole arms, as sketched in Figure 3.65. Also, each loop feed point will need a shunt tuning capacitor, so that the twin line can be effectively utilized as a matching transformer (details given below). Each dipole arm connect through will need to accommodate two independent loops but for three arms a tuning capacitor as well. This imposes fabrication and assembly difficulties.

Segmented loops, which allow larger diameters, require multiple feed lines; a two-segment configuration is sketched in Figure 3.66. Both dipole arms contain twin

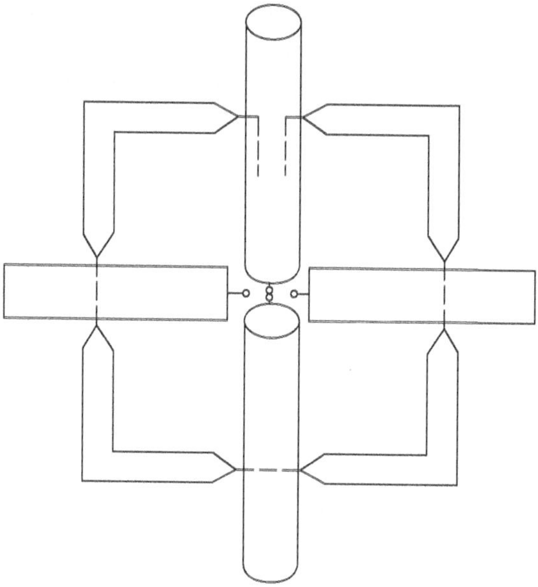

FIGURE 3.65 Coplanar loop and two dipoles.

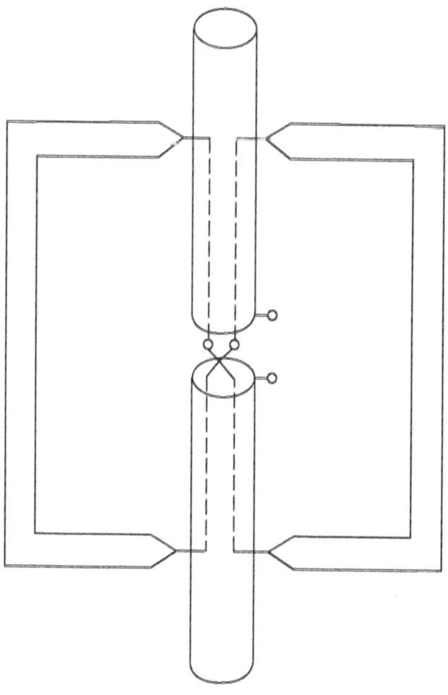

FIGURE 3.66 Coplanar segmented loop and dipole.

line feeds that are paralleled at the final feed point. One pair is reversed, so that the current around the loop is in the same direction.

A vector sensor consists of three loops and three dipoles. Because any single antenna has at least one null direction for a given polarization, use of orthogonal antennas removes nulls. Because all signals will be in the far-field of the antenna, there will be some redundant outputs from a vector sensor. The plane wave response will involve both E and H fields. The principal difference among redundant signals is due to the patterns. For example, the dominant electric field pattern of a dipole, normalized to unity, is

$$E = \frac{\cos(kh \cos\theta) - \cos kh}{\sin\theta} \tag{3.72}$$

where $k = 2\pi/\lambda$, h is the dipole half-length, and θ is measured from the dipole axis. For a loop coplanar with the dipole, the loop normalized electric field is

$$E = J_1(ka \sin\theta) \tag{3.73}$$

where J_1 is the Bessel function, a is the loop radius, and θ is measured from the loop axis. Loops and dipoles small in wavelengths will behave approximately the same.

Orthogonal co-centered dipoles have zero coupling, as the radial component of the near electric field from a dipole is zero in the normal plane through the dipole center. Similarly, orthogonal co-centered loops have zero coupling. A constant current loop has only an H_ϕ, and this flux does not enter the loop. Thus, a vector sensor made of thin orthogonal dipoles and loops should provide six terminals that are decoupled from each other.

Because the determination of Poynting vector direction, either direct or implicitly, involves all six antenna outputs, fabrication and assembly asymmetries must be minimized and the feed-matching circuits must produce sufficiently small phase and amplitude errors. For a specific vector sensor implementation, an error analysis should be made, relating mechanical and electrical errors to error in the Poynting vector. This error function can then be incorporated into the AOA estimation algorithm.

There are many papers on the algorithms for processing vector sensor outputs, such as maximum likelihood, MUSIC, and ESPRIT; see Ko et al. (2002) for references.

3.5 DIELECTRIC RESONATOR ANTENNAS

The dielectric resonator antenna is simply a block of low-loss dielectric placed on a ground plane. A precursor was probably a short monopole surrounded by a centered cylindrical dielectric (pillbox); the monopole has the same height as the pillbox (James and Burrows, 1973). As expected, a short monopole may be resonated by high ε, but the bandwidth is reduced. Long et al. (1983) introduced the DRA, where the dielectric body is resonant. They considered both rectangular slab dielectrics and

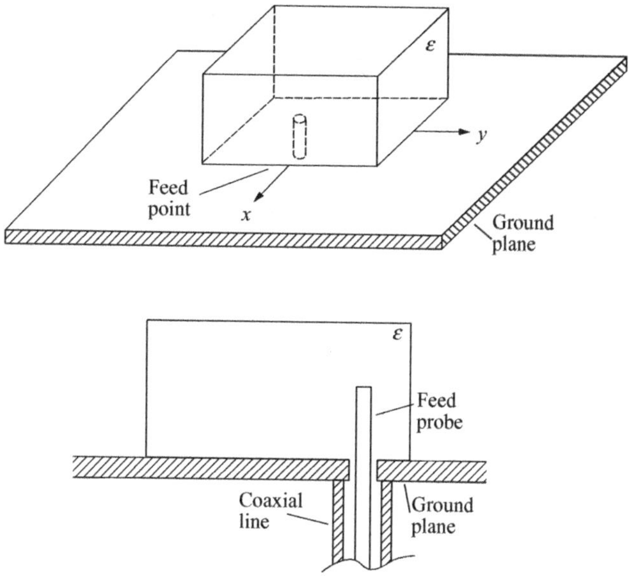

FIGURE 3.67 DRA with probe feed. Courtesy of McAllister et al. (1983).

cylindrical (pillbox) dielectrics. A rectangular block DRA can be fed by a probe (McAllister et al., 1983) as sketched in Figure 3.67, by a microstrip (Mridula et al., 2004), or by a microstrip line with a tee (Bijumon et al., 2005). Circular disk (pillbox) DRA (McAllister and Long, 1984) have been fed by a microstrip excited slot (Leung and To, 1997), an inverted microstrip (Leung et al., 1997a), a waveguide with a slot in the guide top wall–ground plane (Eshrah et al., 2005a), a waveguide with a probe extending into both the guide and the DRA (Eshrah et al., 2005b), or a vertical metal strip on the disk exterior, with the strip connected to a feed line (Leung et al., 2000). Unusual DRA shapes include a hemisphere fed by a slot in the ground plane with microstrip excitation (Leung et al., 1995), a half-cylinder with axis parallel to the ground plane and fed by a probe (Mongia, 1989; Kishk et al., 1999), a cylindrical ring with axis normal to the ground plane and fed by a microstrip (Leung et al., 1997c), and conical dielectric shapes (Kishk et al., 2002). A further reduction in size results from a metal top plate connected to a center grounded rod (Mongia, 1997). A different configuration utilizes a tall, narrow, and thin high-ε dielectric slab normal to the ground plane and fed by a metal strip along the flat side, with a thicker, low-ε block attached (Moon and Park, 2000; see Figure 3.68). Size reduction also occurs when metallic plates connected to the ground plane are placed at the E walls (of a rectangular DRA). This, however, also reduces the bandwidth (Cormos et al., 2003). Some bandwidth can be recovered by adding a grounded metal strip on an H wall, thus exciting also a higher mode (Li and Leung, 2005). Another bandwidth enhancement technique top loads a circular disk with a thin, high-ε disk of larger diameter (Leung et al., 1997b).

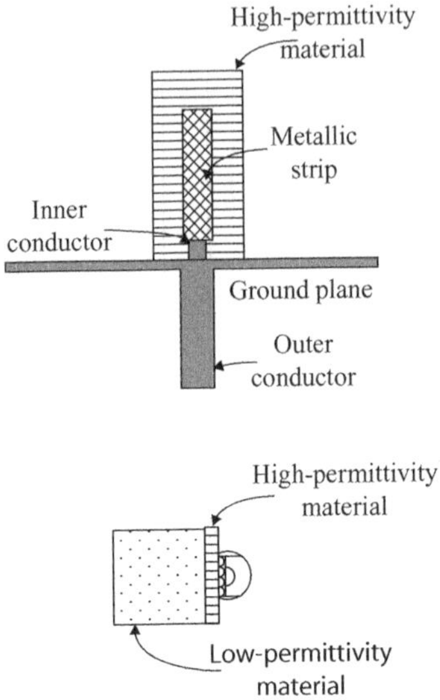

FIGURE 3.68 Rectangular DRA with high-ε vertical strip. Courtesy of Moon (2000).

Because of the needs of mobile communications devices, there has been work on dual-band DRA. Two rectangular blocks placed on either side of a coupling slot produce two narrowband responses (Fan and Antar, 1997). A circular disk fed by a stripline can produce dual-band response if the feed line is properly placed (Kumar et al., 2005). An oval cylindrical disk, by proper choice of diameters, produces dual-band operation (Paul et al., 2004). An alternate geometry uses a circular disk with a concentric circular cavity, fed by microstrip lines at right angles (Sung et al., 2004). Finally, a circular disk of modest ε with a higher ε, smaller diameter, and smaller height disk inside with a probe feed provides dual-band operation (Nannini et al., 2003). All of these dual-band configurations have a few percent (VSWR = 2) bandwidth in each band.

Field distributions inside a cylindrical dielectric were studied via the moment method by Glisson et al. (1983) and by Kajfez et al. (1984). Analyses of resonant frequency and Q of rectangular DRA were made by Mongia (1992) and by Mongia and Ittipiboon (1997). Details on calculations are given below. For cylindrical DRA, Tsuji et al. (1982) and Mongia and Bhartia (1994) derived the relevant equations. Measurements of DRA Q appear to have been made primarily on cylindrical DRA (Mongia et al., 1994a, 1994b). Although a high ε considerably reduces the size of a DRA, metallic loading such as a top plate (Mongia, 1997) or a side plate (Li and Leung, 2005) can further reduce the size. Luk and Leung (2003) have extensively covered the art of DRA.

Most of the analyses and experiments have been made on the cylindrical or spherical configuration, because the wave functions are well-known cylindrical or spherical harmonics, which allow easy satisfaction of the boundary conditions. In contrast, rectangular DRA usually require a numerical method such as the moment method. But for practical antennas, the rectangular DRA is the best choice: it is low profile and easy to fabricate; the three dimensions provide versatility in resonant frequency, polarization, and bandwidth. Thus, only the rectangular DRA is considered in detail here.

It is analogous to the conventional patch antenna, where the patch–dielectric–ground plane forms a cavity supporting the dominant TM mode. Polarization and pattern for the DRA are similar to those of the patch antenna. However, in a rectangular DRA, the length-to-width ratio may need to be adjusted to give an omnidirectional pattern. Circular polarization requires proper location of the feed. Increasing ε_r reduces the size by $\sqrt{\varepsilon_r}$, but the bandwidth is also reduced. Dielectric loss is an important factor as it reduces efficiency and gain.

The analysis given here is from Mongia (1992) and Mongia and Ittipiboon (1997). Let the ground plane be the X–Z plane, with Z the resonant direction, and let the dielectric block have length l (along Z), width w, and height h. Their wave numbers are

$$k_x = \frac{\pi}{w}, \qquad k_y = \frac{\pi}{2h}, \qquad k_z^2 = \varepsilon k^2 - k_x^2 - k_y^2 \tag{3.74}$$

The indicial equation for resonance is

$$k_z \tan\frac{k_z l}{2} = \sqrt{(\varepsilon - 1)k^2 - k_z^2} \tag{3.75}$$

These combine to give an equation for $k_z l$ from which the resonant frequency can be found:

$$k_z^2\left[1 + \varepsilon \tan^2\frac{k_z l}{2}\right] = (\varepsilon - 1)(k_x^2 + k_y^2) \tag{3.76}$$

In terms of convenient design parameters l/h and w/h, Equation 3.76 becomes

$$k_z^2 l^2\left[1 + \varepsilon \tan^2\frac{k_z l}{2}\right] = \frac{\pi^2}{4}(\varepsilon - 1)\left(\frac{l}{h}\right)^2\left[1 + \left(\frac{2h}{w}\right)^2\right] \tag{3.77}$$

Given ε, l/h, and w/h, this equation is solved by this author by a Newton–Raphson rooter (Stark, 1970), a good choice as the required derivative function is easily found. Figure 3.69 shows $k_z l$ versus l/h for values of $w/h = 1$, 2, 5, and 10. The curves are accurate within 1% for ε from 10 to 40, and probably higher ε. It should be noted that $k_z l$ is a function of only ε and of the ratios of dimensions. Because $l/\lambda_g = k_z l/2\pi$, the resonant guide length approaches $\lambda_g/2$ as l/h becomes

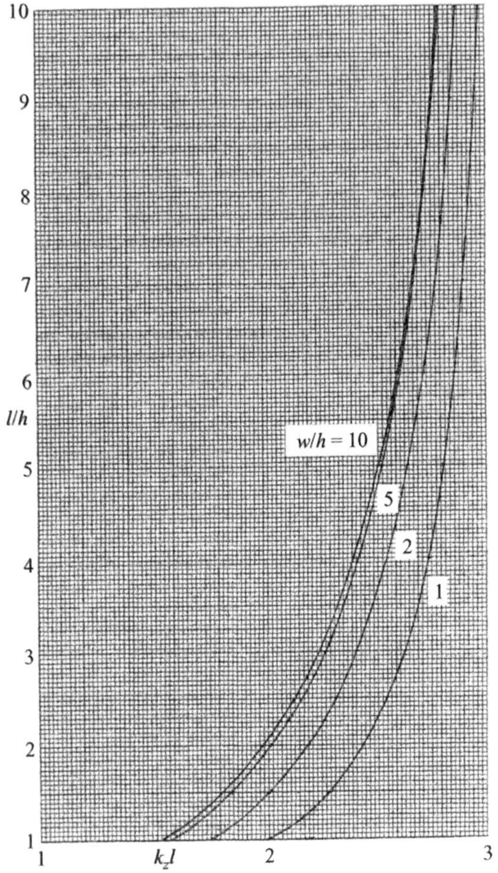

FIGURE 3.69 Propagation constant vs aspect ratio.

large, and as w/h becomes small. The resonant value kdl is now found from

$$k^2l^2 = \frac{1}{\varepsilon}\left[k_z^2l^2 + \frac{\pi^2}{4}\left(\frac{l}{h}\right)^2\left[1+\left(\frac{h}{w}\right)^2\right]\right] \tag{3.78}$$

The resonant frequency f_r is given by

$$f_r = 0.3kl/2\pi \tag{3.79}$$

where f_r is in gigahertz and l is in millimeters.

In free space wavelengths, the square root of Equation 3.78 is divided by 2π to get l/λ. For example, $\varepsilon = 20, l/h = 5, w/h = 5, k_z l = 2.5421, l/\lambda_g = 0.4046$, and the resonant length is $l/\lambda = 0.314$.

Bandwidth for a matched load and for VSWR ≤ 2 is given by

$$\mathrm{BW} = \frac{P}{\sqrt{2}W} \tag{3.80}$$

The radiated power P is given by (Mongia and Ittipiboon, 1997)

$$P = \frac{640k^4\omega^2\varepsilon_0^2(\varepsilon-1)^2\sin^2(k_z l/2)}{k_x^2 k_y^2 k_z^2} \tag{3.81}$$

Energy times 2ω is

$$2\omega W = \frac{\omega\varepsilon_0\varepsilon lh(k_x^2+k_y^2)(1+\mathrm{sinc}\ k_z l)}{16} \tag{3.82}$$

Since $\omega\varepsilon_0 = k/120\pi$, the bandwidth becomes

$$\mathrm{BW} = \frac{128k^5(\varepsilon-1)^2\sin^2(k_z l/2)}{3\sqrt{2}\pi\varepsilon wlhk_x^2k_y^2k_z^2(k_x^2+k_y^2)(1+\mathrm{sinc}\ k_z l)} \tag{3.83}$$

Inserting dimensional ratios,

$$\mathrm{BW} = \frac{512(\varepsilon-1)^2(w/h)^3(kl)^5\sin^2(k_z l/2)}{3\sqrt{2}\pi^7\varepsilon(l/h)^4[1+(w/2h)^2]k_z^2l^2(1+\mathrm{sinc}\ k_z l)} \tag{3.84}$$

It is interesting to note that the bandwidth is a function of only ε and the ratios l/h and w/h. Figure 3.70 shows bandwidth for VSWR = 2, for $\varepsilon = 20$, and for $w/h = 2$, 5, and 10. The range of l/h is 1–10. As expected, longer and wider DRA have more bandwidth. Figure 3.71 shows bandwidth for $\varepsilon = 40$ for the same parameters. The bandwidth is reduced close to the $\varepsilon^{1.5}$ factor that has been suggested for large ε. Finally, Figure 3.72 gives bandwidth for a square DRA, with $\varepsilon = 20$ and 40. These bandwidths are the intrinsic values for the dielectric resonator; the effect of the feed on efficiency and bandwidth can be expected to be important.

A brief comparison with a conventional patch is now made. For the DRA take $\varepsilon = 20$, a square ($l = w$) dielectric, with $l/h = 40$. The value of kl is 1.6637 so that $l/\lambda_0 = 0.2648$, and $h/\lambda_0 = 0.0662$. VSWR = 2 bandwidth is 5.38%. For the square patch, use $\varepsilon = 3$. Patch bandwidth is approximately

$$\mathrm{BW} = \frac{4t}{\sqrt{2}\varepsilon\lambda_0} \tag{3.85}$$

Using the same bandwidth, the patch thickness is $t/\lambda_0 = 0.03295$. The patch length is $l/\lambda_0 = 1/2\sqrt{\varepsilon} = 0.2887$. The DRA and patch lengths are nearly the same,

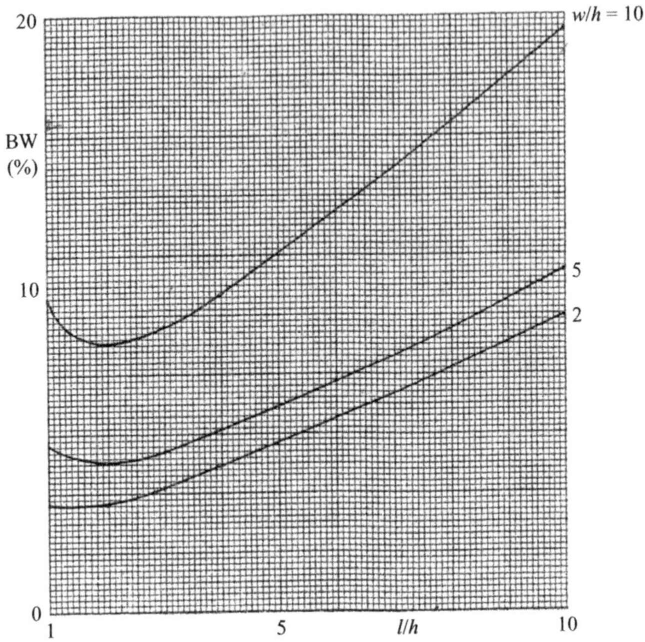

FIGURE 3.70 DRA VSWR = 2 matched bandwidth, $\varepsilon = 20$.

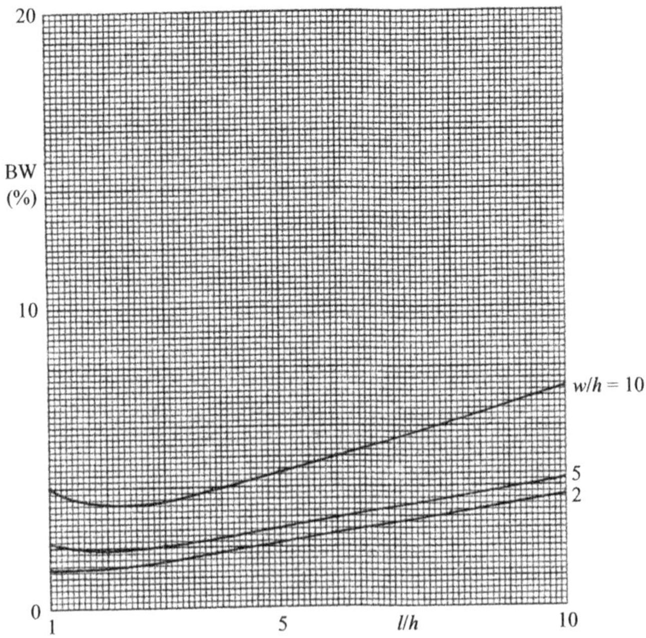

FIGURE 3.71 DRA VSWR = 2 matched bandwidth, $\varepsilon = 40$.

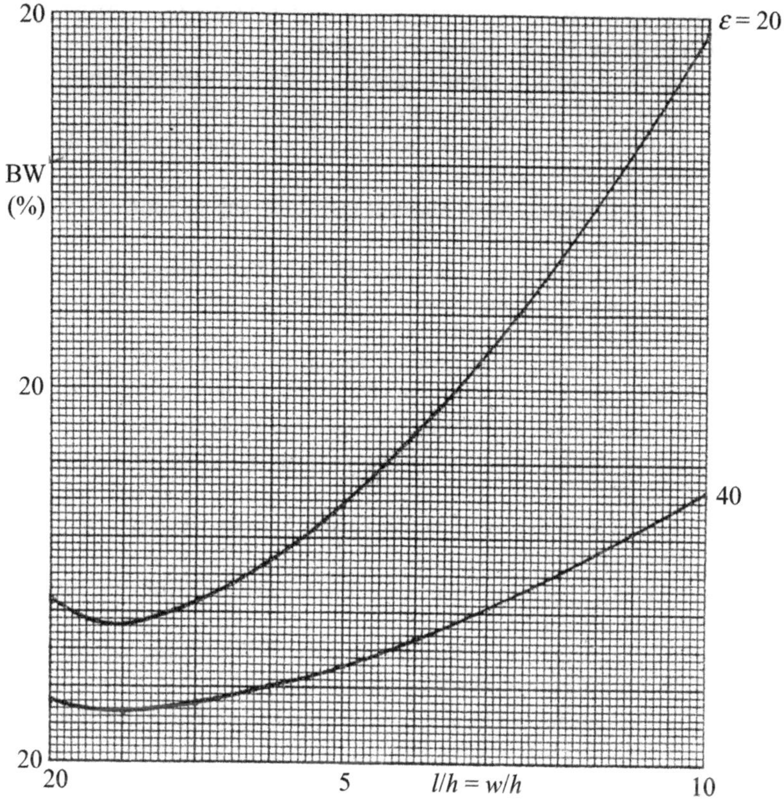

FIGURE 3.72 Square DRA VSWR = 2 matched bandwidth.

with nearly the same resonant frequency, but the DRA thickness is about double that of the patch.

REFERENCES

Abramowitz, M. and Stegun, I. A. *Handbook of Mathematical Functions with Formulas, Graphs, and Mathematical Tables,* NBS Applied Mathematics Series 55, Ninth Printing, November 1970.

Adams, J. J. and Bernhard, J. T. A Class of Electrically Small Spherical Antennas with Near-Minimum *Q. Antenna Applications Symp.*, Allerton, IL, September 2008, pp. 165–175.

Adams, J. J. et al., Conformal Printing of Electrically Small Antennas on Three-Dimensional Surfaces. *Adv. Mater.* 2011, xx, pp. 1–6.

Andreasen, M. G. Scattering from Parallel Metallic Cylinders with Arbitrary Cross Section. *Trans. IEEE*, Vol. AP-12, November 1964, pp. 746–775.

Balanis, C. A. *Antenna Theory*, Wiley, 2005.

Best, S. R. The Radiation Properties of Electrically Small Folded Spherical Helix Antennas. *Trans. IEEE*, Vol. AP-52, April 2004, pp. 953–960.

Best, S. R. Low *Q* Electrically Small Linear and Elliptical Polarized Spherical Dipole Antennas. *Trans. IEEE*, Vol. AP-53, March 2005, pp. 1047–1053.

Best, S. R. A Study of the Performance Properties of Small Antennas. *Antenna Applications Symp.*, Allerton, IL, 2007, pp. 193–219.

Best, S. R. A Low *Q* Electrically Small Magnetic (TE Mode) Dipole. *IEEE Antenn. Wireless Propag. Lett.*, Vol. 8, 2009, pp. 572–575.

Best, S. R. and Morrow, J. D. On the Significance of Current Vector Alignment in Establishing the Resonant Frequency of Small Space-Filling Wire Antennas. *IEEE Antenn. Wireless Propag. Lett.*, Vol. 2, 2003, pp. 201–204.

Bijumon, P. V. et al. T-Strip-Fed High-Dielectric Resonator Antenna for Broadband Applications. *Microwave Opt. Technol. Lett.*, Vol. 47, November 2005, pp. 226–228.

Bulgerin, M. A. and Walters, A. B. *NOLC—Report 154*, 1954, pp. 67–83.

Cebik, L. B.NEC-4: Limitations of Importance to Hams, *QEX*, May–June 1998, ARRL, pp. 3–16.

Chu, L. J. Physical Limitations of Omni-Directional Antennas. *J. Appl. Phys.*, Vol. 19, December 1948, pp. 1163–1175.

Collin, R. E. Minimum *Q* of Small Antennas, *J. Electromagnetic Waves and Applications*, Vol. 12, 1998, pp. 1369–1393.

Cormos, D. et al. Compact Dielectric Resonator Antenna for WLAN Applications. *Electron. Lett.*, Vol. 39, April 2003, pp. 588–590.

Czerwinski, W. P. On Optimizing Efficiency and Bandwidth of Inductively Loaded Antennas. *Trans. IEEE*, Vol. AP-13, September 1965, pp. 811–812.

Czerwinski, W. P. On a Foreshortened Center-Fed Whip Antenna. *Trans. IEEE*, Vol. VC-15, October 1966, pp. 33–40.

Deschamps, G. A. and Sichak, W. Microstrip Microwave Antenna. *Proc. 1953 USAF Antenna R&D Symp.*, Allerton, IL, October 1953.

Eshrah, I. A. et al. Theory and Implementation of Dielectric Resonator Antenna Excited by a Waveguide Slot. *Trans. IEEE*, Vol. 53, January 2005a, pp. 483–494.

Eshrah, I. A. et al. Excitation of Dielectric Resonator Antennas by a Waveguide Probe: Modeling Technique and Wide-Band Design. *Trans. IEEE*, Vol. 53, March 2005b, pp. 1028–1037.

Esselle, K. P. and Stuchly, S. S. Pulse-Receiving Characteristics of Resistively Loaded Dipole Antennas. *Trans. IEEE*, Vol. AP-38, October 1990, pp. 1677–1683.

Fan, Z. and Antar, Y. M. M. Slot-Coupled DR Antenna for Dual-Frequency Operation. *Trans. IEEE*, Vol. AP-45, February 1997, pp. 306–308.

Fanson, P. L. and Chen, K.-M. Modification of Antenna Radiating Characteristics with Multi-Impedance Loading. *Trans. IEEE*, Vol. AP-21, September 1973, pp. 715–721.

Fournier, M. and Pomerleau, A. Experimental Study of an Inductively Loaded Short Monopole Antenna. *Trans. IEEE*, Vol. VT-27, February 1978, pp. 1–6.

Francavilla, L. A. et al. Mode-Matching Analysis of Top-Hat Monopole Antennas Loaded with Radially Layered Dielectric. *Trans. IEEE*, Vol. AP-47, January 1999, pp. 179–185.

Friedman, C. H. Wide-Band Matching of a Small Disk-Loaded Monopole. *Trans. IEEE*, Vol. AP-33, October 1985, pp. 1143–1148.

Guertler, R. J. F., Impedance Transformation in Folded Dipoles, *Proc. IRE*, Vol. 38, Sept. 1950, pp. 1042–1047.

Glisson, A. W., Kajfez, D., and James, J. Evaluation of Modes in Dielectric Resonators Using a Surface Integral Equation Formulation. *Trans. IEEE*, Vol. MTT-31, December 1983, pp. 1023–1029.

Goubau, G. Multi-Element Monopole Antennas. *Proc. Workshop on Electrically Small Antennas*, USARO, Durham, NC, May 6-7, 1976, pp. 63–67, AD-A031–845.

Grimes, D. M. and Grimes, C. A. Bandwidth and Q of Antennas Radiating TE and TM Modes. *Trans. IEEE*, Vol. EMC-37, May 1995, pp. 217–226.

Grimes, D. M. and Grimes, C. A. A Clarification and Extension of "Bandwidth and Q of Antennas Radiating Both TE and TM Modes". *Trans. IEEE*, Vol. EMC-38, May 1996, pp. 201–202.

Hansen, R. C. Formulation of Echelon Dipole Mutual Impedance for Computer. *Trans. IEEE*, Vol. AP-29, November 1972, pp. 780–781.

Hansen, R. C. Efficiency and Matching Tradeoffs for Inductively Loaded Short Antennas. *Trans. IEEE*, Vol. COM-23, April 1975a, pp. 430–435.

Hansen, R. C. Optimum Inductive Loading of Short Whip Antennas. *Trans. IEEE*, Vol. VT-24, May 1975b, pp. 21–29.

Hansen, R. C. Folded and T-Match Dipole Transformation Ratio. *Trans. IEEE*, Vol. AP-30, January 1982, pp. 161–162.

Hansen, R. C. *Phased Array Antennas*, Wiley, 1998, Section 5.4.3.

Hansen, R. C. Q and Bandwidth of Electrically Small Antennas. *Microwave Opt. Technol. Lett.*, Vol. 49, May 2007, pp. 1170–1171.

Hansen, R. C. and Burke, M. Antennas with Magneto-Dielectrics. *Microwave Opt. Technol. Lett.*, Vol. 26, July 2000, pp. 75–78.

Harrington, R. F. Matrix Methods for Field Problems. *Proc. IEEE*, Vol. 55, February 1967, pp. 136–149.

Harrington, R. F. *Field Computation by Moment Methods*, Macmillan, 1968, pp. 10–81.

Harrison, C. W. Monopole with Inductive Loading. *Trans. IEEE*, Vol. AP-11, July 1963, pp. 394–400.

Harrison, C. W. and King, R. W. P. Folded Dipoles and Loops. *Trans. IRE*, Vol. AP-9, March 1961, pp. 171–187.

Hatke, G. F. Conditions for Unambiguous Source Location Using Polarization Diverse Arrays. *IEEE 27th Asilomar Conf.*, 1993, pp. 1365–1369.

Hirasawa, K. and Haneishi, M. *Analysis, Design, and Measurement of Small and Low-Profile Antennas*, Artech House, 1992.

Hong, T. et al. Printed L-Band Monopole Antenna with a Modified Trefoil Knot Structure. *Microwave Opt. Technol. Lett.*, Vol. 52, July 2010, pp. 1585–1588.

Hsiao, F.-R. et al. A Dual-Band Planar Inverted-F Patch Antenna with a Branch Line Split. *Microwave Opt. Technol. Lett.*, Vol. 32, February 2002, pp. 310–312.

James, J. R. and Burrows, R. M. Resonance Properties of Dielectric-Loaded Short Unipoles. *Electron. Lett.*, Vol. 9, July 1973, pp. 300–302.

Jung, J.-H. and Park, I. Electromagnetically Coupled Small Broadband Monopole Antenna. *IEEE Antenn. Wireless Propag. Lett.*, Vol. 2, 2003, pp. 349–351.

KAJFEZ, D., GLISSON, A. W., AND JAMES, J. Computed Modal Field Distributions for Isolated Dielectric Resonators. *Trans. IEEE*, Vol. MTT-32, December 1984, pp. 1609–1616.

KANDA, M. A Relatively Short Cylindrical Broadband Antenna with Tapered Resistive Loading for Picosecond Pulse Measurements. *Trans. IEEE*, Vol. AP-26, May 1978, pp. 439–447.

KANDA, M. Standard Antennas for Electromagnetic Interference Measurements and Methods to Calibrate Them. *Trans. IEEE*, Vol. EMC-36, November 1994, pp. 261–273.

KIM, O. S. Low-Q Electrically Small Spherical Magnetic Dipole Antennas. *Trans. IEEE*, Vol. AP-58, July 2010, pp. 2210–2217.

KISHK, A. A., GLISSON, A. W., AND JUNKER, G. P. Study of Broadband Dielectric Resonator Antennas. *Antenna Applications Symp.*, Allerton, IL, 1999, pp. 45–68.

KISHK, A. A. ET AL. Conical Dielectric Resonator Antennas for Wide-Band Applications. *Trans. IEEE*, Vol. AP-50, April 2002, pp. 469–474.

KO, C. C., ZHANG, J., AND NEHORAL, A. Separation and Tracking of Multiple Broadband Sources with One Electromagnetic Vector Sensor. *Trans. IEEE*, Vol. AES-38, July 2002, pp. 1109–1116.

KUMAR, A. V. P. ET AL. Microstripline Fed Cylindrical Dielectric Resonator Antenna for Dual-Band Operation. *Microwave Opt. Technol. Lett.*, Vol. 47, October 2005, pp. 150–153.

LALLY, J. F. AND ROUCH, D. T. Experimental Investigation of the Broad-Band Properties of a Continuously Loaded Resistive Monopole. *Trans. IEEE*, Vol. AP-18, November 1970, pp. 764–768.

LANGFORD-SMITH, F. *Radiotron Designer's Handbook*, 4th ed., RCA, 1953, Chapter 10.

LEONHARD, J. ET AL. Folded Unipole Antennas. *Trans. IRE*, Vol. AP-3, July 1955, pp. 111–116.

LEUNG K. W. AND TO, M. W. Slot-Coupled Dielectric Resonator Antenna Using a Proximity Feed on a Perpendicular Substrate. *Electron. Lett.*, Vol. 33, September 1997, pp. 1665–1666.

LEUNG, K. W. ET AL. Theory and Experiment of an Aperture-Coupled Hemispherical Dielectric Resonator Antenna. *Trans. IEEE*, Vol. AP-43, November 1995, pp. 1192–1198.

LEUNG, K. W. ET AL. Excitation of Dielectric Resonator Antenna Using a Soldered-Through Probe. *Electron. Lett.*, Vol. 33, February 1997a, pp. 349–350.

LEUNG, K. W. ET AL. Bandwidth Enhancement of Dielectric Resonator Antenna by Loading a Low-Profile Dielectric Disk of Very High Permittivity. *Electron. Lett.*, Vol. 33, April 1997b, pp. 725–726.

LEUNG, K. W. ET AL. Low-Profile Circular Disk DR Antenna of Very High Permittivity Excited by a Microstripline. *Electron. Lett.*, Vol. 33, June 1997c, pp. 1004–1005.

LEUNG, K. W. ET AL. Circular-Polarized Dielectric Resonator Antenna Excited by Dual Conformal Strips. *Electron. Lett.*, Vol. 36, March 2000, pp. 484–486.

LI, B. AND LEUNG, K. W. Strip-Fed Rectangular Dielectric Resonator Antennas with/without a Parasitic Patch. *Trans. IEEE*, Vol. AP-53, July 2005, pp. 220–2207.

LIN, C. J., NYQUIST, D. P., AND CHEN, K. M. Short Cylindrical Antennas with Enhanced Radiation or High Directivity. *Trans. IEEE*, Vol. AP-18, July 1970, pp. 576–580.

LIU, G., GRIMES, C. A., AND GRIMES, D. M. A Time-Domain Technique for Determining Antenna Q. *Microwave Opt. Technol. Lett.*, Vol. 21, June 1999, pp. 395–398.

LO, Y. T. A Note on the Cylindrical Antenna of Noncircular Cross Section. *J. Appl. Phys.*, Vol. 24, October 1953, pp. 1338–1339.

LONG, S. A., MCALLISTER, M. W., AND SHEN, L. C. The Resonant Cylindrical Dielectric Cavity Antenna. *Trans. IEEE*, Vol. AP-31, May 1983, pp. 406–412.

LUK, K. M. AND LEUNG, K. W. *Dielectric Resonator Antennas*, Research Studies Press Ltd., 2003.

LUKE, Y. *Mathematical Functions and Their Approximations*, Academic Press, 1975, Table 4. 4.

LUKE, Y. AND WIMP, J. Polynomial Approximations to Integral Transforms. *Math. Tables Other Aids Comput.*, Vol. 15, Issue 74 1961, pp. 174–178.

MACLEAN, T. S. M. AND SAINI, S. P. S. Efficiency and Signal/Noise Ratio of Short Linear Dipoles. *Electron. Lett.*, Vol. 17, July 1981, pp. 492–494.

MCALLISTER, M. W. AND LONG, S. A. Resonant Hemispherical Dielectric Antenna. *Electron. Lett.*, Vol. 20, August 1984, pp. 657–659.

MCALLISTER, M. W., LONG, S. A., AND CONWAY, G. L. Rectangular Dielectric Resonator Antenna. *Electron. Lett.*, Vol. 19, March 1983, pp. 218–219.

MCLEAN, J. S. A Re-Examination of the Fundamental Limits on the Radiation *Q* of Electrically Small Antennas. *Trans. IEEE*, Vol. AP-44, May 1996, pp. 672–676.

MCLEAN, J. S. AND CROOK, G. E. *Broadband Antenna Incorporating Both Electric and Magnetic Dipole Radiators,* U.S. Patent 6,329,955, December 11, 2001.

MEHDIPOUR, A., ALIAKBARIAN, H. AND RASHED-MOHASSEL, J. A Novel Electrically Small Spherical Wire Antenna with Almost Isotropic Radiation Pattern. *IEEE Antenn. Wireless Propag. Lett.*, Vol. 7, 2008, pp. 396–399.

MEI, K. K. AND VAN BLADEL, J. G. Scattering by Perfectly-Conducting Rectangular Cylinders. *Trans. IEEE*, Vol. AP-11, March 1963, pp. 185–192.

MIDDLETON, W. M. *Reference Data for Engineers*, Newnes–Butterworth–Heinemann, 2002, Section 4. 20.

MONGIA, R. K. Half-Split Dielectric Resonator Placed on Metallic Plane for Antenna Applications. *Electron. Lett.*, Vol. 25, May 1989, pp. 462–464.

MONGIA, R. K. Theoretical and Experimental Resonant Frequencies of Rectangular Dielectric Resonators. *Proc. IEE Part H*, Vol. 139, February 1992, pp. 98–104.

MONGIA, R. K. Reduced Size Metallized Dielectric Resonator Antennas. *IEEE AP Symp. Dig.*, Vol. 35, 1997, pp. 2202–2205.

MONGIA, R. K. AND BHARTIA, P. Dielectric Resonator Antennas—A Review and General Design Relations for Resonant Frequency and Bandwidth. *Int. J. Microwave Millimeter-Wave Comput.-Aided Eng.*, Vol. 4, 1994, pp. 230–247.

MONGIA, R. K. AND ITTIPIBOON, A. Theoretical and Experimental Investigations on Rectangular Dielectric Resonator Antennas. *Trans. IEEE*, Vol. AP-45, September 1997, pp. 1348–1356.

MONGIA, R. K., ITTIPIBOON, A., AND CUHACI, M. Measurement of Radiation Efficiency of Dielectric Resonator Antennas. *IEEE Microwave Guided Wave Lett.*, Vol. 4, March 1994a, pp. 80–82.

MONGIA, R. K. ET AL. Accurate Measurement of *Q*-Factors of Isolated Dielectric Resonators. *Trans. IEEE*, Vol. MTT-42, August 1994b, pp. 1463–1467.

MOON, J.-L. AND PARK, S.-O. Dielectric Resonator Antenna for Dual-Band PCS/IMT-2000. *Electron. Lett.*, Vol. 36, June 2000, pp. 1002–1003.

MRIDULA, S. ET AL. Characteristics of a Microstrip-Excited High-Permittivity Rectangular Dielectric Resonator Antenna. *Microwave Opt. Technol. Lett.*, Vol. 40, February 2004, pp. 316–318.

Nannini, C. et al. A Dual-Frequency Dielectric Resonator Antenna. *Microwave Opt. Technol. Lett.*, Vol. 38, July 2003, pp. 9–10.

Nashaat, D. M., Elsadek, H. A., and Ghali, H. Single Feed Compact Quad-Band PIFA Antenna for Wireless Communication Applications. *Trans. IEEE*, Vol. 53, August 2005, pp. 2631–2635.

Overfelt, P. L. Electric Lines of Force of an Electrically Small Dipole-Loop Antenna Array. *Trans. IEEE*, Vol. AP-46, March 1998, pp. 451–456.

Paul, B. et al. A Compact Very-High-Permittivity Dielectric-Eye Resonator Antenna for Multiband Wireless Applications. *Microwave Opt. Technol. Lett.*, Vol. 43, October 2004, pp. 118–121.

Pinhas, S. and Shtrikman, S. Comparison Between Computed and Measured Bandwidth of Quarter-Wave Microstrip Radiators. *Trans. IEEE*, Vol. AP-36, November 1988, pp. 1615–1616.

Rao, B. L. J., Ferris, J. E., and Zimmerman, W. E. Broadband Characteristics of Cylindrical Antennas with Exponentially Tapered Capacitive Loading. *Trans. IEEE*, Vol. AP-17, March 1969, pp. 145–151.

Rao, S. M., Wilton, D. R., and Glisson, A. W. Electromagnetic Scattering by Surfaces of Arbitrary Shape. *Trans. IEEE*, Vol. AP-30, May 1982, pp. 409–418.

Rashed, J. and Tai, C.-T. A New Class of Resonant Antennas. *Trans. IEEE*, Vol. AP-39, September 1991, pp. 1428–1430.

Richmond, J. H. Scattering by an Arbitrary Array of Parallel Wires. *Trans. IEEE*, Vol. MTT-13, July 1965, pp. 408–412.

Rosa, E. B. and Grover, F. W.Formulas and Tables for the Calculation of Mutual and Self-Inductance. *NBS Circular*, No. 169, December 1916, p. 135.

Rumsey, V. H. Reaction Concept in Electromagnetic Theory. *Phys. Rev.*, Vol. 94, 1954, pp. 1483–1491.

Salonen, P., Keskilammi, M., and Kivikoski, M. Single-Feed Dual-Band Planar Inverted-F Antenna with U-Shaped Slot. *Trans. IEEE*, Vol. AP-48, August 2000, pp. 1262–1264.

Shen, L.-C. An Experimental Study of the Antenna with Nonreflecting Resistive Loading. *Trans. IEEE*, Vol. AP-15, September 1967, pp. 606–611.

Shen, L.-C. and Wu, T. T. Cylindrical Antenna with Tapered Resistive Loading. *Radio Sci.*, Vol. 2, February 1967, pp. 191–201.

Sim, C.-Y.-D. and Cheng, Y.-T. Quad-Band PIFA Design for WLAN/Satellite Applications. *Microwave Opt. Technol. Lett.*, Vol. 52, October 2010, pp. 2331–2336.

Smith, D. L. The Trap-Loaded Cylindrical Antenna. *Trans. IEEE*, Vol. AP-23, January 1975, pp. 20–27.

Smith, G. S. Proximity Effect in Systems of Parallel Conductors. *J. Appl. Phys.*, Vol. 43, May 1972, pp. 2196–2203.

Stark, P. A. *Introduction to Numerical Methods*, Macmillan, 1970, p. 130.

Stuart, H. R. Eigenmode Analysis of Small Multielement Spherical Antennas. *Trans. IEEE*, Vol. AP-56, September 2008, pp. 2841–2851.

Stuart, H. R. and Tran, C. Subwavelength Microwave Resonators Exhibiting Strong Coupling to Radiation Modes. *Appl. Phys. Lett.*, Vol. 87, 2005, 151108.

Stuart, H. R. and Tran, C. Small Spherical Antennas Using Arrays of Electromagnetically Coupled Planar Elements. *IEEE Antenn. Wireless Propag. Lett.*, Vol. 6, 2007, pp. 7–10.

STUART, H. R., BEST, S. R., AND YAGHJIAN, A. D. Limitations in Relating Quality Factor to Bandwidth in a Double Resonance Small Antenna. *IEEE Antenn. Wireless Propag. Lett.*, Vol. 6, 2007, pp. 460–463.

SUNG, Y., AHN, C. S., AND KIM, Y. S. Microstripline Fed Dual-Frequency Dielectric Resonator Antenna. *Microwave Opt. Technol. Lett.*, Vol. 42, September 2004, pp. 388–390.

TAYLOR, C. D. Cylindrical Transmitting Antenna: Tapered Resistivity and Multiple Impedance Loadings. *Trans. IEEE*, Vol. AP-16, March 1968, pp. 176–179.

TERMAN, F. E. *Radio Engineers' Handbook*, McGraw-Hill, 1943, Section 2.

TILSTON, M. A. AND BALMAIN, K. G. A Multiradius, Reciprocal Implementation of the Thin-Wire Moment Method. *Trans. IEEE*, Vol. AP-38, October 1990, pp. 1636–1644.

TSUJI, M. ET AL. Analytical and Experimental Considerations on the Resonant Frequency and the Quality Factor of Dielectric Resonators. *Trans. IEEE*, Vol. MTT-11, November 1982, pp. 1952–1958.

VILLEGER, S. ET AL. Dual-Band Planar *Inverted*-F Antenna. *Microwave Opt. Technol. Lett.*, Vol. 38, July 2003, pp. 40–42.

WAIT, J. R. Receiving Propagation Properties of a Wire Loop with a Spheroidal Core. *Can. J. Technol.*, Vol. 31, January 1953a, pp. 9–14.

WAIT, J. R. The Receiving Loop with a Hollow Prolate Spheroidal Core. *Can. J. Technol.*, Vol. 31, June 1953b, pp. 132–137.

WALSER, R. M. Electromagnetic Metamaterials. In *Complex Mediums II: Beyond Linear Isotropic Dielectrics*, Lakhtakia, A., Werner, S. W., and Hodgkinson, I. J., Eds., *Proc. SPIE*, Vol. 4467, 2001, pp. 1–15.

WEGSTEIN, J. *Algorithm 2. CACM*, Vol. 3, Issue 2 1960, p. 74.

WHEELER, H. A. Fundamental Limitations of Small Antennas. *Proc. IRE*, Vol. 35, December 1947, pp. 1479–1484.

WHEELER, H. A. Transmission-Line Propagation Properties of Parallel Strips Separated by a Dielectric Sheet. *Trans. IEEE*, Vol. MTT-13, March 1965, pp. 172–185.

WONG, K.-L. *Compact and Broadband Microstrip Antennas*, Wiley, 2002.

WONG, K.-L. *Planar Antennas for Wireless Communications*, Wiley, 2003.

WONG, J. L. AND KING, H. E. Height-Reduced Meander Zigzag Monopoles with Broad-Band Characteristics. *Trans. IEEE*, Vol. AP-34, May 1986, pp. 716–717.

WU, T. T. AND KING R. W. P. The Cylindrical Antenna with Nonreflecting Resistive Loading. *Trans. IEEE*, Vol. AP-13, May 1965, pp. 369–373;correction November 1965, p. 998.

YAGHJIAN, A. D. AND BEST, S. R. Impedance, Bandwidth, and *Q* of Antennas. *Trans. IEEE*, Vol. AP-53, April 2005, pp. 1298–1324.

CHAPTER 4

CLEVER PHYSICS, BUT BAD NUMBERS

4.1 CONTRAWOUND TOROIDAL HELIX ANTENNA

The first example of an ESA concept that was clever and sound, but has unfortunate characteristics (tight tolerances, low radiation resistance, etc.), is the contrawound toroidal helix antenna (CWTHA). This was invented and patented by Corum (1986, 1988), who started with a coil of many turns bent into a ring or toroid. Inexplicably, some years later the patent office granted patents for the CWTHA to former lab assistants of Corum! Applying a voltage at the terminals produces a ring of current, like a single-turn loop. Then a second toroidal coil is added in the same volume, but wound in the opposite direction. If the second winding is excited 180° out of phase with the first, the loop currents cancel. Figure 4.1 is a sketch of a CWTHA; the loop lines represent lines of magnetic flux. The feeds are at the origin. However, the currents around the turns add, and a vertical (dipole) electric field is created. The CWTHA was heavily promoted by others without any careful measurements or calculations. Maclean and Rahman (1978) showed that the CWTHA coordinates could be exactly written in terms of a single spherical coordinate variable. A ring-bar analysis, adapted from TWT work, was performed by Hansen and Ridgley (1999); this was followed by an exact analysis (Hansen and Ridgley, 2001) using exact geometry. The exact vector potential integrals were integrated numerically. It was shown that any combination of dimensions that gave an omnidirectional pattern (in the plane of the toroid) produced a very small ratio of dipole/loop fields. Thus, the mechanical and excitation tolerances are very tight. The corresponding radiation resistance is low, where the reactance is modest. When the dimensions produce resonance, the radiation resistance is again low and the bandwidth is narrow

Small Antenna Handbook, Robert C. Hansen and Robert E. Collin.

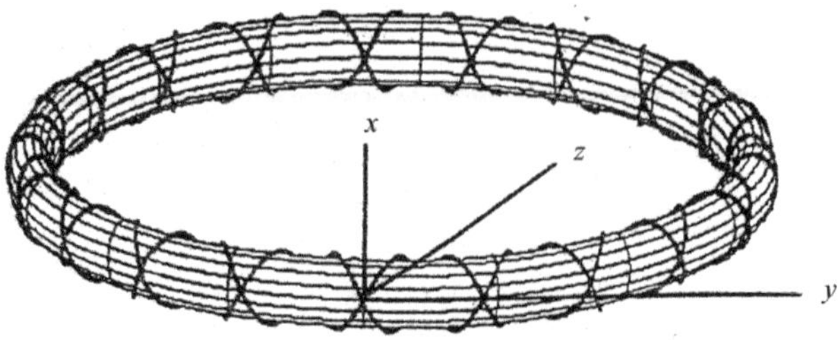

FIGURE 4.1 Contrawound toroidal helix (on foam core). Courtesy of R. C. Hansen, and R. D. Ridgley.

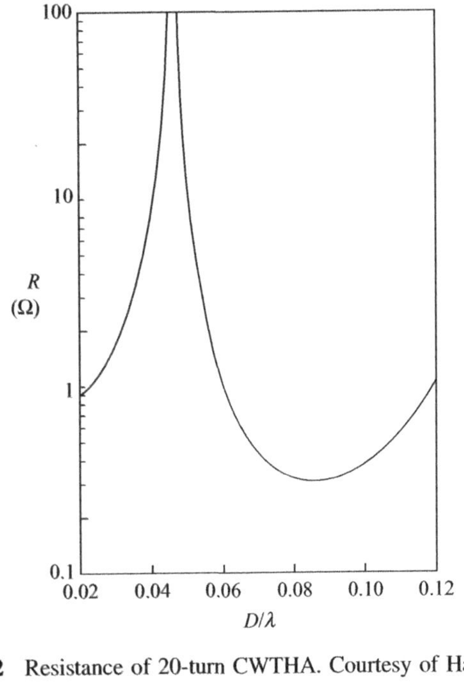

FIGURE 4.2 Resistance of 20-turn CWTHA. Courtesy of Hansen (2001).

(Hansen, 2001). Figures 4.2 and 4.3 show resistance and reactance for a 20-turn model, with the ratio of toroid diameter to turn diameter of 10. Figure 4.4 shows the azimuth pattern. Even a two-turn CWTHA, which can be implemented with printed flat strips, is narrowband (Hansen, 2002). Other approximate analyses are done by Miron (2001) and Pertl et al. (2005); the second paper is about patterns and has no information about the critical areas of tolerances, radiation resistance, impedance, and so on. See also Hansen (2005). So the CWTHA was a good idea, but the very tight tolerances and low radiation resistances doomed it.

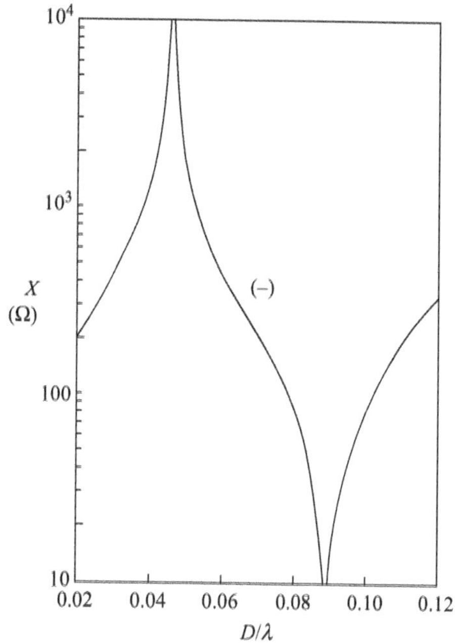

FIGURE 4.3 Reactance of 20-turn CWTHA. Courtesy of Hansen (2001).

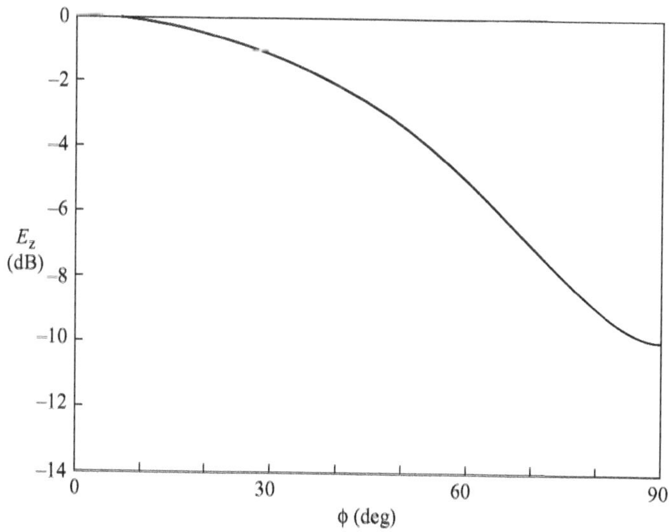

FIGURE 4.4 Azimuth pattern of 20-turn CWTHA. Courtesy of Hansen (2001).

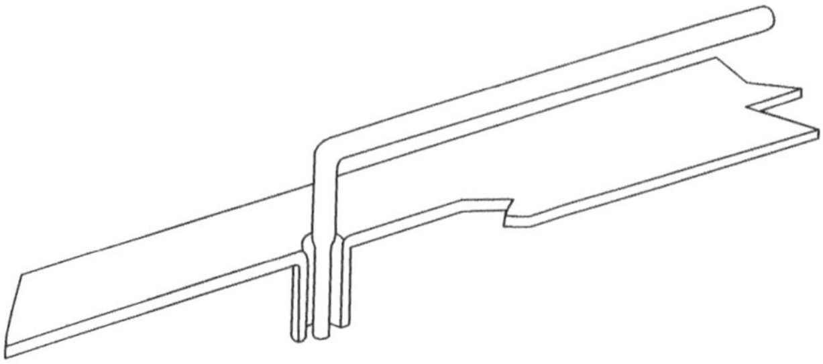

FIGURE 4.5 Transmission line antenna. Courtesy of King et al. (1960).

4.2 TRANSMISSION LINE ANTENNAS

The transmission line antenna, sketched in Figure 4.5, is simply a wire or strip over a ground plane. The wire length is usually $\lambda/4$ or less, and the height above the ground plane is much less. A shunt stub or capacitor may be employed at the end. This antenna is essentially a vertical monopole loaded by a wire parallel to the ground plane (King et al., 1960; Prasad and King, 1961; Guertler, 1977). The loading may allow the monopole to have a nearly uniform current, with a nearly quadrupling of the monopole-alone radiation resistance. In that case, the radiation resistance would be $40k^2h^2$, where h is the monopole height. Thus, the limitation on the transmission line antenna is the low radiation resistance.

4.3 HALO, HULA HOOP, AND DDRR ANTENNAS

A small loop antenna can be folded, as sketched in Figure 4.6. This has been called the halo loop (Harrison and King, 1961). It could also be the result of bending a folded dipole into the halo shape. A narrow bandwidth results. The closely spaced half-loop conductors carry opposing currents, resulting in a radiation resistance reduction of 5:1, and a reactance increase of 2:1, over the folded dipole, for half-wave size. For smaller antennas, the performance is even poorer. Radiation resistance is roughly $30k^2a^2$, with a the loop radius.

The hula hoop antenna is simply a short monopole with capacitive loading bent into a hoop to save space (Boyer, 1963; Nakano et al., 1993; see Figure 4.7). The wire is $\lambda/4$ in length. It was observed by Burton and King (1963) that such an antenna behaves essentially the same when the wire is straight or in a hoop, and thus the radiation resistance is again $30k^2h^2$. Although the silhouette is low, the resistance is very low. If the wire length is electrically short, the reactance becomes very high and the resistance even lower. Thus, this antenna experienced only a brief flurry of interest before being consigned to history.

The simplest top loading of a short monopole is with a top hat, a metal plate at the monopole end, which acts as a capacitive load. A large hat might produce a nearly

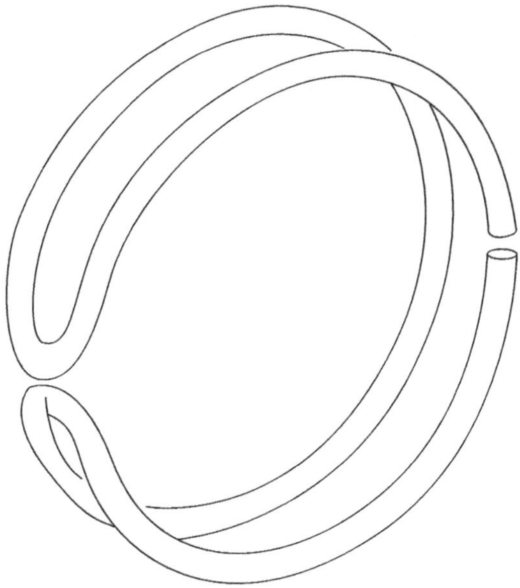

FIGURE 4.6 Halo loop.

constant current in the monopole, with nearly quadruple the radiation resistance, but the hat diameter would be larger than the monopole length. A moment method analysis of a monopole with large top disk was given by Simpson (2004), who found that the load could produce resonance, but the resulting bandwidth was narrow.

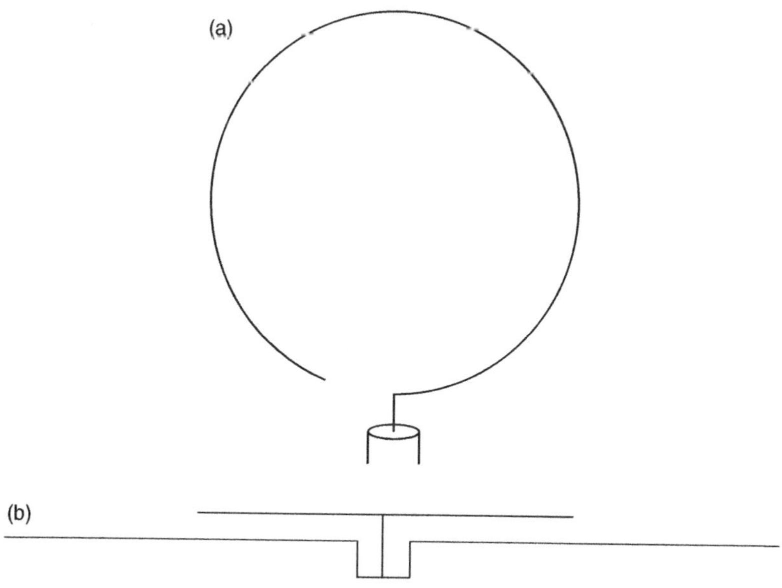

FIGURE 4.7 Hula hoop antenna: (a) top view; (b) side view.

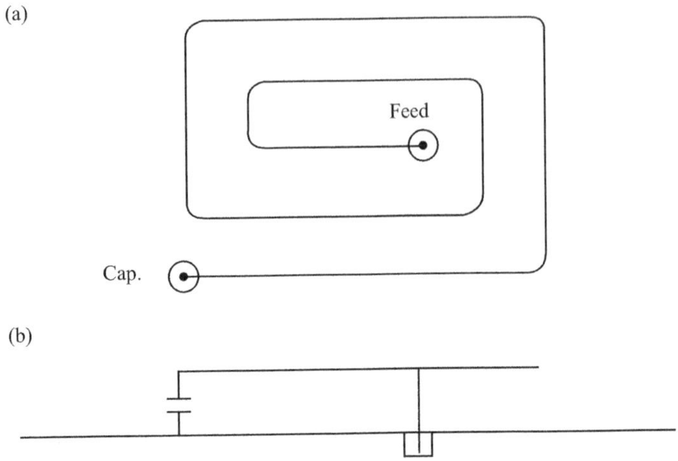

FIGURE 4.8 DDRR antenna: (a) top view; (b) side view.

More elaborate transmission line antennas were developed by Fenwick (1965). Here the single capacitive wire above ground plane is replaced by a meandering wire, by a planar strip spiral, or by a partially counterwound planar strip spiral. The end opposite from the feed may be grounded or connected to a resistor. These configurations tend to be less effective than a top plate. See also Hallbjörner (2004). The transmission line may be zigzagged, to reduce the length (Lee and Mei, 1970).

When a simple planar wire spiral, with the outside end connected to the ground plane by a capacitor, is used as the top load (Wanselow and Milligan, 1966), the radiation resistance is improved somewhat, but the resonant circuit Q reduces the bandwidth. This configuration was called the direct driven resonant radiator (DDRR); the name was more impressive than the performance (see Figure 4.8).

Top loading schemes for MF and HF monopoles are described and evaluated by Trainotti (2001).

4.4 DIELECTRIC-LOADED ANTENNAS

The first attempt in this roughly chronological chronicle was to encapsulate a short dipole in dielectric, in the hope that the result would be similar to that produced by loading a loop with a high-permeability core. Unfortunately, the loop and dipole are not analogs; there are no magnetic currents.

In gross terms, dielectric loading of antennas is undesirable; the current is decreased by $1/\varepsilon_r$ (Wheeler, 1947; Schelkunoff and Friis, 1952). Dipoles or monopoles with dielectric sleeves have been investigated for over 40 years. Polk (1959) analyzed a biconical dipole embedded in a dielectric sphere, using the Schelkunoff transmission line approach. For a short biconical dipole, the reactance was reduced by the dielectric, but the bandwidth was also reduced. Grimes (1958) showed that a dielectric sleeve around a dipole would affect the impedance due to ε, but not from μ. Galejs (1962, 1963) used a spherical mode approach for a short dipole in a dielectric sphere. Using the lowest mode, he showed that the efficiency $\times$ bandwidth product varies as $9\varepsilon/(\varepsilon+2)^2$. For $\varepsilon = 1$ this factor is 1, of course, and for large ε it approaches $9/\varepsilon$. The maximum occurs for $\varepsilon = 2$, and is 1.125. This contradicts the results of Wheeler (1958), who assumed a dielectric surface tangential to E. The results of Galejs were recapitulated by Chatterjee (1985). Measurements were made of short probes with dielectric sleeves by Birchfield and Free (1974): the sleeves improved the impedance match, but data are sparse. A moment method analysis of dielectric-coated dipoles was made by Richmond and Newman (1976). Resonant conductance increased roughly linearly with $(\varepsilon-1)/\varepsilon \, \ell n \, b/a$, where a is the dipole radius and b is the sleeve radius. Bandwidth is reduced, but no data are given. See also Popovic et al. (1981). Smith (1977) in a brief note indicates that the efficiency $\times$ bandwidth product is reduced by the addition of the dielectric, based on the work of Galejs. A cylindrical dielectric or ferrite sleeve on a dipole was treated as a resonator by James et al. (1974) and James and Henderson (1978). See also Fujimoto et al. (1987). The external field has a continuous eigenvalue spectrum, and the cavity has discrete modes. Their perturbation approach ensures that the cavity fields are only slightly changed by loss and by radiation. A variational method is then used to find the fields. Results show decrease in efficiency (due to material losses) and decrease in bandwidth. The degradation appears less when $\mu = \varepsilon$ for the dielectric. Reduced dipole length, for resonance, may not compensate adequately for the lower efficiency and smaller bandwidth. Returning now to subresonant cladding, King and Smith (1981) formulated an integral equation for the currents; only a few results are given. Sinha and Saoudy (1990) did a Weiner–Hopf analysis of a dielectric-coated dipole; the asymptotic approximation of current shows unexplained small changes in impedance with ε. An FDTD approach was used by Bretones et al. (1994), for the case where sleeve thickness equals wire radius, with $\varepsilon = 3.2$. A small shift in impedance peaks was observed. Francavilla et al. (1999) added a circular dish top hat to a monopole, with dielectric between hat and ground plane. Cylindrical wave functions and mode matching were used. As ε was increased, the resonant frequency decreased, and the bandwidth decreased, both as expected. See also Janapsatya and Bialkowski (2004). A different use of dielectric loading uses a permeable bead placed on a long monopole, to make the monopole act as a $\lambda/4$ monopole (Kennedy et al., 2003).

Dielectric coatings on monopoles or dipoles can produce resonance at lengths well below $\lambda/4$, but the lower efficiency and lower bandwidth are generally a poor trade. Better results are obtained by inserting a series inductor as mentioned in Section 3.2.2.

4.5 MEANDERLINE ANTENNAS

The resonant size of a dipole can be reduced by meandering the conductor, as in Figure 4.9 (Rashed and Tai, 1991). For resonance, the wire length is roughly half-wave. These antennas can be used over a ground plane at quarter-wave spacing, or as

FIGURE 4.9 Meanderline antenna. Courtesy of Rashed and Tai (1991).

a patch. A meanderline monopole for a notebook computer has been described by Lin et al. (2005). Meander patches have been used by the HTS community; see Section 7.3.2. Probe feeding was used by Chang and Kuo (2005). A less effective arrangement inserts a folded dipole crosswise in the center of a monopole (Altshuler, 1993).

Although the meander monopole antenna can be resonant in a width (or length) below $\lambda/4$, there are several significant disadvantages (Best, 2003). The closely spaced wires store reactive energy, which reduces bandwidth. Radiation resistance tends to be that of a monopole with the physical length, not the wire length; thus, it is often low. And the wire length introduces nonnegligible loss resistance. Comparisons of meanderline and zigzag antennas are made in Section 3.2.4.

4.6 CAGE MONOPOLE

A novel monopole loading scheme uses a cage of four monopoles with a loading wire at the top of each; the monopoles are independently fed (Breakall et al., 2002, 2003). Each monopole is connected to a transmission line transformer to raise the low radiation resistance; then the four transformer coax are connected in parallel (see Figure 4.10). Moment method simulations have been concerned with modeling the ground wires and the earth. A simple understanding of how this antenna works is provided by a model of four closely spaced dipoles, all much shorter than $\lambda/4$. Unlike the monopoles, where the top loads make the vertical current nearly constant, the short dipoles have a triangular current, but for understanding impedance and performance the cage of dipoles is adequate. If one dipole alone has impedance Z_{11}, the mutual impedances Z_{12}, Z_{13}, and Z_{14} will be very close to Z_{11}, but with somewhat less reactance, as the dipoles are closely spaced. Then the impedance of each dipole in the cage is closely $4Z_{in}$. The impedance transformation process does not change this basic result. Thus, the Q of the cage is essentially that of one monopole as fat as the cage. Better performance (easier matching) would result from

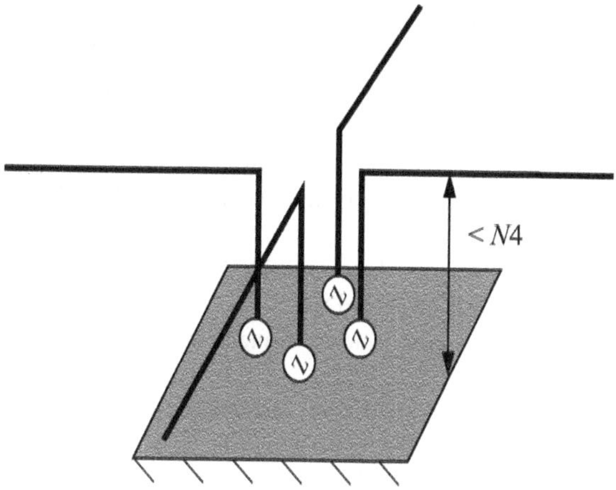

FIGURE 4.10 Cage monopole.

using a single fat monopole (four or more wires connected together at the feed point) with an equivalent top load. Thus, the separate feeding is of no value as the monopoles are eventually connected together.

REFERENCES

ALTSHULER, E. E. A Monopole Antenna Loaded with a Modified folded Dipole. *Trans. IEEE*, Vol. AP-41, July 1993, pp. 871–876.

BEST, S. R. A Discussion on the Significance of Geometry in Determining the Resonant Behavior of Fractal and Other Non-Euclidian Wire Antennas. *IEEE AP Mag.*, Vol. 45, June 2003, pp. 9–28.

BIRCHFIELD, J. L. AND FREE, W. R. Dielectrically Loaded Short Antennas. *Trans. IEEE*, Vol. AP-22, May 1974, pp. 471–472.

BOYER, J. M. Hula-Hoop Antennas: A Coming Trend? *Electronics*, January 1963, pp. 44–46.

BREAKALL, J. K. ET AL. A Novel Short AM Monopole Antenna with Low-Loss Matching System. *Proc. IEEE Broadcast Technology Symp.*, Washington, DC, 2002.

BREAKALL, J. K. ET AL. Testing and Results of a New, Efficient Low-Profile AM Medium Frequency Antenna System. *Proc. NAB Broadcast Engineering Conf.*, 2003.

BRETONES, A. R. ET AL. Time Domain Analysis of Dielectric-Coated Wire Antennas and Scatterers. *Trans. IEEE*, Vol. AP-42, June 1994, pp. 815–819.

BURTON, R. W. AND KING, R. W. P. Theoretical Considerations and Experimental Results for the Hula-Hoop Antenna. *Microwave J.*, Vol. 6, November 1963, pp. 89–90.

CHANG, T.-N. AND KUO, C.-C. Meander Antenna with Backside Tuning Stubs. *Trans. IEEE*, Vol. AP-53, April 2005, pp. 1274–1277.

CHATTERJEE, R. *Dielectric and Dielectric-Loaded Antennas*, Wiley, 1985, Chapter 7.

CORUM, J. F. *Toroidal Antennas*, U.S. Patent No. 4,622,558, November 11, 1986.

CORUM, J. F. *Electromagnetic Structure and Method*, U.S. Patent No. 4,751,515, June 14, 1988.

Fenwick, R. C. A New Class of Electrically Small Antennas. *Trans. IEEE*, Vol. AP-13, May 1965, pp. 379–383.

Francavilla, L. A. et al. Mode-Matching Analysis of Top-Hat Monopole Antennas Loaded with Radially Layered Dielectric. *Trans. IEEE*, Vol. AP-47, January 1999, pp. 179–185.

Fujimoto, K. et al. *Small Antennas*, Research Studies Press, Wiley, 1987, Chapter 3.

Galejs, J. Dielectric Loading of Electric Dipole Antennas. *J. Res. NBS*, Vol. 66D, September–October 1962, pp. 557–562.

Galejs, J. Small Electric and Magnetic Antennas with Cores of a Lossy Dielectric. *J. Res. NBS*, Vol. 67D, July–August 1963, pp. 445–451.

Grimes, D. M. Miniaturized Resonant Antenna Using Ferrites. *J. Appl. Phys.*, Vol. 29, March 1958, pp. 401–402.

Guertler, R. J. F. Isotropic Transmission-Line Antenna and Its Toroid-Pattern Modification. *Trans. IEEE*, Vol. AP-25, May 1977, pp. 386–392.

Hallbjörner, P. Electrically Small Unbalanced Four-Arm Wire Antenna. *Trans. IEEE*, Vol. AP-52, June 2004, pp. 1424–1428.

Hansen, R. C. Resonant Contrawound Toroidal Helix Antenna. *Microwave Opt. Technol. Lett.*, Vol. 29, June 2001, pp. 408–410.

Hansen, R. C. Two-Turn Contrawound Toroidal Helix Antenna. *Microwave Opt. Technol. Lett.*, Vol. 33, May 2002, pp. 251–252.

Hansen, R. C. Comments on "An Analysis of the CWTHA's Resonance Currents and the Resulting Radiation Shapes". *Trans. IEEE*, Vol. AP-53, June 2005, p. 2132.

Hansen, R. C. and Ridgley, R. Modes of the Contrawound Toroidal Helix Antenna. *Microwave Opt. Technol. Lett.*, Vol. 23, December 1999, pp. 354–368.

Hansen, R. C. and Ridgley, R. D. Fields of the Contrawound Toroidal Helix Antenna. *Trans. IEEE*, Vol. AP-49, August 2001, pp. 1138–1141.

Harrison, C. W. and King, R. W. P. Folded Dipoles and Loops. *Trans. IRE*, Vol. AP-9, March 1961, pp. 171–187.

James J. R. and Henderson, A. Electrically Short Monopole Antennas with Dielectric or Ferrite Coatings. *Proc. IEE*, Vol. 125, September 1978, pp. 793–803.

James, J. R., Schuler, A. J., and Binham, R. F. Reduction of Antenna Dimensions by Dielectric Loading. *Electron. Lett.*, Vol. 10, June 1974, pp. 263–265.

Janapsatya, J. and Bialkowski, M. E. Reducing the Height of a Circular Array of Monopoles Using Top Hats and Dielectric Coatings. *Radio Sci.*, Vol. 39, RS3004, 2004.

Kennedy, T. F., Long S. A., and Williams, J. T. Modification and Control of Currents on Monopole Antennas Using Magnetic Bead Loading. *IEEE Antenn. Wireless Propag. Lett.*, Vol. 2, 2003, pp. 208–211.

King, R. W. P. and Smith, G. S. *Antennas in Matter—Fundamentals, Theory, and Applications*, MIT Press, 1981, Chapter 8.

King, R. W. P. et al. Transmission-Line Missile Antennas. *Trans. IRE*, Vol. AP-8, January 1960, pp. 88–90.

Lee, S. H. and Mei, K. K. Analysis of Zigzag Antennas. *Trans. IEEE*, Vol. AP-18, November 1970, pp. 760–764.

Lin, C. C., Kuo, S. W., and Chuang, H. R. A 2.4-GHz Printed Meander-Line Antenna for USB-WLAN with Notebook-PC Housing. *IEEE Microwave Wireless Component Lett.*, Vol. 15, September 2005, pp. 546–548.

Maclean, T. S. M. and Rahman, F. Small Toroidal Antennas. *Electron. Lett.*, Vol. 14, May 1978, pp. 339–340.

Miron, D. B. A Study of the CTHA Based on Analytical Models. *Trans. IEEE*, Vol. AP-49, August 2001, pp. 1130–1137.

Nakano, H. et al. A Curl Antenna. *Trans. IEEE*, Vol. AP-41, November 1993, pp. 1570–1575.

Pertl, F. A., Smith, J. E., and Nutter, R. S. An Analysis of the CTHA's Resonance Currents and the Resulting Radiation Shapes. *Trans. IEEE*, Vol. AP-53, January 2005, pp. 377–385.

Polk, C. Resonance and Supergain Effects in Small Ferromagnetically or Dielectrically Loaded Biconical Antennas. *Trans. IRE*, Vol. AP-7, December 1959, pp. S414–S423.

Popovic, B. D., Djordejevic, A. R., and Kircanski, N. M. Simple Method for Analysis of Dielectric-Coated Wire Antennas. *Radio Electron. Eng.*, Vol. 51, March 1981, pp. 141–145.

Prasad, S. and King, R. W. P. Experimental Study of Inverted L-, T-, and Related Transmission Line Antennas. *J. Res. NBS*, Vol. 65D, September–October 1961, pp. 449–454.

Rashed, J. and Tai C.-T. A New Class of Resonant Antennas. *Trans. IEEE*, Vol. AP-39, September 1991, pp. 1428–1430.

Richmond, J. H. and Newman, E. H. Dielectric Coated Wire Antennas. *Radio Sci.*, Vol. 11, January 1976, pp. 13–20.

Schelkunoff, S. A. and Friis, H. T. *Antennas—Theory and Practice*, Wiley, 1952, Section 10.14.

Simpson, T. L. The Disk Loaded Monopole Antenna. *Trans. IEEE*, Vol. AP-52, February 2004, pp. 542–550.

Sinha, B. P. and Saoudy, S. A. Rigorous Analysis of Finite Length Insulated Antenna in Air. *Trans. IEEE*, Vol. AP-38, August 1990, pp. 1253–1258.

Smith, M. S. Properties of Dielectrically Loaded Antennas. *Proc. IEE*, Vol. 124, October 1977, pp. 837–839.

Trainotti, V. Short Medium Frequency AM Antennas. *Trans. IEEE Broadcasting*, Vol. 47, September 2001, pp. 263–284.

Wanselow, R. D. and Milligan D. W. A Compact, Low Profile, Transmission Line Antenna—Tunable over Greater than Octave Bandwidth. *Trans. IEEE*, Vol. AP-14, November 1966, pp. 701–707.

Wheeler, H. A. Fundamental Limitations of Small Antennas. *Proc. IRE*, Vol. 35, December 1947, pp. 1479–1484.

Wheeler, H. A. The Spherical Coil as an Inductor, Shield, or Antenna. *Proc. IRE*, Vol. 46, September 1958, pp. 1595–1602.

CHAPTER 5

PATHOLOGICAL ANTENNAS

Claims on these antennas typically have performance characteristics that violate the physical laws we work under.

5.1 CROSSED-FIELD ANTENNA

The crossed-field antenna (CFA) was conceived by Hately and Kabbary in the late 1980s. It was based on several "new" principles: that E and H fields could be created independently, that Maxwell's displacement current produces magnetic fields, and that the near-fields could be avoided. The basic geometry is shown in Figure 5.1: Two horizontal circular metallic disks form a capacitor that is excited by the transmitter, but it is alleged that the displacement current between the plates produces an azimuthal magnetic field. Two hollow metallic circular cylinders are stacked vertically, and above the horizontal plates. The transmitter also excites the cylinders, but 90° out of phase with the plate excitation. An electric field is produced by the cylinders as sketched. Because of the Poynting vector, the H and E fields radiate a wave into space. Several of these CFAs have been built, at MF. Later versions have incorporated a large flare structure into the upper cylinder, presumably to increase the electric field in Figure 5.1. To quote the authors, "E and H fields are produced from separate electrodes," "the two fields are compelled to cross at right angles in the interaction zone," "any CFA is capable of radiation over a decade of frequency," and "Moment Method cannot be used to model CFA, as it only models currents" (see Kabbary et al., 1989, 1997, 1999; Hately et al., 1991; see also U.K. Patent 2215524 and U.S. Patent 5155495). It is significant that peer-reviewed antenna journals have not accepted papers on CFA.

Small Antenna Handbook, Robert C. Hansen and Robert E. Collin.

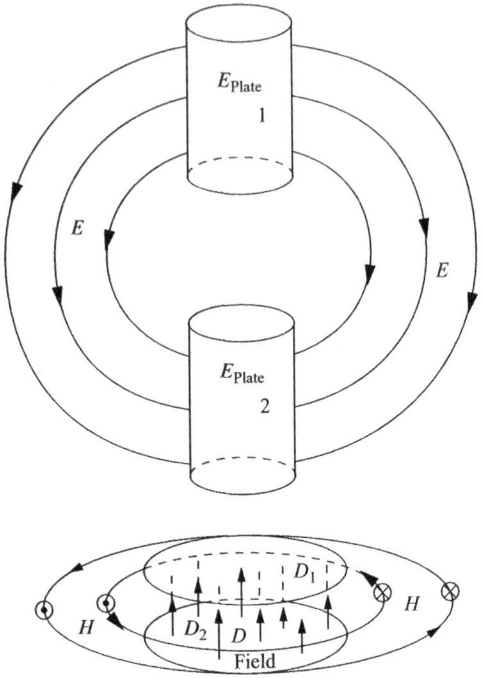

FIGURE 5.1 CFA geometry.

A critical analysis was made by Smith (1992). He performed a simple circuit analysis on one of the CFA antennas, with the result that the efficiency is below 50% and the Q is roughly 20. This value is due to the "fatness" of the CFA monopole. Belrose (2000a, 2000b) did both a moment method analysis and experiments on a CFA (Figure 5.2). This experimental model closely follows a CFA built at Tanta, Egypt. Both the E cylinders and the H plates had high Q (low radiation resistance), and the 90° phasing resulted in power oscillating between the cylinders and plates, further increasing loss. The NEC simulation and measurements both showed that, unlike the claims of Kabbary, (1) the relative voltage drives of the disk and cylinder are not critical; (2) in the vicinity of the CFA "electrodes," the wave impedance is much greater than $120\pi\ \Omega$; and (3) the CFA gain is not sensitive to the relative phase of the two voltage drives. The moment method calculations agreed well with measured data.

Hatfield (2000) calculated that for a typical CFA to maintain the E/H ratio of 120π, the plates would require a voltage of more than 1 MV. Kabbary and Hately apparently do not understand that both the H plates and the E cylinders produce near-fields, that the impedance of these near-fields is not 120π, and that it varies with distance. Furthermore, the near-fields are not orthogonal.

All measurements made by Kabbary et al. are doubtful for the following reasons: Almost all CFAs were mounted on top of buildings, where the building had at least a

FIGURE 5.2 CFA built and tested by Belrose.

partial metal skin; almost all CFA measurements were made at radio transmitting sites where extensive ground wire systems existed. In at least one case, the measurement inspectors were denied access to the phasing and matching cabinet; in at least one case, a hidden ground cable was discovered; in general, the tuning and adjustment of the CFA has proved difficult, and all performances have been narrowband. A variant of the CFA is the EH antenna (see Section 5.3), but the "principles" are the same. Foolishness leads to fraud? (Robert Park).

5.2 INFINITE EFFICIENCY ANTENNA

Kabbary Antenna Technologies (KAT) now claims to have designed and built an improved crossed-field antenna, called the Superpower Positive Feedback Antenna. KAT claims that this antenna can radiate at least five times its input power, thus almost infinite efficiency. Among the statements by KAT to justify the new antenna are the following: Ampere's law does not exist; and the fourth equation of Maxwell needs major correction. Not everyone understands that the Kabbary claims are absurd: Isle of Man International Broadcasting is trying to retrieve the £300,000 advance payment paid to KAT.

5.3 E–H ANTENNA

This antenna, patented by Hart (2002), is a fat dipole, consisting of two collinear fat cylinders slightly separated so that each can be connected to a "phasing and matching network." Claims are that the phasing network allows the *E*- and *H*-field components to be in phase. Of course, the far-field E_z and H_ϕ components radiated by a dipole are in phase. But the near-field relative phases depend upon both distance and angles. No feed network can alter this basic electromagnetics. The patent also claims that the EH antenna does not accept noise.

A more recent variant is the "Super C" antenna. This VHF–UHF antenna consists of a wire (mesh) basket 6 ft × 6 ft × 6 ft that surrounds a monopole. All this is over a slightly larger counterpoise. The usual claims are made: *E* and *H* fields are independent; the antenna is extremely wideband. The performance is probably due to radiation from the long vertical and horizontal feed cable. This is apparently based on the "high-efficiency compact antenna" of Henf (1997). Don't we ever learn?

5.4 TE–TM ANTENNA

It has been recognized since the work of Chu that exciting an ESA with both TM_1 and TE_1 modes would improve the bandwidth as much as twofold. That is because the Chu Q of the TM_1 and TE_1 modes are equal. However, the work of Thal (see Section 1.4), which included energy storage in the radian sphere, revealed that the TE_1 mode Q is twice that of the TM_1 mode. This, of course, limits the bandwidth of a dual-mode antenna.

Dockon, Inc. (Reno, NV) is building dual-mode antennas consisting of a single-turn loop and a coplanar dipole whose length roughly equals the loop diameter. They cite papers by Grimes and Grimes that purport to show that a dual-mode antenna has lower Q than a TM_1 mode antenna. In one paper, Grimes and Grimes (1999) announce a compound antenna with $Q = 0$. All of the Grimes papers (1995, correction 1996, 2001) have basic errors in calculating stored energy. Collin (1998) showed that Grimes calculated stored energy from the integral of the time-dependent Poynting vector over a spherical surface without considering the detailed energy density expressions. Another Grimes approach in calculating energy density used the sum of traveling and standing energy density. The continuity equation in space and time was applied, and differentiated with respect to time. But as pointed out by Collin, there is no Lorentz frame corresponding to uniform motion in the radial direction at velocity *c*. Alas, $Q = 0$ is a mirage. Their errors were admitted (Grimes and Grimes, 2005). The Grimes and Grimes work is now widely discredited.

The second problem is concerned with loop and dipole dimensions. The total length of wire in the loop must be roughly less than $\lambda/10$ in order to avoid pattern breakup and impedance swings. Thus, a single-turn square loop as shown in Figure 5.3 has a maximum side length of $\lambda/40$. The radiation resistance will be

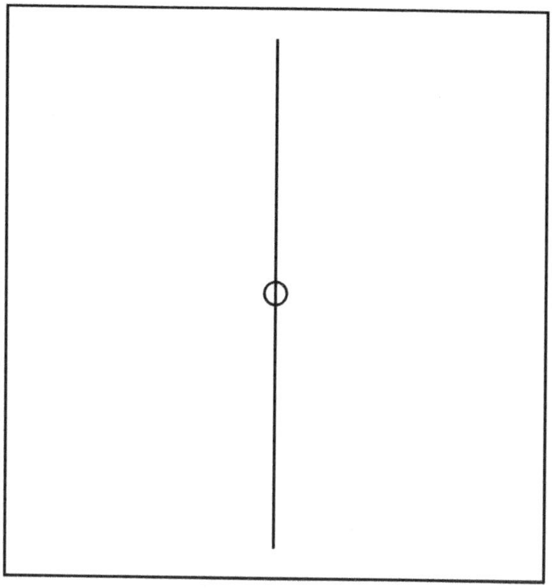

FIGURE 5.3 Dockon loop and dipole.

a maximum of a few ohms and this will reduce the efficiency somewhat. A dipole on the other hand with length of the order $\lambda/40$ has bandwidth of roughly 1% (Hansen, 2007). To achieve significant bandwidth, say 10% half-power bandwidth, the dipole needs to be roughly $\lambda/4$ long. Of course, such a long dipole can be packaged with a much smaller loop but at best the Q is reduced by one-half. Of course, losses can increase the bandwidth at the cost of efficiency. If the copper loss of the loop is comparable to its radiation resistance, the efficiency will be of the order 50% and the loop bandwidth will be doubled. The pattern problem is that a vertical dipole has an omnidirectional azimuth pattern. A loop in the same plane has a figure-eight azimuth pattern. A second loop could be added with appropriate feed circuitry but this increases the volume of the antenna significantly. There is no hope that this technology will provide a wideband electrically small antenna.

5.5 CROSSED DIPOLES

An idea for a wideband dipole uses two dipoles roughly at right angles, with the shorter dipole roughly half the length of the longer dipole (Kuo et al., 2010). The concept is simple: The orthogonal dipoles have zero mutual coupling, and the resonant impedance peaks are staggered in frequency, thereby providing a wideband impedance. There are two major flaws in this configuration. First, the polarizations are orthogonal, or nearly so. Second, at frequencies where either dipole is just over a wavelength in length, the pattern breaks up and splits.

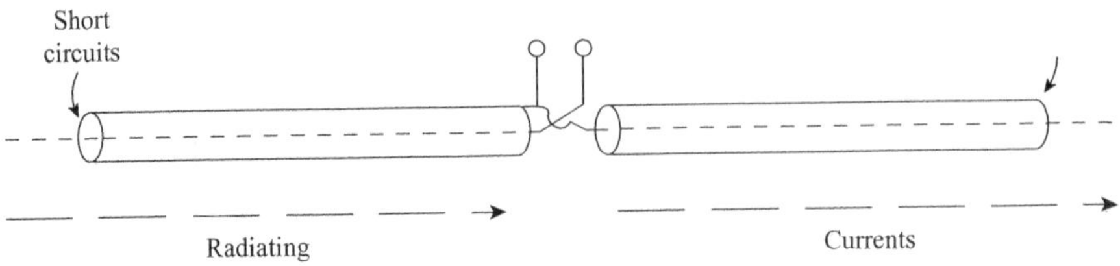

FIGURE 5.4 Snyder dipole.

5.6 SNYDER DIPOLE

A dipole with built-in coaxial matching stubs was invented by Snyder (1984a, 1984b); wideband performance was claimed. The Snyder dipole is sketched in Figure 5.4; the radiating currents are indicated in the figure. This configuration can be decomposed into the parallel combination of three structures: a thin wire dipole and two short-circuited transmission line stubs (see Figure 5.5). Thus, an equivalent circuit for this antenna is simply composed of dipole and transmission line components. Note that the coaxial structure shields the stub currents, preventing them from radiating, as long as the stubs are near resonant length. Call the dipole length L, the length of each stub S, and the diameter D; the wall thickness is assumed to be negligible. It is convenient to normalize the electrical stub length by the resonant electrical stub length and the stub diameter by the dipole length:

$$\alpha = \frac{\sqrt{\varepsilon} kS}{\pi/2} \quad \text{and} \quad \beta = D/L \tag{5.1}$$

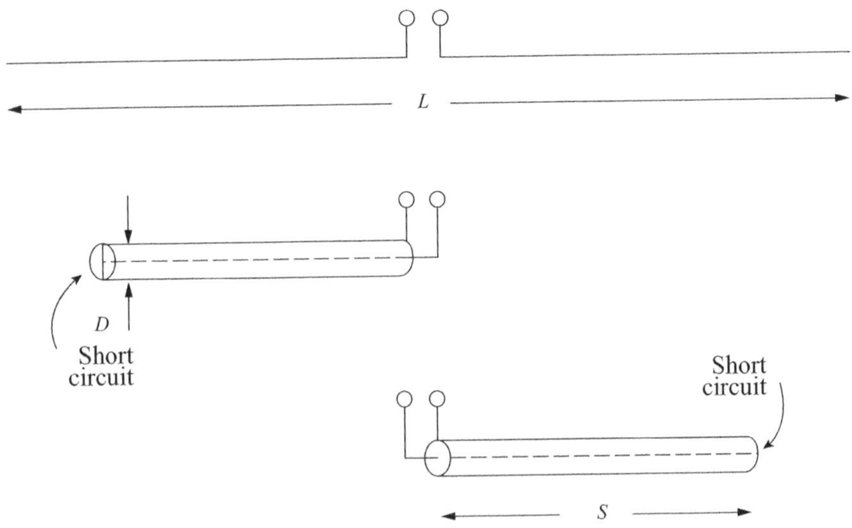

FIGURE 5.5 Snyder dipole components.

ε is the stub transmission line dielectric constant, and the characteristic impedance is given by

$$Z_{0\ell} = \frac{60}{\sqrt{\varepsilon}} \ell n\, D/2a \tag{5.2}$$

where a is the wire radius. The input impedance of one stub, where $k = 2\pi/\lambda$, is

$$Z_s = jZ_{0\ell} \tan \alpha\, \pi/2 \tag{5.3}$$

If the dipole impedance is called Z_d, the input impedance of the Snyder dipole will be

$$Z_{in} = \frac{Z_s Z_d}{Z_s + 2Z_d} \tag{5.4}$$

Note that this is different from that of a folded dipole, whose impedance is given by (Stutzman and Thiele, 1981)

$$Z_{in} = \frac{4Z_s Z_d}{Z_0 + 2Z_d} \tag{5.5}$$

This folded dipole result is four times the Snyder formula; both contain the parallel combination of dipole and two stubs.

The Snyder dipole is used with a matching transformer, and approximate value for its impedance Z_0 can be indicated by a simple analysis. An admittance $G + jB$ connected to a circuit of characteristic admittance $Y_0 = 1/Z_0$ has a VSWR given implicitly by

$$Y_0 G(V^2 + 1) = V(Y_0^2 + G^2 + B^2) \tag{5.6}$$

For VSWR = 2, this reduces to

$$Y_0 = \frac{5G}{4} \pm 1/(4\sqrt{9G^2 - 16B^2}) \tag{5.7}$$

When a simple resonant matching circuit is used optimally, the VSWR has a peak of 2 near center frequency and rises at band edges. Thus, near resonance, the dipole susceptance B_d and the stub susceptance B_s are both small, giving the approximate result that Y_0 should be $2G_d$ or $0.5G_d$. It will appear that the second is a good choice.

Snyder (1984b) uses as comparison a thin wire dipole with no matching transformer. This is obviously unfair. The comparison dipole used here uses a resonant tuning circuit as well as a transformer. This makes a fair comparison between the Snyder dipole, where the tuning is provided by coaxial stubs, and a

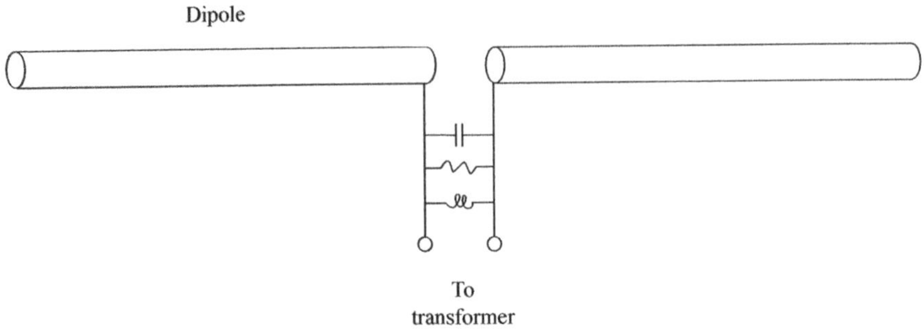

FIGURE 5.6 Dipole with matching circuit.

dipole where the tuning is provided by a lumped resonant circuit at the feed terminals. A parallel resonant circuit is used in parallel with the feed terminals (see Figure 5.6). The admittance of this circuit is given by

$$Y_{\mathrm{ckt}} = (1 + jQH)/R \tag{5.8}$$

where $H = f/f_0 - f_0/f$, $Q = 2\pi f_0 RC$, and $f_0^2 = 1/(4\pi^2 LC)$. It is convenient to use Q and R/Q as input parameters. Q should be selected very large so that the matching circuit loss does not increase the bandwidth. The parameter R/Q can be determined for maximum bandwidth. The matching transformer ratio, or, equivalently, the characteristic impedance of the dipole plus circuit, is selected to make VSWR $= 2$ at the resonant peak.

The Snyder dipole and the comparison dipole model have been implemented by this author in computer programs for ease of determining optimum results. A wire diameter of 0.005 times dipole length is used.

For the comparison dipole with matching resonant circuit, the circuit Q is set at 1000, a value sufficiently large so as to have little effect on bandwidth. The matching circuit is tuned to dipole resonance. It may be possible to improve the bandwidth somewhat by staggering the dipole and circuit resonance frequencies, but this is not done here. As occurred for the Snyder dipole, the optimum match impedance is a little less than twice the resonant resistance; $Z_0 = 126\ \Omega$. To find the optimum pair of values of R/Q and Z_0, a number of values of R/Q are used. For each R/Q, value of Z_0 is adjusted to make the VSWR peak around resonance equal to 2. Finally, the bandwidth of each pair is evaluated, and the largest is chosen. Note that R/Q must be carefully chosen; it controls the circuit reactance that offsets the dipole reactance. The value used was 30. Use of lower Q matching circuit components will increase bandwidth; a value of $Q = 100$ significantly improves bandwidth but lowers efficiency. The bandwidth was 21.0%.

For the Snyder dipole, taking $\alpha = 1$, which is for a stub that is resonant at the dipole resonant frequency, the maximum VSWR $= 2$ bandwidth occurs for $Z_{0\ell} = 45\ \Omega$ and $Z_0 = 127\ \Omega$; the normalized bandwidth is 20.7%. Calculations were made by picking values of α and $Z_{0\ell}$, then adjusting Z_0 to make the VSWR peak around center frequency equal to 2. This allows the low and high VSWR = 2 frequencies to be determined, hence the bandwidth. When the stub is tuned off the dipole resonance, the bandwidth decreases; this is typical of matching of high-Q circuit (the dipole) with a high-Q resonant circuit. Maximum bandwidth for $\alpha = 0.95$ occurs for $Z_{0\ell} = 42\,\Omega$ and $Z_0 = 110\,\Omega$; the normalized bandwidth is 18.3%.

At center frequency where the coaxial stubs are resonant, the radiating currents on the Snyder dipole are just those of an equivalent fat plain dipole. Thus, the basic gain is expected to be that of a dipole: 1.64 or 2.15 dB. In the center of the band and at band edges, where the VSWR = 2, there is a mismatch loss of 0.51 dB. In addition, the gain is reduced by the loss in the coaxial stubs, due to dissipation both in the coaxial dielectric and in the metallic conductors. A further gain reduction occurs away from center frequency, where the stubs are no longer resonant. This results in currents flowing from the inside of the stub around the open end to the outside. These exterior currents will radiate, and in some parts of the frequency band their radiation will oppose the primary radiation; thus, there will be a gain reduction. So the bandwidth calculations presented here are optimistic, in that the dipole gain diminished by the mismatch loss will not be realized over the entire bandwidth.

One might consider replacing the Snyder dipole with twin conductor stubs, but this configuration is not equivalent to the coaxial stub antenna. In the Snyder dipole, the return stub current is shielded, whereas in the twin conductor stubs the currents nearly cancel.

In summary, the Snyder dipole performance is no better than a fat dipole of the same diameter as the stub diameter. At band edges, it is worse because of stub transmission line currents that flow on the outside of the stubs away from resonance; this was not included in the simple analysis used here.

5.7 LOOP-COUPLED LOOP

In this configuration, a large tuned parasitic loop is excited by a coplanar, but eccentric, small loop (Dunlavy, 1971; see Figure 5.7). The large loop couples loss to the small loop, thereby increasing bandwidth and decreasing efficiency, but the gain is low for the large size. This antenna is also dispersive (Barrick, 1986). For some years, this loop was sold commercially by a U.S. company.

Because both loops are small in wavelengths, the performance can be analyzed by simple circuit theory. Figure 5.8 shows the equivalent circuit of the two loops. For the large single-turn loop, R_r is the radiation resistance, R_ℓ is the loss resistance, L is the inductance, and C is the tuning capacitor. For the small loop, the loss resistance is $R_{\ell 1}$, the inductance is L_1, and the load resistance is R_0. The mutual inductance is X_m. The circuit equations are

$$\begin{aligned} V &= I_1 \underbrace{[R_r + R_\ell + j(X_c + X_\ell + X_m)]}_{A} - I_2 \underbrace{[R_0 + R_{\ell 1} + j(X_{\ell 1} + X_m)]}_{B} \\ O &= I_1 \underbrace{j X_m}_{C} - I_2 \underbrace{[R_0 + R_{\ell 1} + j(X_{\ell 1} + X_m)]}_{D} \end{aligned} \tag{5.9}$$

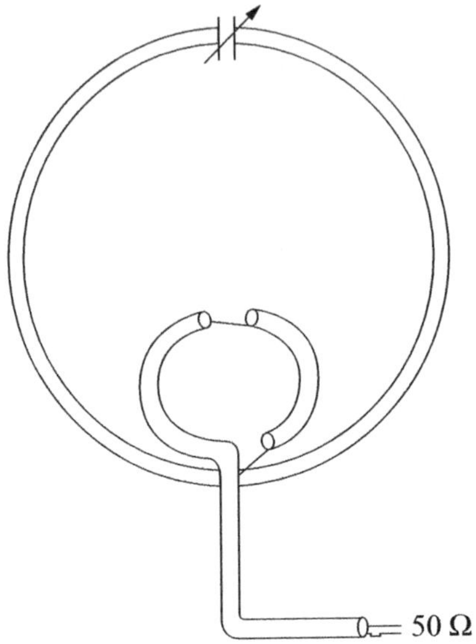

FIGURE 5.7 Loop-coupled loop.

The complex brackets are called A and B for brevity. The solution for the two currents is

$$I_1 = \frac{V}{A-C}, \quad I_2 = \frac{VC}{B(A-C)} \tag{5.10}$$

The power delivered to the load resistor as a result of the induced voltage is

$$P_1 = I_2 I_2^* R_0 = V^2 R_0 \left[\frac{C}{B(A-C)}\right]\left[\frac{C}{B(A-C)}\right]^* \tag{5.11}$$

To maximize this power, the quantity $B(A-C)$ times its conjugate should be minimized. If the loop dimensions are fixed, the minimization occurs when $X_c = X_\ell$; that is, the big loop is tuned for the desired frequency. This gives received power as a

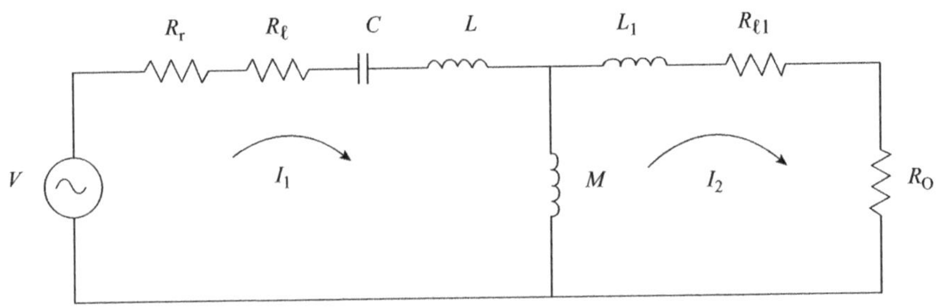

FIGURE 5.8 Loop-coupled loop equivalent circuit.

function of the resistances, the mutual inductance, and the small loop inductance:

$$P_1 = \frac{V^2 R_0 X_{\mathrm{m}}^2}{[R_0+R_{\ell 1}+j(X_{\ell 1}+X_{\mathrm{m}})][R_0+R_{\ell 1}+j(X_{\ell 1}+X_{\mathrm{m}})]^* [R_{\mathrm{r}}+R_\ell+jX_{\mathrm{m}}][R_{\mathrm{r}}+R_\ell+jX_{\mathrm{m}}]^*} \tag{5.12}$$

Because the maximum diameter of the large loop is roughly 0.1λ to avoid pattern breakup and impedance swings, the large loop reactance is significant while the radiation resistance and loss resistance are typically much smaller than 1 Ω. Thus, the power can be approximated with the result

$$P_1 = \frac{V^2 R_0}{(X_{\ell 1}+X_{\mathrm{m}})^2} \tag{5.13}$$

In comparison, the large loop by itself, when matched, has a power delivered to the load of

$$P = \frac{V^2 R_0}{(R_{\mathrm{r}}+R_\ell+R_0)^2} \tag{5.14}$$

The efficiency of the loop-coupled loop system is now found from the ratio of its power to the single-loop power just derived. This ratio is given by

$$\eta_{\mathrm{t}} = \frac{P_1}{P} = \frac{(R_{\mathrm{r}}+R_\ell+R_0)^2}{(X_{\ell 1}+X_{\mathrm{m}})^2} \tag{5.15}$$

Overall performance of the loop-coupled loop system now requires this efficiency to be multiplied by that of the single large loop, which is

$$\eta = \frac{R_{\mathrm{r}}}{R_{\mathrm{r}}+R_\ell} \tag{5.16}$$

It is immediately clear that the resistances in the numerator of Equation 5.15 are all very small, whereas the reactances in the denominator, although less than those for the large loop, are still much larger than the resistances. Thus, the transfer efficiency is very small. The formulas for the approximate mutual inductance between single-turn coplanar but noncoaxial loops are complicated (Grover, 1946). Very roughly, the ratio of mutual inductance to self-inductance varies as $\sqrt{d/D}$, where d is the diameter of the small loop and D is the diameter of the large loop. It appears that the mutual inductance term in Equation 5.15 may be comparable to the self-inductance of the small loop. In any case, the transfer efficiency is very small. It might be thought that the load resistance in Equation 5.15 could be increased in order to increase the transfer impedance. However, this would violate the assumptions made in reducing Equation 5.12 to Equation 5.13. It is expected that if the load resistance value were raised, and the exact formulas used, the transfer impedance would still be very small. The single-loop efficiency, which multiplies the transfer efficiency, may or may not be small. For the 0.1λ-diameter loop, the radiation resistance is $1.923\ \Omega$, and it should be easy to build a loop of a large conductor with high efficiency. However, most loops are much smaller than 0.1λ, and for these the efficiency may be poor, as the radiation resistance varies as the diameter in wavelengths to the fourth power. Thus, in all cases the efficiency of the loop-coupled loop system is small. Belrose (2004, 2005a, 2005b) has modeled the loop-coupled loop with the NEC-4 moment method code and has performed measurements on a model that he constructed. Because he was interested in modeling the effect of the earth on the antenna, a moment method analysis was appropriate. His results show in detail that the efficiency of the antenna is very poor, and that the bandwidth is very narrow, because of the high Q of the large loop. The conclusion is that if a loop antenna is needed for a given application, it should be a single loop tuned with a capacitor. Whether the loop is comprised of single turn or multiple turns will depend on a variety of other factors.

5.8 MULTIARM DIPOLE

A multiarm dipole represents an idea that has reappeared, phoenix like, many times from the ashes of critical engineering evaluation (Turner and Richard, 1968). The concept is simple. In Figures 5.9 and 5.10, each pair of arms is cut to a different frequency, so the ensemble should be broadband. However, whether the arms are resonant or short, the ensemble acts like a single fat antenna because of mutual coupling among the closely spaced arms. The result is again a narrowband antenna resulting from the lack of basic understanding of antennas and mutual coupling.

5.9 COMPLEMENTARY PAIR ANTENNA

In the complementary pair antenna (Schroeder, 1964, 1969; Schroeder and Soo Hoo, 1976), two antennas are connected to the side arms of a 180° hybrid junction (see Figure 5.11). The concept is that the reactances are cancelled because of the π phase, whereas the radiation resistances are added. Hybrids with highly mismatched (and usually not quite equal) loads give only a partial reactance cancellation. The cancellation may also vary with frequency. Furthermore, the environment of the two

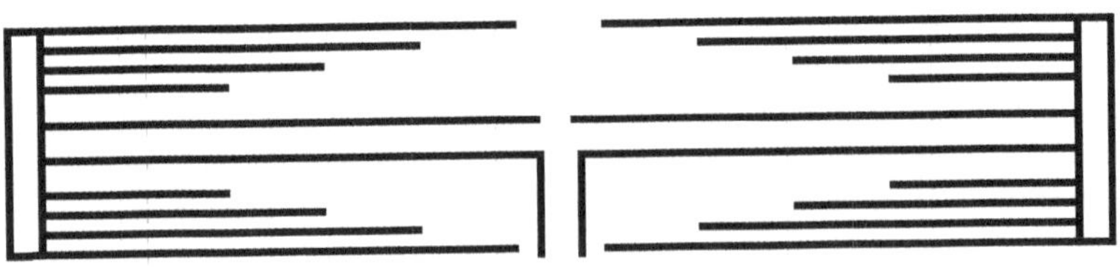

FIGURE 5.9 Multiarm dipole.

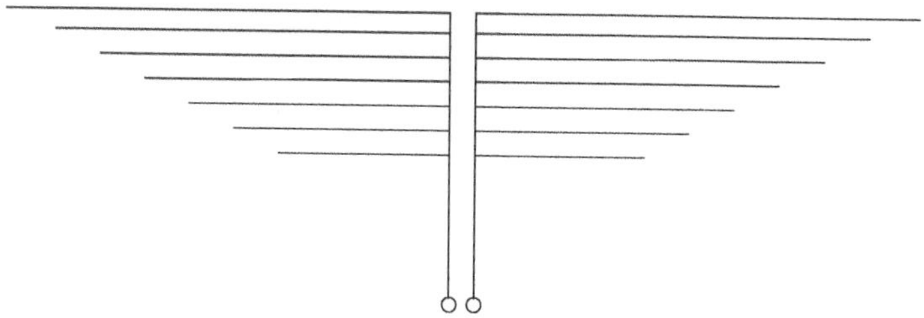

FIGURE 5.10 Multiarm dipole. Courtesy of Turner and Richard (1968).

antennas may make the reactances somewhat different. Thus, the reactance cancellation may be only partial. Mutual coupling between the two antennas will also affect the hybrid performance. When the volume occupied by the two antennas is utilized by a single, fatter antenna, the complementary pair advantage disappears. The single antenna is simpler.

5.10 INTEGRATED ANTENNA

This is the name given by Turner and Meinke to an electrically small antenna that has a semiconductor element, usually a transistor, connected in the interior of an antenna. The first discussion of the integrating semiconductor electronics into antennas was given by Frost, who conceived a parametric amplifier using portions of the antenna as resonant circuits (Frost, 1960, 1964). This work was followed at OSU, where an amplifier was integrated with the antenna, although in a more conventional sense (Copeland et al., 1964). For the integrated antennas, although many configurations have been devised, the most common is a folded monopole with capacitive "top hat." The transistor is inserted at the top hat junction (see Figure 5.12). Other connections of the transistor have been tried; this is the broadband configuration (Flachenecker and Meinke, 1967). The transistor can be located just below the top hat or can be near

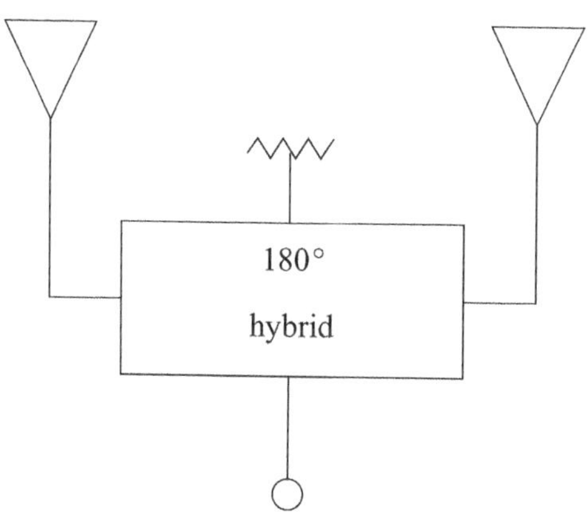

FIGURE 5.11 Complementary pair antenna.

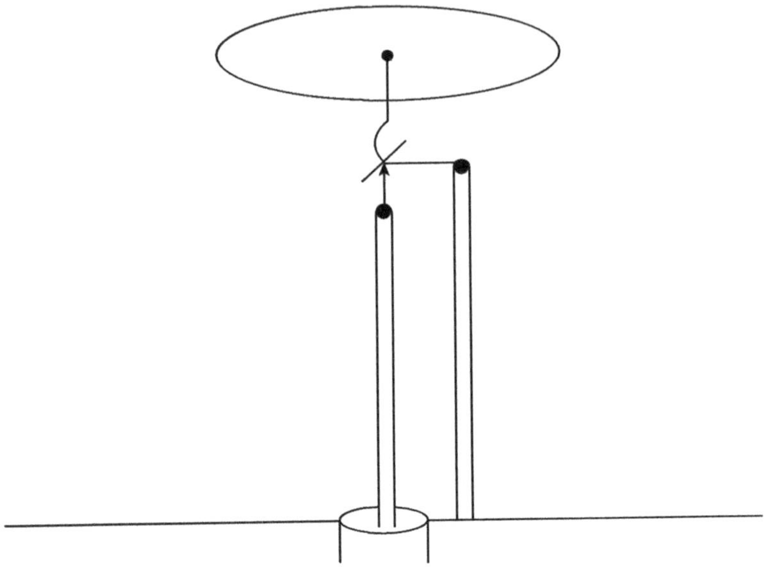

FIGURE 5.12 Integrated antenna. Courtesy of Plachenccker and Meinke (1967).

the feed point, with the location affecting the frequency of minimum VSWR. Operation of this scheme can be understood by considering only the example of Figure 5.12, with the transistor just below the top plate. Effectively, all the induced voltage appears from base to emitter, as the folded monopole operating as a loop will yield a much smaller voltage. Thus, the transistor acts as a variable resistance inserted at the loop of the folded dipole, with the resistance value depending on the bias level. Such a loaded folded antenna was discussed above. The resistance makes the folded monopole broadband, with no increase in output voltage. The transistor may also be used as a variable reactance circuit to add capacitance to tune the monopole inductance. However, this application and the broadband application are only partly compatible, with the relative effectiveness of each depending on the transistor drive phase angle. The integrated antenna has an effective length like that of the top hat monopole alone, but the transistor loss broadens the bandwidth, reduces efficiency, and introduces appreciable noise (Maclean and Ramsdale, 1975). Instabilities incurred in active loading of dipoles were examined by Fanson and Chen (1973). After years of high-level hype, the consensus was reached that the optimum placement of active devices is not in the antenna, but at the antenna terminals. Subsequent amplification can be accomplished with appropriate band limiting for noise factor control.

5.11 $Q = 0$ ANTENNA

As discussed in Chapter 1, an antenna radiating both TM and TE modes should realize a Q lower than a single-mode antenna. In their 1999 *Radio Science* paper,

Grimes and Grimes show that an antenna consisting of four dipoles, x and y directed electric dipoles and x and y directed magnetic dipoles, can have $Q = 0$!

As mentioned in Section 5.5, Collin (1998) showed that they made errors in the calculation of stored energy, and that their definition of Q is not in agreement with all earlier investigators. In their later paper (Grimes and Grimes, 1997), a time-domain calculation of stored energy is performed, and then a time and space continuity equation is applied to obtain the stored energy. But as Collin points out, there is no Lorentz frame corresponding to uniform motion in the radial direction at velocity c. The moral is: If it sounds too good to be true, it probably is not true.

5.12 ANTENNA IN A NIM SHELL

In another attempt to beat the fundamental limits on bandwidth of an electrically small antenna, such as a short dipole, Ziolkowski and colleagues (Ziolkowski and Kipple, 2003; Ziolkowski and Erentok, 2006, 2007; Arslanagic et al., 2007) placed a thin shell around a short dipole, with the shell diameter also small in wavelengths (see Figure 5.13). When the shell has the ideal properties of a negative index metamaterials (NIM), it can act as an impedance transformer over that part of the dispersion curve versus frequency where the NIM property exists. It is claimed that the shell can greatly increase radiation resistance and bandwidth of the dipole and

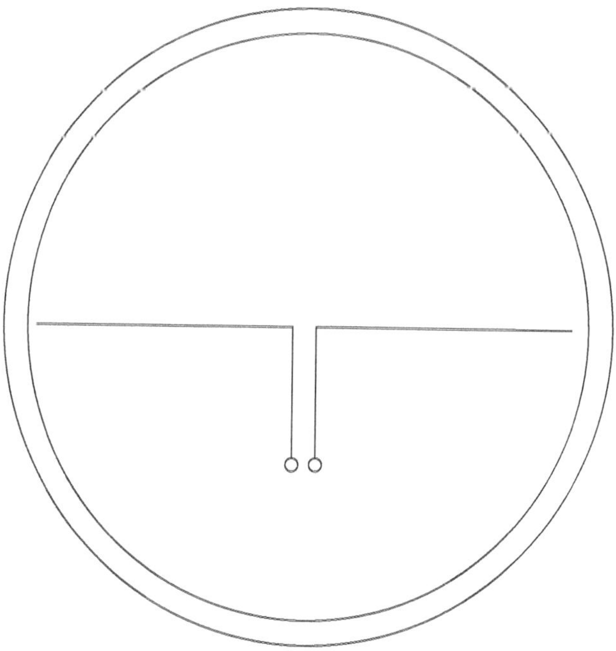

FIGURE 5.13 NIM shell and dipole.

decrease reactance. And the Wheeler–Chu–McLean limit is violated by a significant amount.

However, in these papers the dipole with shell and the dipole in free space are both fed by 1 A of current. This results in a comparison of radiation resistance for the two cases (Kildal, 2006). As this is not directly related to Q, the low Q values calculated are not valid. Other examples in the Ziolkowski papers use 1 W of available power for both dipole cases. This results in a comparison of antenna mismatch factors for the two dipoles, but not a comparison of Q values. The papers' evaluation of Q calculates total complex power at some point $r = a$, and then divides it by the square of current to get input impedance. This is not correct as the complex power must be evaluated at the location of the source. For the Hertzian dipole, it would be infinite.

An attempt was made to demonstrate experimentally the NIM shell effects. The design and construction of the shell and the very small loop antenna was so poorly done that the results were useless (Holloway, 2007). A subsequent software simulation (Greegor et al., 2009) used Floquet unit cell analysis; this simulated an infinite array of shells instead of a single shell.

Alas, the electrically small, thin NIM shell is made of that well-known material, unobtainium. NIM require at least one of several awkward features: an array of long (in wavelengths), closely spaced wires parallel to the E field; a conducting ground plane not small in wavelengths (for mushroom NIM); or conductive waveguide walls that allow circuit elements. None of these configurations fits into an electrically small, thin shell. Closely spaced and long wires (in wavelengths) greatly attenuate the field transmitted through, although several papers have shown the field without attenuation. The reviewers should be ashamed.

A further difficulty is that the metallic inclusions that are integral to any NIM are small in wavelengths and are close to the dipole wire in wavelengths. Just as a plethora of higher order wire dipoles result in one lowest order wire dipole, because of mutual coupling, the NIM inclusions have their induced currents and scattering properties altered with the resulting loss of NIM properties.

In addition to the calculation errors in the Ziolkowski papers, there is a catastrophic problem. Inclusion of realistic dispersion in the shell removes the NIM properties. Because this information is so critical for antenna R&D, the paper (Karawas and Collin, 2008) that contains a thorough analysis is included as Appendix C.

To quote Robert Park, "It never pays to underestimate the human capacity for self-deception."

5.13 FRACTAL ANTENNAS

The science of fractals was developed by Mandelbrot and published in seminal books (Mandelbrot, 1977, 1982). Fractal antennas have incurred a large interest and a large number of papers in recent years. Only the most significant are mentioned here. This activity arose because there is very little new in antennas, and fractal antennas are a new frontier. In addition, the mathematics is fairly simple, and the antenna forms are neat. For ESA, the forms are either monopoles (dipoles) or loops, and there are five general categories, each with its initiator and generators. These are shown in the various stages:

- von Koch (1870–1924) snowflake, 1904 (Puente et al., 1998a; Baliarda et al., 2000a). Figure 5.14 shows stages and Figure 5.15 shows a segment that could be a monopole.

FIGURE 5.14 von Koch stages.

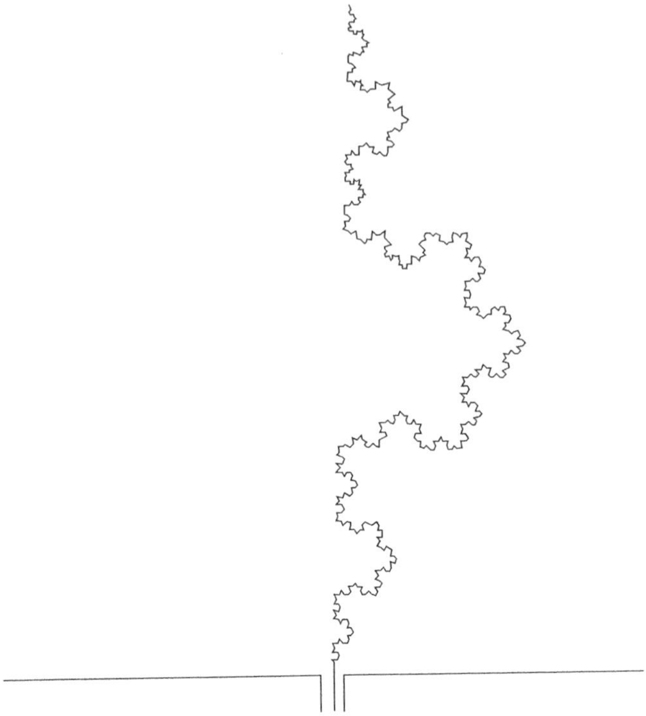

FIGURE 5.15 von Koch monopole.

- Sierpinski (1882–1969) gasket (Puente et al., 1996, 1998b; Baliarda et al., 2000b). Figure 5.16 shows stages and Figure 5.17 shows a segment that could be a monopole.
- Hilbert (1862–1943) curve (Vinoy et al., 2001; Anguera et al., 2003). Figure 5.18 shows monopoles.
- Minkowski (1864–1909) island, circa 1890 (Cohen, 1995; Best and Morrow, 2003). Figure 5.19 shows monopoles and Figure 5.20 shows loops.
- Peano (1858–1932) curves (Zhu and Engheta, 2004). Figure 5.21 shows monopoles.

Most of the papers concern Sierpinski monopoles.

Fractals occur in many places in nature, and in biology in particular. Ferns, leaves, and coral are among the many examples. Fractals may be useful in scattering from rough surfaces or ocean waves (Jaggard, 1990). However, there is nothing in Maxwell's equations that indicates any particular spatial periodicity in performance parameters, so one must look to the world of practical antennas to see whether fractal mathematics would improve performance. Wideband antennas such as the spiral, the log-periodic dipole array (LPDA), and the TEM horn have in common the frequency-independent principle: The feed point excites a small geometry first. If this geometry is a resonant active region, it radiates. If not, the exciting currents are passed on

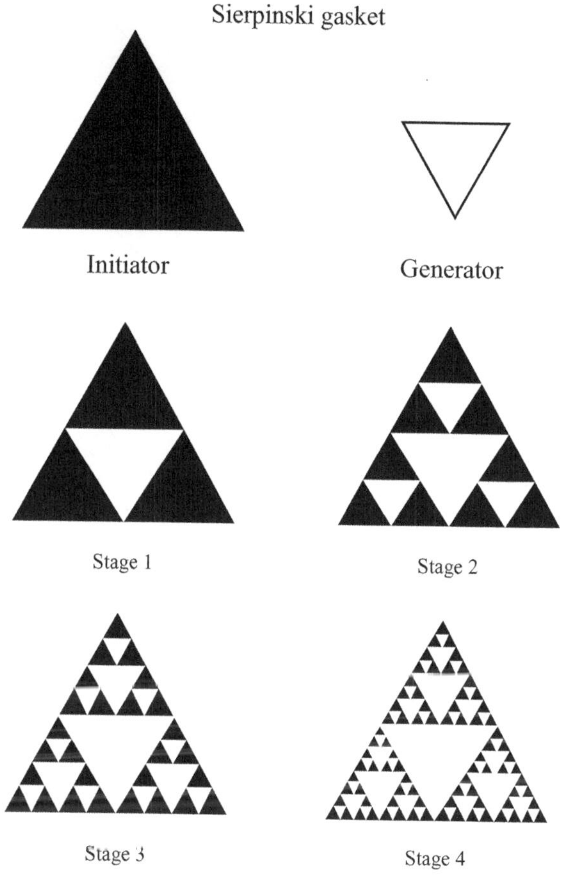

FIGURE 5.16 Sierpinski stages.

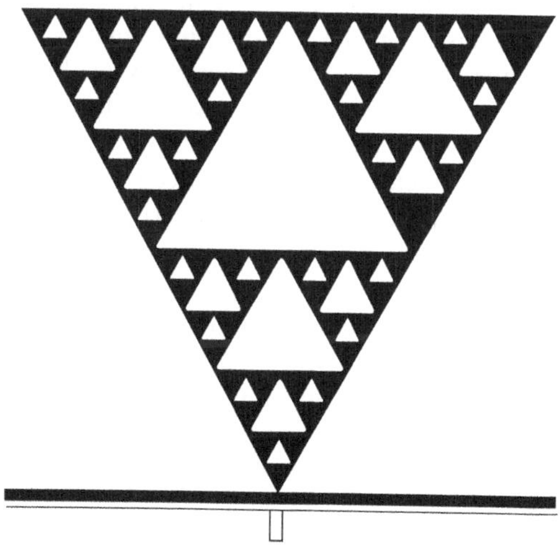

FIGURE 5.17 Sierpinski monopole.

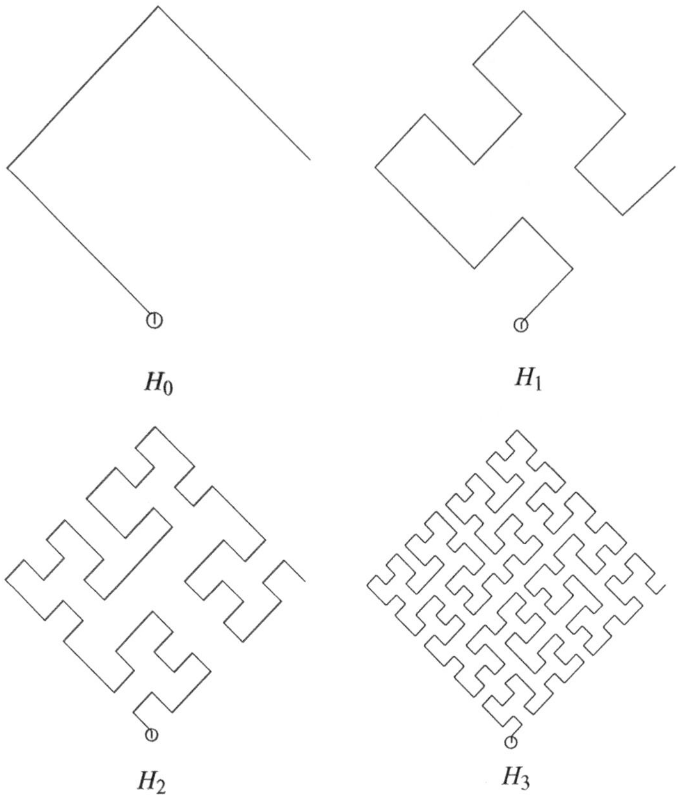

FIGURE 5.18 Hilbert monopoles.

without significant change to a subsequent larger active region. The active regions start small at the feed point and become large at the end of the antenna. In the spiral antenna, the active region is an annulus that starts immediately around the feed point and as frequency decreases it moves to the periphery of the spiral at the lowest operating frequency. Similarly, the LPDA has a high-frequency resonant region consisting of short dipoles near the feed and longer, lower frequency dipoles situated along the boom. The active region for the lowest frequency contains long dipoles at the big end of the LPDA (see Figure 5.22). Similarly, the TEM horn radiation is primarily from the throat region at high frequencies and primarily from the mouth region at low frequencies.

Fractals, in contrast, tend to behave in just the opposite manner. The classic Mandelbrot diagram (see Figure 5.23) starts with a large figure at the origin (feed point); smaller replicas are then clustered at strategic points around the large figure, and even small figures are clustered around each of these medium-sized figures, and so on. Thus, the feed point occurs at the large structure instead of at the small active region; the currents to excite the small structures must travel through the large structure, clearly violating the principles of broadband radiation.

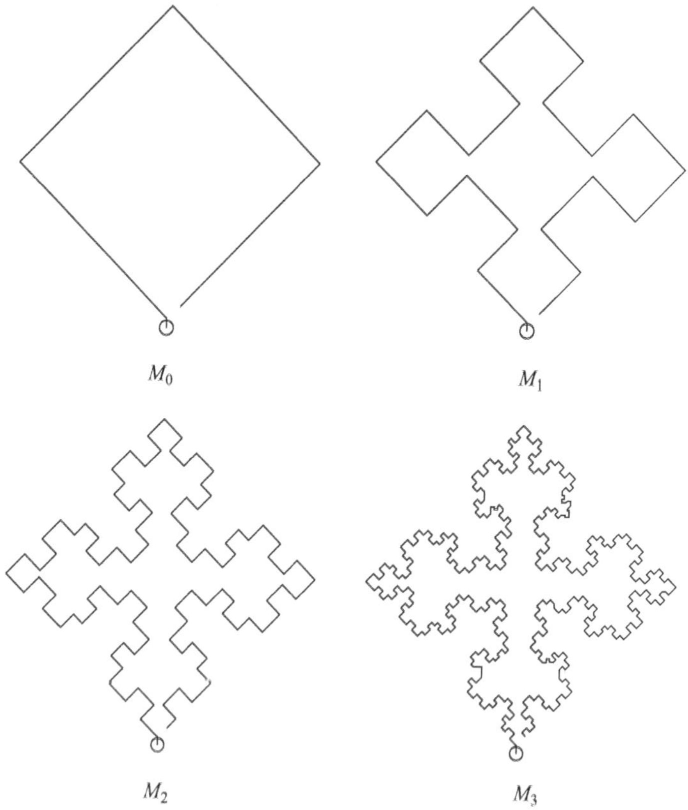

FIGURE 5.19 Minkowski monopoles.

Best and Morrow (2002, 2003) have shown that, for the von Kock, Hilbert, Minkowski, and Peano fractal antennas, although their space-filling nature reduces resonant frequency, the strong coupling between parallel segments with opposite currents reduces the antenna effective length. Also, reactance is increased over that of simple antennas such as dipoles and bowties, thereby reducing bandwidth. Hilbert fractal antennas with narrow bandwidth at resonance are shown by Zhu et al. (2003) and by Guterman et al. (2004). Similarly, Koch fractal antennas have narrow bandwidth at resonance (Best, 2002). The Sierpinski holes in the bowtie shape produce higher frequency resonances, but these make a multiple narrowband antenna rather than a broadband antenna (see Puente, 1998b; Liang and Chia, 1999; Soler et al., 2002). The resonant frequency of wire monopoles such as Hilbert, Minkowski, and meander is controlled primarily by wire length (Best and Morrow, 2002, 2003; Best, 2002). Self-resonant ESA because their input resistance is not small (Best, 2005) do not need the matching circuit loss enhancement described in Section 2.4. Thus, we may conclude that in general fractal antennas should not be expected to contribute performance improvement in size or bandwidth.

FIGURE 5.20 Minkowski loops.

Electrically small planar monopoles, broadly speaking, have only two design parameters besides height. These are the fatness and the wire length (for a wire monopole). It is well known that fat monopoles, such as bowtie monopoles, have much wider bandwidth than wire monopoles. A wire monopole can become resonant even if the overall height is small in wavelengths. Best (2002, 2003) compared ESA resonant wire antennas of meander, Koch fractal, meander helix, Hilbert fractal, and Minkowski fractal. The wire length-controlled resonance and the bandwidth were essentially the same. The increase in resistance usually did not affect bandwidth significantly because of a large radiation resistance. For loops, the wire length affects efficiency because of a small radiation resistance. González-Arbesú et al. (2003) showed that space-filling monopoles, such as meander, Hilbert fractal, and Peano fractal, store a lot of energy in the near-field and have more loss. Thus, the bandwidth and efficiency are both diminished. If wire length is used to produce resonance in an ESA, the wires should be disposed to minimize cross-polarization and to maximize radiation resistance. Generally, higher order fractal designs have less bandwidth and lower efficiency. A closely spaced meanderline monopole is

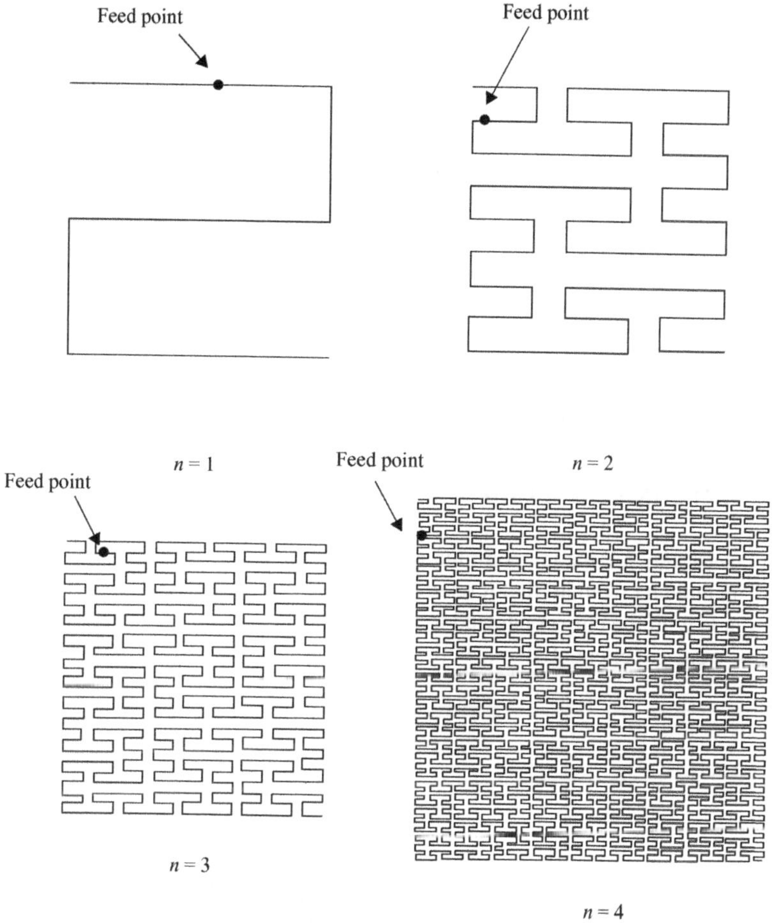

FIGURE 5.21 Peano curves.

better than Koch and Hilbert fractal monopoles, but fat bowties or top-loaded fat dipoles are better.

The Sierpinski monopole is slightly poorer in efficiency and thus slightly better in bandwidth, because of slightly higher loss resistance, than a solid bowtie. The Sierpinski holes in the bowtie shape produce higher frequency resonance, but these make a multiple narrowband antenna rather than a broadband antenna.

Fractal array designs, such as Cantor set arrays, are outside the scope of this book, but they offer no advantages and have high sidelobes; there are much better nonuniformly spaced array designs available. An array design related to fractals uses difference sets (a brand of combinatorial mathematics) to formulate a thinned array with constant sidelobe envelope (Leeper, 1999).

A careful review of the many papers on fractal antennas shows that they offer no advantages over fat dipole, loaded dipoles, and simple loops with or without magnetic core. Nonfractals are always better.

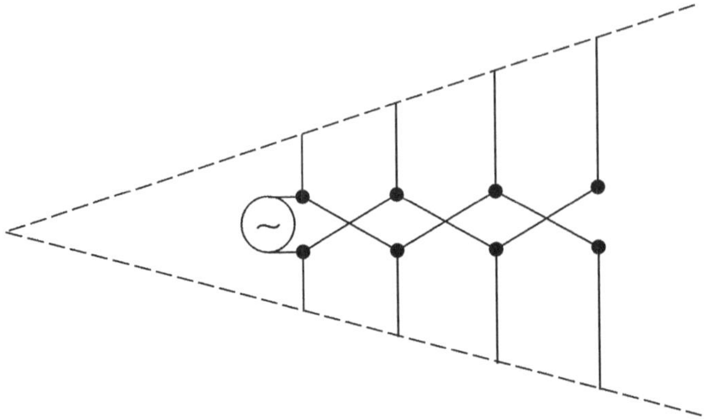

FIGURE 5.22 Log-periodic dipole array.

5.14 ANTENNA ON A CHIP

A few years ago, a large telecommunications company announced an "antenna on a chip," an antenna so small that it resided on the printed circuit board. Measurements were made with the chip connected to a network analyzer by a small diameter coax. Results were excellent. Later, when the cable was removed and the antenna was

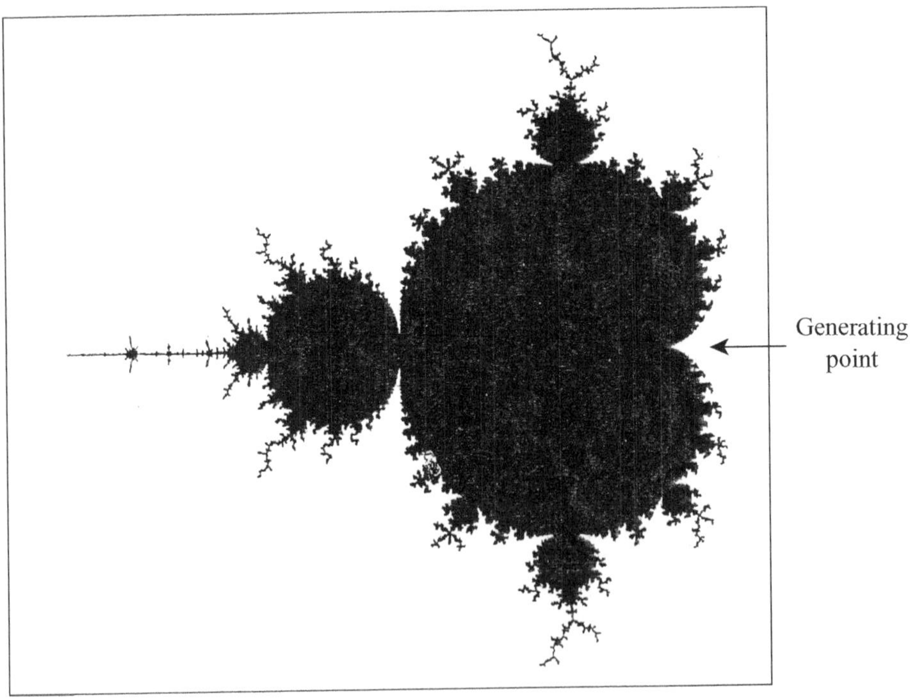

FIGURE 5.23 Original Mandelbrot fractal.

activated by the circuit board, the antenna did not operate. This was another case of an unbalanced antenna connected to a coax cable; the cable often makes an excellent radiator! All announcements on the chip antenna ceased. Similar problems occurred in most of the CWTHA tests.

5.15 RANDOM SEGMENT ANTENNAS

It has become fashionable to design wire antennas with some type of optimizer program, almost independent of good physics or high-quality performance. The results sometimes have wire segments in all directions; see Figure 5.24 for an example. A long total wire length may achieve resonance in a small volume, but there are several disadvantages. If Z is the normal monopole direction, the X currents tend to cancel, as do the Y currents. However, in certain directions the cross-polarized field may not be negligible. Longer total wire length increases loss resistance, reduces efficiency, and increases reactance. And generally the bandwidth is narrow. Examples are Altshuler and Linden (2004), Choo et al. (2005), Altshuler (2005), and Best (2002, 2003). Use of fractals and meanderlines to fill space (González-Arbesú et al., 2003; Best and Morrow, 2002) suffers from the same problems.

"Do not confuse inexperience with creativity" (Linda Whittaker) is appropriate here.

5.16 MULTIPLE MULTIPOLES

The fundamental limitations on antennas make it clear that the maximum bandwidth of an ESA occurs when both the electric dipole mode and the magnetic dipole mode

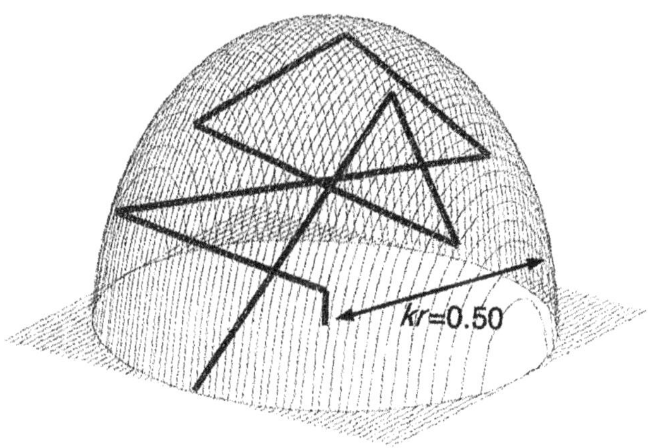

FIGURE 5.24 Random segment antenna. Courtesy of Choo et al. (2005).

are excited. For many years, an effort has been made to show that higher order modes, or multipoles, could also be excited within the imaginary enclosing small sphere, and thereby invalidate Chu's results. Grimes and Grimes (1995, 1996) improperly applied Poynting's theorem to individual modes and "derived" a non-Maxwellian equation for the conservation of power. Grimes and Grimes (1997, 1999) produced a "standing energy density" (reactive power) that cannot be derived from electrodynamics. It was stated that ESA Q could be as low as zero! It is well known from antenna theory that a dipole along the axis of a loop, with the dipole center in the loop plane, has zero mutual impedance to the loop. McLean (1996), using the moment method, showed this but, more importantly, showed that a coplanar dipole and loop were coupled and that the dipole and loop energy densities were interrelated. Notwithstanding, Grimes et al. (2000a, 2000b) claimed that two loop–dipole pairs used "energy that returns from the radiation field to the antenna," and that when the pairs are excited for CP, "large near-field energy is not supported." In a review paper (Grimes and Grimes, 2001), it was stated again that $Q = 0$ was possible. Collin (1998) gives a detailed critique of how Grimes has bent the laws of electrodynamics.

The several Grimes papers only vaguely explain how to produce higher order multipoles in a small volume. Of course, for the lowest order TE and TM modes, a dipole along the loop axis suffices. In a patent (Grimes and Grimes, 1989), it is suggested that the lowest order mode may be produced by a dipole, the next mode may be produced by two parallel dipoles fed 180° out of phase, and so on, and similarly for the loops (see Figure 5.25). Of course, what happens is that the mutual coupling of three closely spaced dipoles is very strong, with the result that all the dipoles revert to a single fatter dipole, but of course with reactance changed. This is much like the multiarm dipole of Section 5.8. Similar effects can be expected from the loops. Also damaging is that the dipoles and loops must be connected to the external world by wire segments inside the enveloping sphere. These wires cross-couple the multipoles and become part of the antenna. It is not

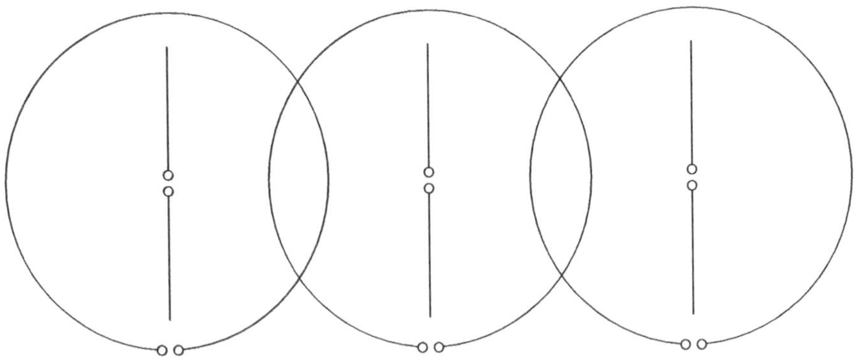

FIGURE 5.25 Multipoles, after Grimes and Grimes (1989).

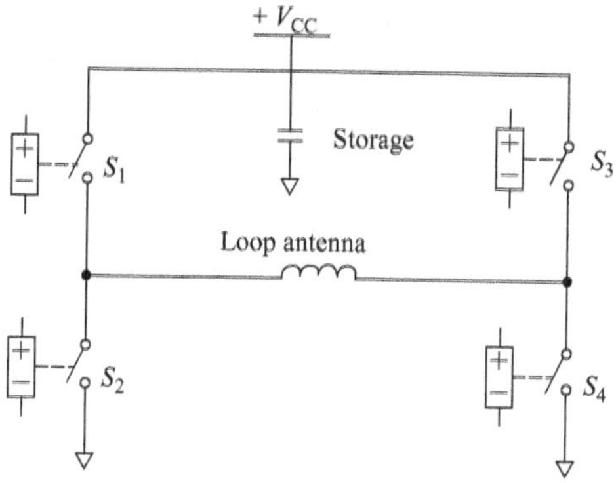

FIGURE 5.26 Merenda switching circuit.

surprising that there are no credible measurements of antennas that beat the Chu limit.

As Robert Park states in *Voodoo Science: The Road from Foolishness to Fraud*: "Where accepted laws of electrodynamics must be changed to give the researcher the results he wants, he is almost certainly wrong."

5.17 SWITCHED LOOP ANTENNAS

The switching of energy between a capacitor and a loop to provide wideband operation was invented by Merenda (2001). As shown in Figure 5.26, an energy storage device (capacitor) and a loop are connected with a DC supply by four switches. Figure 5.27 depicts the cycle: Current flows through the loop for the first 60° of cycle, building up to a peak value. During the next 60°, the loop is short-circuited; the current decays

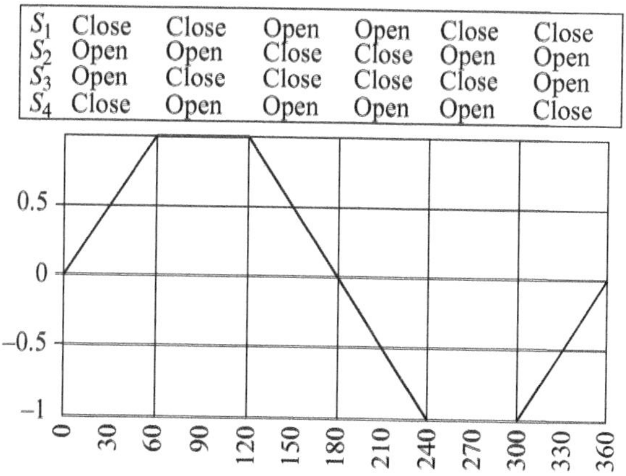

FIGURE 5.27 Merenda switching cycle.

somewhat. For the next 120°, the loop excitation is reversed; the current decays to zero and builds up in the opposite direction. From 240° to 300°, the loop is again shorted, thus allowing a slow current decay. Finally, from 300° to 360°, the current decays to zero and starts the buildup in the next cycle. The capacitor and power supply are connected to the loop during the current increasing cycles. The waveform of Figure 5.27 approximates a sine wave, the RF carrier.

The question arises: How does this switching scheme differ from simply connecting a capacitor in parallel with the loop and connecting both to an RF source? The switching circuit effectively generates the RF carrier. Disadvantages of the Merenda circuit are that the waveform of Figure 5.27 contains, in addition to the RF carrier frequency, some higher harmonics, which cause interference, corrupt the signal modulation, and reduce efficiency somewhat, and that the switching circuit transistors will reduce the DC to RF carrier efficiency and introduce noise. In both cases, the antenna is basically a loop with a shunt capacitor; the bandwidth will be limited by the Q of the tuned loop.

"The absolute ingenuity of this idea almost blinds one to its utter worthlessness" (Air Marshall Tedder) is appropriate here.

5.18 ELECTRICALLY SMALL FOCAL SPOTS

A complementary problem to that of the ESA is the production of an electrically small focal spot. An early work by Carlin (1964) used the Heisenberg uncertainty principle to show that a spot size roughly half that of the normal pattern could be produced. This was refined using a uniformly excited linear array by Hansen (1965); the minimum spot size available is roughly 0.35λ.

In 2000, Pendry showed that Veselago-type negative refraction could produce a "perfect lens" (Pendry, 2000). His calculations indeed showed a subwavelength focal spot! Using thermodynamic arguments, Markel (2008) showed that negative refraction is not possible in Pendry's case. Many readers may not be comfortable with thermodynamic calculations. Collin (2010) showed, in electromagnetic detail, that in addition to Pendry's solution there will exist resonant surface wave modes and other modes that preclude a coherent reconstruction of the object in the image plane. The Pendry lens does produce a small focal spot, just as a focused antenna does. But there is no useable modulation transfer function.[1] The lengthy but important Collin paper is included as Appendix D.

5.19 ESA SUMMARY

Although there have been many attempts to make electronically small antennas that outperform the fundamental limitations, none has succeeded. The Wheeler–

[1] MTF is the spatial frequency amplitude response, in lines/mm or pixels/mm^2 (RCA Contributors, 1968).

Chu–McLean fundamental limitations are absolute, regardless of government needs or expenditures. What has changed over the years is that the attempts have become more sophisticated, although in each case, basic principles were ignored.

Clever antenna design can make the most of available space; a reasonable objective is to get close to the limits. In general, dipole-type ESA should use as much area (and volume) as feasible, for example, bowtie, to take advantage of "free" bandwidth. Judicious inductive loading is also productive. Bandwidth can be improved further through use of clever matching networks. HTS materials will affect these conclusions only by allowing more efficient matching sections to be used. Non-Foster matching circuits will become very important.

The reader is urged to read the fascinating *Voodoo Science* by Robert Park (2000): "Foolishness, if pursued can lead to denial, and that sometimes leads to fraud."

REFERENCES

Altshuler, E. E. A Method for Matching an Antenna Having a Small Radiation Resistance to a 50-ohm Coaxial Line. *Trans. IEEE*, Vol. AP-53, September 2005, pp. 3086–3089.

Altshuler, E. E. and Linden, D. S. An Ultrawide-Band Impedance-Loaded Genetic Antenna. *Trans. IEEE*, Vol. AP-52, November 2004, pp. 3147–3150.

Anguera, J. et al. The Fractal Hilbert Monopole: A Two-Dimensional Wire. *Microwave Opt Technol. Lett.*, Vol. 36, January 2003, pp. 102–104.

Arslanagic, S., Ziolkowski, R. W., and Breinbjerg, O. Analytical and Numerical Investigation of the Radiation from Concentric Metamaterial Spheres Excited by an Electric Hertzian Dipole. *Radio Sci.*, Vol. 42, 2007, RS6S16.

Baliarda, C. P., Romeu, J., and Cardama, A. The Koch Monopole: A Small Fractal Antenna. *Trans. IEEE*, Vol. AP-48, November 2000a, pp. 1773–1781.

Baliarda, C. P. et al. An Iterative Model for Fractal Antennas: Application to the Sierpinski Gasket Antenna. *Trans. IEEE*, Vol. AP-48, May 2000b, pp. 713–719.

Barrick, D. Miniloop Antenna Operation and Equivalent Circuit. *Trans. IEEE*, Vol. AP-34, January 1986, pp. 111–114.

Belrose, J. S. Characteristics of the CFA Obtained by Numerical and Experimental Modeling. *Proc. IEEE BTS Symp.*, 2000a.

Belrose, J. S. The Crossed Field Antenna-Analyzed by Simulation and Experiment. *ICAP–JINA Conf. on Antenna and Propagation*, Davos, 2000b.

Belrose, J. S. Electrically Small Transmitting Loops—Part 1. *Radcom*, June 2003, pp. 65–67; Part 2, July 2004, pp. 88–90.

Belrose, J. S. Electrically Small Transmitting Loops. *IEEE APS Symp. Dig.*, 2005a.

Belrose, J. S. Performance Analysis by Experiment and Simulation of Small Tuned Transmitting Loop Antennas. *IEEE APS Symp. Dig.*, 2005b.

Best, S. R. A Comparison of the Performance Properties of the Hilbert Curve Fractal and Meander Line Monopole Antennas. *Microwave Opt. Technol. Lett.*, Vol. 35, November 2002, pp. 258–262.

Best, S. R. A Comparison of the Resonant Properties of Small Space-Filling Fractal Antennas. *IEEE Antenn. Wireless Propag. Lett.*, Vol. 2, 2003, pp. 197–200.

Best, S. R. A Discussion on the Quality Factor of Impedance Matched Electrically Small Wire Antennas, *Trans. IEEE*, Vol. AP-53, Jan. 2005, pp. 502–508.

Best, S. R. and Morrow, J. D. The Effectiveness of Space-Filling Fractal Geometry in Lowering Resonant Frequency. *IEEE Antenn. Wireless Propag. Lett.*, Vol. 1, 2002, pp. 112–115.

Best, S. R. and Morrow, J. D. On the Significance of Current Vector Alignment in Establishing the Resonant Frequency of Small Space-Filling Wire Antennas. *IEEE Antenn. Wireless Propag. Lett.*, Vol. 2, 2003, pp. 201–204.

Carlin, P. W. The Minimum Spot Size for a Focused Laser and the Uncertainty Relation. *Proc. IEEE*, Vol. 52, November 1964, p. 1371.

Choo, H., Rogers, R. L., and Ling, H. Design of Electrically Small Wire Antennas Using a Pareto Genetic Algorithm. *Trans. IEEE*, Vol. AP-53, March 2005, pp. 1038–1046.

Cohen, N. Fractal Antennas—Part 1. *Commun. Q.*, Summer 1995, pp. 7–22.

Collin, R. E. Minimum *Q* of Small Antennas. *J. Electromagn. Waves Appl.*, Vol. 12, 1998, pp. 1369–1393.

Collin, R. E. Frequency Dispersion Limits in Veselago Lens. *Prog. Electromagn. Res. B*, Vol. 19, 2010, pp. 233–261.

Copeland, J. R. et al. Antennafier Arrays. *Trans. IEEE*, Vol. AP-12, March 1964, pp. 227–233.

Dunlavy, J. H. Jr.Wide Range Tunable Transmitting Loop Antenna, U.S. Patent 3,588,905, June 1971.

Fanson, P. L. and Chen, K.-M. Modification of Antenna Radiating Characteristics with Multi-Impedance Loading. *Trans. IEEE*, Vol. AP-21, September 1973, pp. 715–721.

Flachenecker, G. and Meinke, H. R. Active Antennas with Transistors. *Can. Int. Electron. Conf. Rec.*, September 1967, pp. 142–143.

Frost, A. D. Parametric Amplifier Antenna. *Proc. IRE*, Vol. 48, June 1960, pp. 1163–1164.

Frost, A. D. Parametric Amplifier Antenna. *Trans. IEEE*, Vol. AP-12, March 1964, pp. 234–235.

González-Arbesú, J. M., Blanch, S., and Romeu, J. Are Space-Filling Curves Efficient Small Antennas? *IEEE Antenn. Wireless Propag. Lett.*, Vol. 2, 2003, pp. 147–150.

Greegor, R. B. et al. Demonstration of Impedance Matching Using a mu-Negative (MNG) Metamaterial. *IEEE Antenn. W ireless Propag. Lett.*, Vol. 8, 2009, pp. 92–95.

Grimes, C. A. and Grimes, D. M. A Clarification and Extension of Bandwidth and *Q* of Antennas Radiating Both TE and TM Modes. *Trans. IEEE*, Vol. EMC-38, May 1996, pp. 201–202.

Grimes, C. A. et al. Time-Domain Measurement of Antenna *Q*. *Microwave Opt. Technol. Lett.*, Vol. 25, April 2000a, pp. 95–100.

Grimes, C. A. et al. Characterization of a Wideband, Low-*Q*, Electrically Small Antenna. *Microwave Opt. Technol. Lett.*, Vol. 27, October 2000b, pp. 53–58.

Grimes, D. M. and Grimes, C. A. *Resonant Antenna,* U.S. Patent, 4,809,009, February 28, 1989.

Grimes, D. M. and Grimes, C. A. Bandwidth and *Q* of Antennas Radiating TE and TM Modes. *Trans. IEEE*, Vol. EMC-37, May 1995, pp. 217–226.

Grimes, D. M. and Grimes, C. A. Power in Modal Radiation Fields: Limitations of the Complex Poynting Theorem and the Potential for Electrically Small Antennas. *J. Electromagn. Waves Appl.*, Vol. 11, 1997, pp. 1721–1747.

Grimes, D. M. and Grimes, C. A. Radiation Q of Dipole-Generated Fields. *Radio Sci.*, Vol. 34, March–April 1999, pp. 281–296.

Grimes, D. M. and Grimes, C. A. Minimum Q of Electrically Small Antennas: A Critical Review. *Microwave Opt. Technol. Lett.*, Vol. 28, February 2001, pp. 172–177.

Grimes, D. M. and Grimes, C. A. *The Electromagnetic Origin of Quantum Theory of Light*, World Scientific, 2005, Chapter 3.

Grover, F. W. *Inductance Calculations*, Dover, 1946.

Guterman, J., Moreira, A. A., and Peixeiro, C. Microstrip Fractal Antenna for Multistandard Terminals. *IEEE Antenn. Wireless Propag. Lett.*, Vol. 3, 2004, pp. 351–354.

Hansen, R. C. Minimum Spot Size of Focused Apertures. *URSI Electromagnetic Wave Theory Symposium*, Pergamon Press, 1965, pp. 661–667.

Hansen, R. C. Q and Bandwidth of Electrically Small Antennas. *Microwave Opt. Technol. Lett.*, Vol. 49, May 2007, pp. 1170–1171.

Hart, R. T. *E-H Antenna,* US Patent 6,486,846B1, November 26, 2002.

Hately, M. C., Kabbary, F. M., and Khattab, M. *An Operational MF Broadcast Antenna Using Poynting Vector Synthesis. ICAP*, York, IEE Conf. Publ. 333, 1991, pp. 645–648.

Hatfield, J. B. Magnetic Fields from Displacement Current Densities Generated by the Crossed Field Antenna. *IEEE Broadcast Technol. Soc. 50th Broadcast Symp.*, September 28, 2000, pp. 1–3.

Henf, G. *High Efficiency Compact Antenna,* U.S. Patent 5,796,369, August 18, 1998.

Holloway, C. L. Metamaterials and Metafilms: Overview and Applications. *IEEE EMC Symp. Rec.*, 2007.

Jaggard, D. L. On Fractal Electrodynamics. In *Recent Advances in Electromagnetic Theory*, Kritikos, H. N. and Jaggard, K. L.Eds., Springer, 1990, Chapter 6.

Kabbary, F. M., Hately, M. C., and Stewart, B. G. Maxwell's Equations and the Crossed-Field Antenna. *Electron. Wireless World*, March 1989, pp. 216–218.

Kabbary, F. M., Khattab, M., and Hately, M. C. Extremely Small High Power MW Broadcasting Antennas. *Int. Broadcasting Convention*, IEE Conf. Publ. 447, September 12–16, 1997, pp. 421–430.

Kabbary, F. M. et al. Four Egyptian MW Broadcast Crossed-Field-Antennas. Proc. NAB Conf., Las Vegas, NV, April 1999, pp. 235–241.

Karawas, G. K. and Collin, R. E. Spherical Shell of ENG Material Surrounding a Dipole Antenna. *IEEE MILCOM*, 2008, pp. 1–15.

Kildal, P.-S. Comments on "Application of Double Negative Materials to Increase the Power Radiated by Electrically Small Antennas". *Trans. IEEE*, Vol. AP-54, February 2006, p. 766.

Kuo, F.-Y. et al. A Novel Dipole Antenna Design with an Over 100% Operational Bandwidth. *Trans. IEEE*, Vol. AP-58, August 2010, pp. 2737–2741.

Leeper, D. G. Isophoric Arrays—Massively Thinned Phased Arrays with Well-Controlled Sidelobes. *Trans. IEEE*, Vol. AP-47, December 1999, pp. 1825–1835.

Liang, X. and Chia, M. Y. W. Multiband Characteristics of Two Fractal Antennas. *Microwave Opt. Technol. Lett.*, Vol. 23, November 20, 1999, pp. 242–245.

Maclean, T. S. M. and Ramsdale, P. A. Signal/Noise Ratio for Short Integrated Antennas. *Electron. Lett.*, Vol. 11, February 1975, pp. 62–63.

MANDELBROT, B. B. *Fractals: Form, Chance and Dimension*, W. H. Freeman, 1977.

MANDELBROT, B. B. *The Fractal Geometry of Nature*, W. H. Freeman, 1982.

MARKEL, V. A. Correct Definition of the Poynting Vector in Electrically and Magnetically Polarizable Medium Reveals That Negative Refraction is Impossible. *Opt. Express*, Vol. 16, November 2008, pp. 19152–19168.

MCLEAN, J. S. A Re-Examination of the Fundamental Limits on the Radiation *Q* of Electrically Small Antennas. *Trans. IEEE*, Vol. AP-44, May 1996, pp. 672–676.

MERENDA, J. T. *Radiation Synthesizer Systems and Methods,* U.S. Patent 6,229,494, May 8, 2001.

PARK, R. L. *Voodoo Science*, Oxford University Press, 2000.

PENDRY, J. B. Negative Refraction Makes a Perfect Lens. *Phys. Rev. Lett.*, Vol. 85, No. 18, October 2000, pp. 3966–3969.

PUENTE, C. ET AL. Fractal Multiband Antenna Based on the Sierpinski Gasket. *Electron. Lett.*, Vol. 32, January 1996, pp. 1–2.

PUENTE, C. ET AL. Small but Long Koch Fractal Monopole. *Electron. Lett.*, Vol. 34, January 1998a, pp. 9–10.

PUENTE, C. ET AL. On the Behavior of the Sierpinski Multiband Fractal Antenna. *Trans. IEEE*, Vol. AP-46, April 1998b, pp. 517–524.

RCA Contributors. *Electro-Optics Handbook*, RCA Commercial Engineering, 1968.

SCHROEDER, K. G. The Complementary Pair—A Broadband Element Group for Phased Arrays. *IEEE AP Conv. Rec.*, 1964, pp. 128–133.

SCHROEDER, K. G. *Complementary Pair Antenna Element Groups,* U.S. Patent 3,449,751, June 10, 1969.

SCHROEDER, K. G. AND SOO HOO, K. M. Electrically Small Complementary Pair (ESCP) with Interelement Coupling. *Trans. IEEE*, Vol. AP-24, July 1976, pp. 411–418.

SMITH, M. S. Conventional Explanation for "Crossed Field Antenna". *Electron. Lett.*, Vol. 28, February 1992, pp. 360–361.

SNYDER, R. D. *Broadband Antennae Employing Coaxial Transmission Line Sections*, U.S. Patent 4479130, October 23, 1984a.

SNYDER, R. D. The Snyder Antenna. *RF Des.*, September–October 1984b, pp. 49–51.

SOLER, J., PUENTE, C., AND PUERTO, A. A Dual-Band Bidirectional Multilevel Monopole Antenna. *Microwave Opt. Technol. Lett.*, Vol. 34, September 2002, pp. 445–448.

STUTZMAN, W. L. AND THIELE, G. A. *Antenna Theory and Design*, Wiley, 1981.

TURNER, E. M. AND RICHARD, D. J. Development of an Electrically Small Broadband Antenna. *Proc. 18th Symp. on USAF Antenna Research and Development Program*, Allerton, IL, October 1968.

VINOY, K. J. ET AL. Hilbert Curve Fractal Antenna: A Small Resonant Antenna for VHF/UHF Applications. *Microwave Opt. Technol. Lett.*, Vol. 29, May 2001, pp. 215–219.

ZHU, J. AND ENGHETA, N. Peano Antennas. *IEEE Antenn. Wireless Propag. Lett.*, Vol. 3, 2004, pp. 71–74.

ZHU, J., HOORFAR, A., AND ENGHETA, N. Bandwidth, Cross-Polarization, and Feed-Point Characteristics of Matched Hilbert Antennas. *IEEE Antenn. Wireless Propag. Lett.*, Vol. 2, 2003, pp. 2–5.

Ziolkowski, R. W. and Erentok, A. Metamaterial-Based Efficient Electrically Small Antennas. *Trans. IEEE*, Vol. AP-54, July 2006, pp. 2113–2130.

Ziolkowski, R. W. and Erentok, A. At and Below the Chu Limit: Passive and Active Broad Bandwidth Metamaterial-Based Electrically Small Antennas. *IET Microwaves Antenn. Propag.*, Vol. 1, January 2007, pp. 116–128.

Ziolkowski, R. W. and Kipple, A. D. Application of Double Negative Materials to Increase the Power Radiated by Electrically Small Antennas. *Trans. IEEE*, Vol. AP-51, October 2003, pp. 2626–2640.

CHAPTER 6

SUPERDIRECTIVE ANTENNAS

6.1 HISTORY AND MOTIVATION

A useful operational definition of antenna array superdirectivity (formerly called supergain) is directivity higher than that obtained with the same array length and elements uniformly excited (constant amplitude and linear phase). Superdirectivity applies in principle to ESA, to apertures, to arrays of isotropic elements, and to actual antenna arrays composed of nonisotropic elements such as dipoles, slots, and patches. Small dipoles and loops are superdirective; their directivity remains at 1.5 as size decreases, but their efficiency decreases. Excessive array superdirectivity inflicts major problems in low radiation resistance (hence low efficiency), sensitive excitation and position tolerances, and narrow bandwidth. It is important to distinguish between directivity and gain. Gain as used in this book follows the antenna industry definition: Gain includes the effects of both losses (conductor and dielectric) and impedance mismatch. The IEEE and textbook definition where only loss is included is unrealistic and of little use; directivity includes neither loss nor impedance match. Thus, the gain of a superdirective antenna may be low.

Taylor (1948) was one of the first to use the term "superdirectivity"; supergain should include efficiency, which in many cases would negate the directivity increase. He proposed a physical explanation of superdirectivity in terms of spherical modes and their cutoff due to dimensions. Taylor (1955) defined a most useful parameter: superdirective ratio (SDR). This is the ratio of array (or aperture) directivity to the directivity that would be obtained if the amplitude(s) were uniform, and the phase

Small Antenna Handbook, Robert C. Hansen and Robert E. Collin.

constant or progressive. It is the ratio of radiated power plus reactive power to radiated power, which for a broadside line source is

$$\mathrm{SDR} = \frac{\int_{-\infty}^{\infty} [f(u)]^2 du}{\int_{-\pi L/\lambda}^{\pi L/\lambda} [f(u)]^2 du} \tag{6.1}$$

where $u = (L/\lambda)\sin\theta$ and $f(u)$ is the aperture distribution.

For other than broadside, the limits in the denominator change. For a long uniform-amplitude line source, the broadside SDR = 1, whereas for endfire SDR = 2. Values of SDR for 10λ uniform-amplitude line sources are given by Stutzman and Thiele (1998):

	SDR
Broadside	1.01
Endfire	2.01
Hansen–Woodyard	8.23

Because Q may be expressed as the ratio of reactive power to radiated power,

$$Q = \frac{\int_{-\infty}^{-\pi L/\lambda} [f(u)]^2 du + \int_{\pi L/\lambda}^{\infty} [f(u)]^2 du}{\int_{-\pi L/\lambda}^{\pi L/\lambda} [f(u)]^2 du} \tag{6.2}$$

These two equations give $\mathrm{SDR} = 1 + Q$.

Superdirectivity is a mature technical area, but many newer researchers have not been exposed to its capabilities and limitations. Thus, this chapter aims at providing a broad reference framework, as well as some more recent results. Probably, the first work on this subject was due to Oseen, in 1922. The next work that appeared was dated 1938. A flurry of papers appeared around the World War II era, up to roughly 1960. Another burst of activity occurred from 1964 to 1974. Since then, only a few papers have appeared. Details of these advances are given below.

6.2 MAXIMUM DIRECTIVITY

6.2.1 Apertures

The question arises whether a continuous aperture with amplitude and phase chosen appropriately can have an infinite directivity. Oseen (1922) discussed forming an arbitrarily narrow beam, analogous to the quantum mechanical "needle" radiation of Einstein, and the possibilities of superdirectivity. See Bloch et al. (1953, 1960) for a list of early references. Another early contributor was Franz (1943). Schelkunoff, in a classic paper (1943) on linear arrays, discussed, among other topics, array spacings less than $\lambda/2$, showing how equal spacing of the array polynomial zeroes over that portion of the unit circle represented by the array gives superdirectivity. The field received wide publicity when La Paz and Miller (1943) purported to show that a given aperture would allow a maximum directivity. However, Bouwkamp and De Bruijn (1945/1946) showed that they had made mathematical errors and that there was no limit on theoretical directivity. Thus, the important theorem: A fixed aperture size can achieve (in theory) any desired directivity value. This theorem is now widely recognized, but the practical implications are less well known. Bloch et al. (1960) say that the theorem has been rediscovered several times; the practical limitations of superdirectivity occur as a surprise to systems engineers and others year after year!

Reid (1946) generalized the Hansen–Woodyard endfire directivity as $d \to 0$. The Bouwkamp and De Bruijn work was extended to a two-dimensional aperture by Riblet (1948). Kyle (1959) discussed transforming a linear superdirective distribution to a cylindrical distribution. Superdirective aperture design thus requires a constraint; see Section 6.3.

6.2.2 Arrays

6.2.2.1 Broadside Arrays of Fixed Spacing An array with fixed length and number of elements represents a determinate problem. Clearly, a maximum directivity exists. Uzkov (1946), and later Gilbert and Morgan (1955), showed that in the limit of zero element spacing the maximum directivity is

$$D = \sum_{n=0}^{N-1} (2n+1)[P_n(\sin\theta)]^2 \tag{6.3}$$

The array has N elements, and broadside is $\theta = 0$. P_n is the Legendre polynomial of order n. The broadside case is discussed next and the endfire case in Section 6.2.2.2.

Tai (1964) developed broadside array results starting with the mutual resistance series and then optimizing by setting derivatives to zero and evaluating the resulting matrix. He plotted directivity for $0 \le d/\lambda \le 2$ for isotropes, parallel dipoles, and collinear dipoles. For broadside in the $d = 0$ limit, the Uzkov result is obtained:

$$D = \sum_{n=0}^{N-1} (2n+1)[P_n(0)]^2 \tag{6.4}$$

The Legendre polynomial of argument 0 can be written as a product of factors (Burington, 1973); the result for maximum directivity is a double sum. It is simpler to use $P_n(0)$ from Jahnke and Emde. Since $P_{odd}(0) \equiv 0$, thus the following result: three- and four-element arrays have the same limiting value, five- and six-element arrays have the same value, and so on. This maximum directivity is plotted in Figure 6.1.

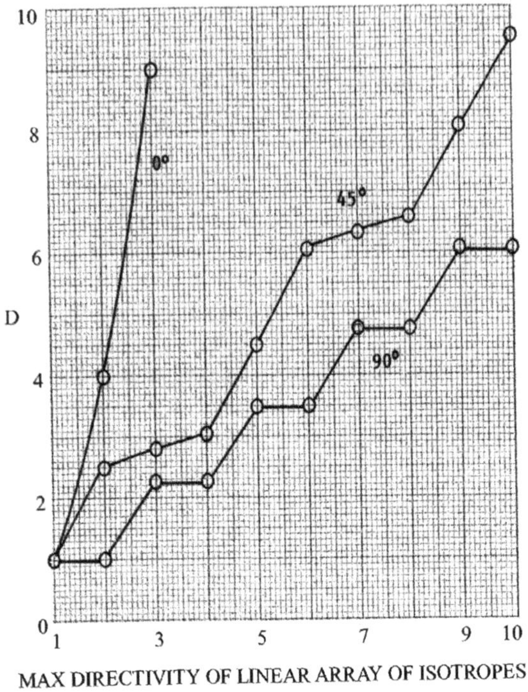

FIGURE 6.1 Maximum broadside directivity in zero spacing limit.

Bloch et al. (1953) used a formulation involving mutual resistances (see Hansen, 1998; Section 2.4.3), but only very limited calculations were presented.

The maximum value of directivity can be found by using the Lagrange multiplier method (Sokolnikoff and Sokolnikoff, 1941). The directivity D of an array of isotropes at broadside can be written as

$$D = \frac{\left(\sum_{n=1}^{N} A_n\right)^2}{\sum_{n=1}^{N}\sum_{m=1}^{N} A_n A_m \,\mathrm{SINC}[(n-m)2\pi d/\lambda]} \tag{6.5}$$

Here it is assumed that N array elements are isotropic and equally spaced by d. In the formulation in Equation 6.6, the array amplitudes are A_n and the SINC function $[(\sin x)/x]$ represents the mutual impedance between isotropic elements (Hansen, 1983), which is 120 SINCkd. Although the directivity expression could be maximized directly, it is convenient to constrain the sum of the coefficients to unity, and then to minimize the denominator. The Lagrangian equations are

$$\begin{aligned} &2\sum_{n=1}^{N} A_n \mathrm{SINC}[(n-i)2\pi d/\lambda] - \beta = 0, \quad i = 1, 2, 3, \ldots, N \\ &\sum_{n=1}^{N} A_n = 1 \end{aligned} \tag{6.6}$$

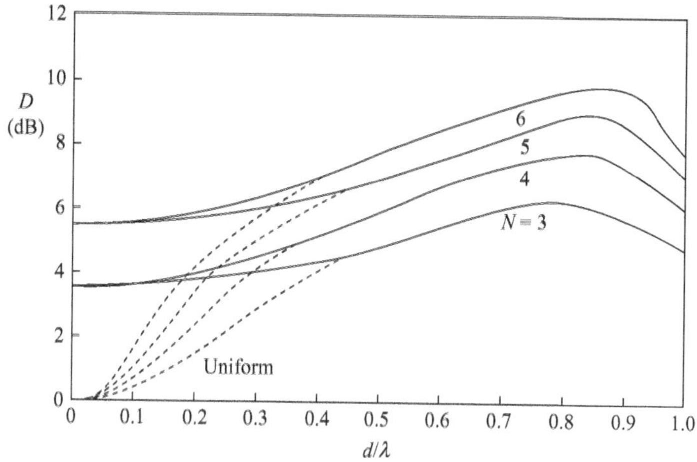

FIGURE 6.2 Maximum directivity for fixed spacing.

where β is the Lagrangian multiplier. Solving the first equation for $i = 1$ for β and substituting gives N equations in the unknown coefficients for an N-element array. These are easily solved by using simultaneous equation computer subroutines with multiple precision as needed. These were solved by hand for $N = 3$, 5, and 7 by Pritchard (1953). Hansen (1983) compares the maximum directivity of small arrays with the uniform-amplitude directivity versus element spacing (see Figure 6.2). Above $d/\lambda = 0.5$, the two are very close. Also, some minor oscillations in the directivity curves have been smoothed out, as they are not important here. The coalescing of pairs of curves at zero spacing occurs because arrays of $2N$ and $2N - 1$ elements have the same number of degrees of freedom as previously mentioned.

To give an idea of the coefficients, input resistances, and pattern of a small array with modest superdirectivity, Table 6.1 shows the amplitude coefficients for an array of seven elements with quarter-wave spacing. Directivity is 5.21, and the pattern is the solid line in Figure 6.3. Also shown (dashed line) is the pattern of the same length array with half-wave spacing. Directivity of a corresponding uniform-amplitude array is 3.64. Three of the input resistances in the table are negative; these elements have power flow in the reverse direction, a phenomenon not unusual in superdirective arrays.

TABLE 6.1 Seven-Element Superdirective Array ($d = \lambda/4$)

Element Number	Amplitude	Resistance (Ω)
1	1.443	0.13299
2	−3.933	−0.04879
3	7.122	0.002694
4	−8.264	−0.02322
5	7.122	0.02694
6	−3.933	−0.04879
7	1.443	0.13299

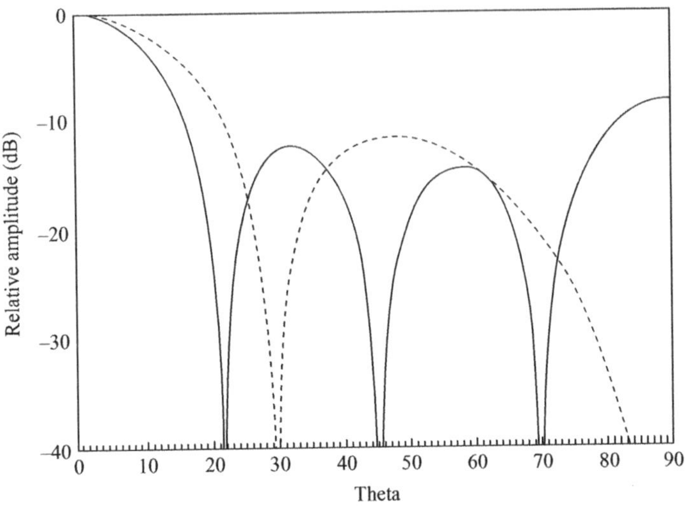

FIGURE 6.3 Maximum directivity array, $N = 7$, $d = \lambda/4$.

An extreme example of a superdirective array was computed by Yaru (1951); this was a nine-element broadside array of isotropes with an overall length of $\lambda/4$. The design was a nominal Chebyshev pattern with 25 dB SLR. Because of the limitations of desk calculators circa 1950, the currents calculated by Yaru contain small errors, which were corrected (Jordan and Balmain, 1968). Correct values are

$$\begin{aligned} I_1 &= I_9 = 260{,}840.2268 \\ I_2 &= I_8 = 2{,}062{,}922.9994 \\ I_3 &= I_7 = 7{,}161{,}483.1266 \\ I_4 &= I_6 = 14{,}253{,}059.7032 \\ I_5 &= 17{,}787{,}318.7374 \end{aligned}$$

The net broadside current is 0.039! The superdirective ratio is 17.3!

Almost all papers have neglected the network used to feed the array elements, but Harrington (1965) sets up a matrix representing array element self- and mutual impedances. Directivity is a ratio of products of current and impedance matrices, which is then differentiated to find maximum D. An eigenvalue equation results with one nonzero eigenvalue. Unfortunately, no examples or calculations accompanied this work.

6.2.2.2 Endfire Arrays

The Uzkov–Gilbert–Morgan result for maximum endfire directivity as element spacing goes to zero is

$$D = \sum_{n=0}^{N-1} (2n+1)[P_n(1)]^2 \qquad (6.7)$$

Because $P_n(1) = 1$, the directivity limit is

$$D = \sum_{n=0}^{N-1} (2n+1) = N^2 \tag{6.8}$$

Although calculations are scarce, this result was validated by Tai (1964) and by Stearns (1961). Another validation, both theoretical and experimental, for two-element monopole arrays has been given by Altshuler et al. (2005).

Endfire superdirectivity is produced by interference whereby the main beam is scanned into the invisible region where $|u| > \pi L/\lambda$ or $|\cos\theta_0| > 1$. This causes energy to be stored in the near-field, resulting in a large antenna Q.

Hansen and Woodyard (1938) developed an endfire line source with modest superdirectivity. This is of interest because the distribution can be sampled to get array excitations and because the amplitude is constant, a feature that is attractive for arrays. They observed that if the free space phase progression along the aperture was increased, the space factor power integral decreased faster than the peak value; thus, the directivity increases up to a point. The endfire pattern, for a source of length L, is

$$f(\theta) = \mathrm{SINC}\left(\frac{L}{2}(k\sin\theta - \beta)\right) \tag{6.9}$$

where β is the wave number over the aperture. Inverse directivity is proportional to

$$\begin{aligned} \frac{1}{D} &\propto \frac{1}{\mathrm{SINC}^2[(L/2)(k-\beta)]}\int \mathrm{SINC}^2\left(\frac{L}{2}(k\sin\theta - \beta)\right)\cos\theta\, d\theta \\ &= \frac{1}{\mathrm{SINC}^2(\phi/2)}\left(\frac{\pi}{2} + \mathrm{Si}\,\phi + \frac{\cos\phi - 1}{\phi}\right) \end{aligned} \tag{6.10}$$

Here Si is the sine integral, and $\phi = L(k-\beta)$, the additional phase along the aperture (in addition to the progressive endfire phase). Maximum directivity of $7.2143L/\lambda$ was determined to occur for $\phi = 2.922$ rad. In many books, it is carelessly stated that π extra radians of phase are needed, but there is no physical reason for this; a better approximation to 2.922 is 3. Directivity increase over normal endfire is 2.56 dB, and the sidelobe ratio is 9.92 dB. The distribution is suitable for long arrays; for short arrays, a computer optimization of phase is recommended; see Hansen (1992). A modest improvement was made by Goward (1947) by adjusting the endfire source amplitude.

The Hansen–Woodyard distribution is endfire. In general, the maximum directivity does not occur there. The most general solution of uniform amplitude would allow any element phase needed to maximize directivity. Such a solution could be formally realized for a given number of elements and spacing, but the equations would require a numerical solution. A slightly simpler problem was worked by Bach (1970); he started with a uniform-amplitude array of isotropic elements that was phased to produce a main beam at θ_0. The interelement phase factor is $\delta = kd \sin\theta_0$, and the directivity is

$$D = \frac{N}{1 + (2/N)\sum_{n=1}^{N-1}(N-n)\,\mathrm{SINC}(nkd)\cos(n\delta)} \tag{6.11}$$

Calculations were made for 2-, 3-, 4-, and 10-element arrays for all beam angles, and for spacings up to λ. Figure 6.4 shows the results for 2- and 3-element arrays, and Figure 6.5 for 4- and 10-element arrays. Figure 6.5 is striking in that high directivity occurs along a line roughly for $kd + \delta = 0$, or $\theta_0 = -\pi 2$, with peak directivity near endfire at λ spacing. Along the line, roughly for $kd = \delta + 2\pi$ directivity is changing

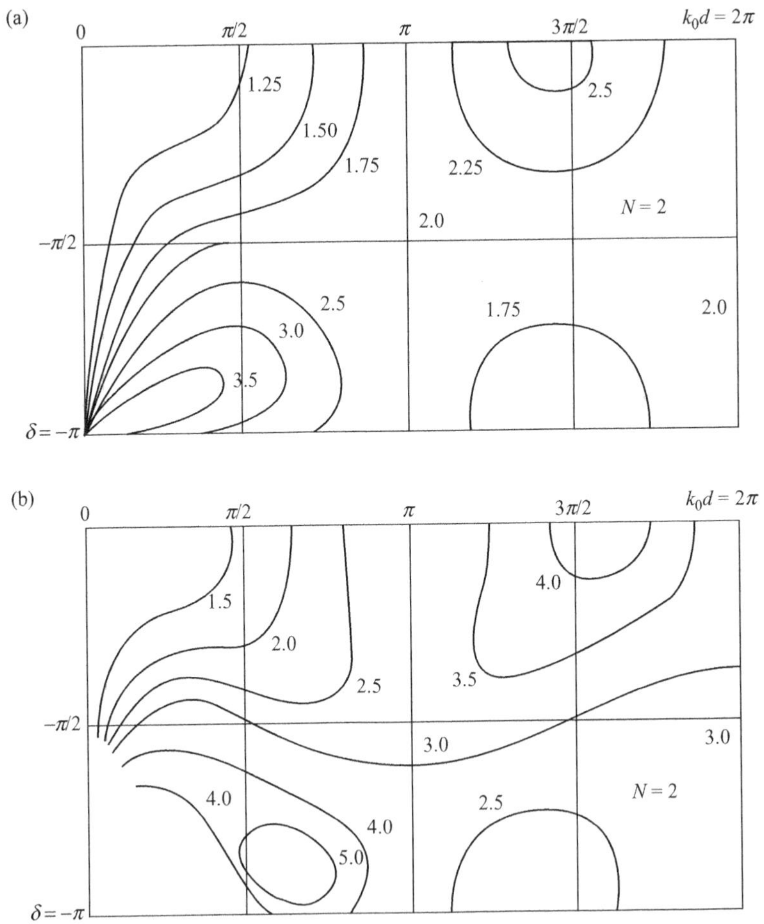

FIGURE 6.4 Uniform array directivity, $N = 2$ and 3. Courtesy of Bach (1970).

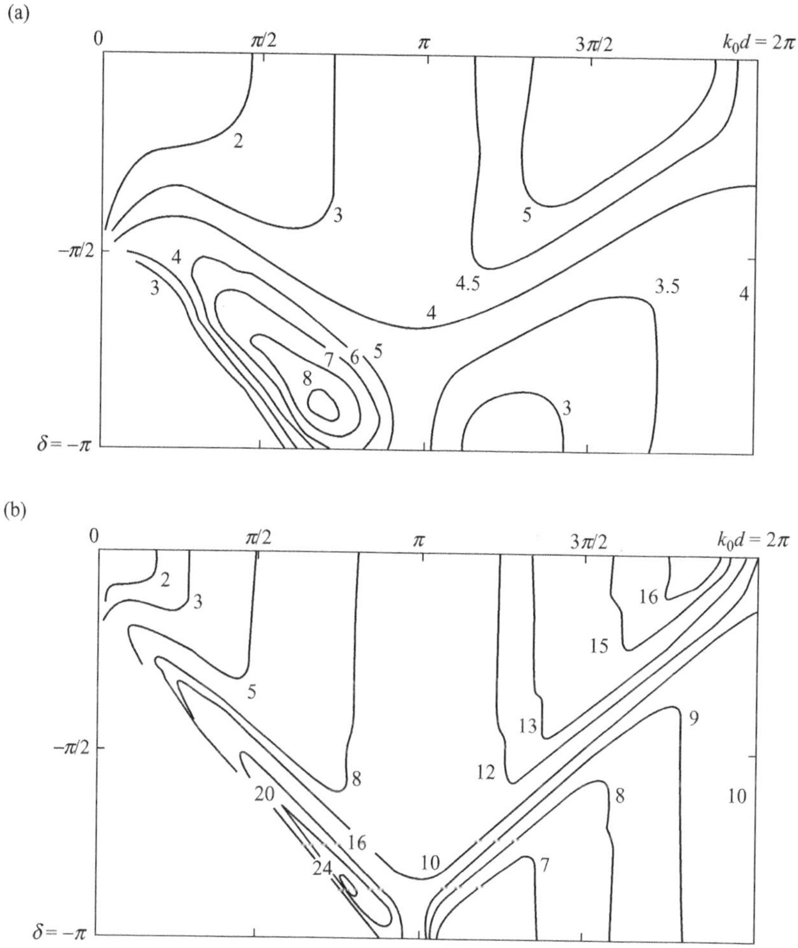

FIGURE 6.5 Uniform array directivity, $N = 4$ and 10. Courtesy of Bach (1970).

rapidly, perhaps owing to the appearance of another lobe. Directivity values are shown at θ_0, but in some cases a "sidelobe" may have higher amplitude. Thus, even for uniform amplitude an array is complex.

The simplicity of linear arrays of isotropic elements allows the directivity to be exactly written in closed form when the excitation coefficients are given. A three element linear array of isotropes is used as an example. Let the phase parameter ψ be:

$$\psi = kd(\cos\theta - \cos\theta_0) \tag{6.12}$$

Then the array excitation coefficients and function are a, b, c, d, e, f, and F.

$$F = a + jb + (c + jd)\exp(\psi) + (e + jf)\exp(2\psi) \tag{6.13}$$

The directivity is given by:

$$D = \frac{2|F_0|^2}{\int_0^\pi |F(\theta)|^2 \sin\theta d\theta} \tag{6.14}$$

When the magnitude squared of the array function is expanded and inserted into the integral above, the resulting integration reduces to integrals of two basic types, where kd and $2kd$ are the parameters. The scan angle θ_0 is zero for endfire, and 90 deg. for broadside. The two key integrals are:

$$\begin{aligned}\int_0^\pi cos[kd(\cos\theta - \cos\theta_0)]\sin\theta d\theta &= cos(kd\cos\theta_0)\int_0^\pi \cos(kd\cos\theta)\sin\theta d\theta \\ &+ \sin(kd\cos\theta_0)\int_0^\pi \sin(kd\cos\theta)\sin\theta d\theta\end{aligned} \tag{6.15}$$

Fortunately both of the integrals on the right hand side can be evaluated (Grobner and Hofreiter, 1958). These integrals are:

$$\begin{aligned}\int_0^\pi sin(kd\cos\theta)\sin\theta d\theta &= 0 \\ \int_0^\pi cos(kd\cos\theta)\sin\theta d\theta &= 2sinc\, kd\end{aligned} \tag{6.16}$$

Now the directivity can be written in closed form using the given array excitation coefficients. The directivity is given by:

$$D = \frac{(a+c+e)^2 + (b+d+f)^2}{DENOM} \tag{6.17}$$

The steering factor is γ:

$$\gamma = kd\cos\theta_0 \tag{6.18}$$

Now the denominator can be written in exact closed form.

$$\begin{aligned}DENOM = {} & 4(ac+bd+ce+df)cos(\gamma)sinc\, kd \\ & + 4(bc-ad-cf+de)sin(\gamma)sinc\, kd \\ & + 4(ae+bf)\cos 2\gamma\, sinc\, 2kd + 4(be-af)\sin 2\gamma\, sinc\, 2kd \\ & + a^2+b^2+c^2+d^2+e^2+f^2\end{aligned} \tag{6.19}$$

There are eighteen terms in the denominator resulting from the six excitation coefficients. This formula has been validated for broadside arrays of three istoropes using results from Ma (1974) and for endfire using coefficients found from minimization codes described below.

Lo et al. (1966) formulated maximum endfire directivity by setting derivatives of the array polynomial to zero. They compare D for a 10-element array of isotropes for four cases: D(U)—uniform-amplitude, progressive endfire phase; D(OC)—amplitude-optimized, progressive endfire phase; D(HW)—Hansen–

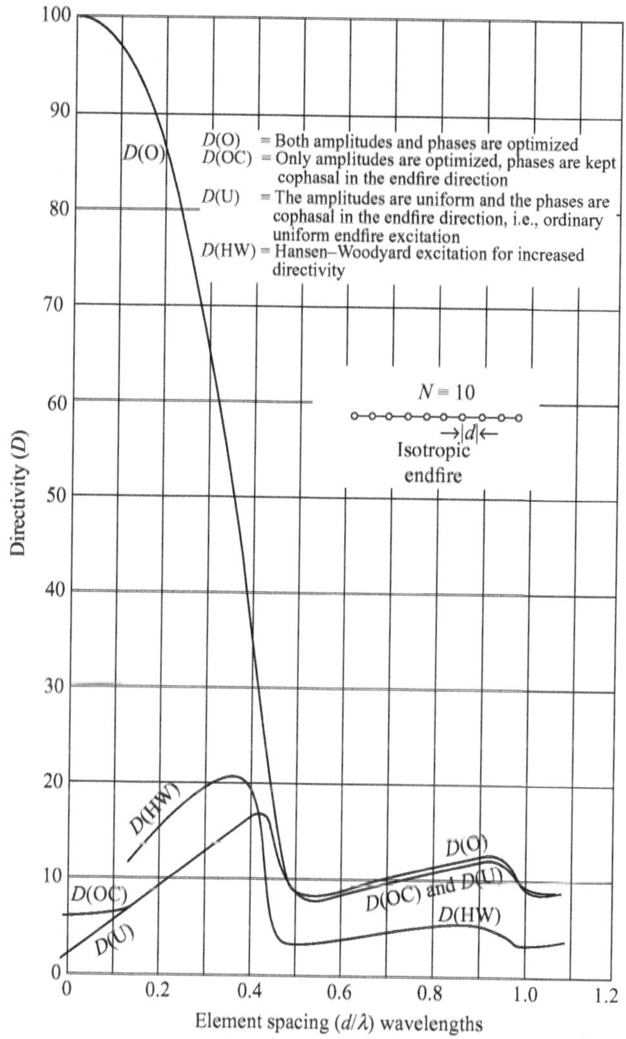

FIGURE 6.6 Linear endfire array directivity; various excitations. (Courtesy of Lo (1988).

Woodyard; $D(\mathrm{O})$—amplitude and phase optimized. Figure 6.6 shows results; for $d \leq \lambda/4$, which is the normal endfire range, the Hansen–Woodyard case gives better directivity than the uniform-amplitude or the optimized-amplitude case. Of course, the fully optimized case gives directivity equal to N^2 in the $d = 0$ limit. Note that Figure 6.6 (Lo, 1988) is a correction of Figure 4 of Lo et al. (1966). With the high directivity of $D(\mathrm{O})$ goes severe bandwidth and tolerance problems; see Section 6.4.1.

Cheng and Tseng (1965) optimized directivity by using the Hermitian quadratic forms ratio; their calculations for an eight-element endfire array are not useful as the element spacing was $> \lambda/4$, thereby producing additional main lobes.

A special case of an endfire dipole array was investigated by Bacon and Medhurst (1969). Here four unequally spaced dipoles were used, with one dipole fed. Mutual resistances were used, with the spacings optimized. Array lengths of 0.5λ, 0.6λ, and 0.7λ were used. It was necessary to couple dipole 2 to dipole 4 (dipole 1 is fed). This configuration gave a maximum directivity of 19.7, with an overall length of $\lambda/2$. From the feed end, spacings were 0.05λ, 0.25λ, and 0.20λ. A typical Yagi–Uda array of 0.5λ length will have a directivity of about 5 (Hansen, 2002). Hansen–Woodyard directivity would be $1.8L/\lambda = 3.6$. The superdirectivity ratio of 10.7 is accompanied by tight tolerances and a small input resistance.

Seeley (1963a, 1963b) investigated arrays of two and three loops; phasing was used to steer nulls. As expected, tolerances were tight.

6.2.2.3 Minimization Codes

Another method for maximizing directivity of a linear array is to minimize $1/D$. The reader is warned that this process is full of pitfalls: trendy codes such as genetic algorithm and simulated annealing often lock up on subsidiary minima and are hard to shift. Recommended codes are the Fletcher variable metric (quasi-Newton). Versions are available that do not require explicit derivatives (Fletcher, 1987). Calculations described below utilize double precision on a 64-bit workstation but the large slopes in superdirectivity may require more that double precision as the element spacing becomes very small.

Of particular interest is how the excitation coefficients (currents) behave as the spacing between elements approaches zero. For the broadside case the currents are real, but an equal set of imaginary currents gives the same results. For $N=3$, broadside, as the spacing approaches zero the currents approach A, $-2A$, A. As d/λ approaches zero, A approaches infinity. For $d/\lambda = .01$, $A = 885$. For a four-element broadside array the coefficients for small spacing are A, $-A$, $-A$, A. Again as spacing approaches zero A approaches infinity. For $d/\lambda = .01$, $A = 470$.

For the endfire case for $N=3$ as spacing approaches zero, the real part of the currents are $-A$, $2A$, $-A$ where for $d/\lambda = .02$, $A = 496$ plus a small imaginary part. For $N=4$ the coefficients are $-A$, $2A$, $-2A$, A. For $d/\lambda = .05$, $A = 300$ plus a small imaginary part. These results have been validated for three-elements using the exact formulation and also using Romberg numerical integration, both using the coefficients from the minimization codes.

6.2.2.4 Resonant Endfire Arrays

The classic superdirective arrays have been electrically small, using small elements and small element spacings. A new class of superdirective arrays uses electrically small resonant elements of the type discussed in Sections 3.2.4 and 3.2.5. This allows the isolated elements to be designed for a desired radiation resistance, say 50 Ω. The array may then have radiation resistance of a few ohms, instead of the usual milliohm values of superdirective arrays. And so the efficiency will be much higher than that for conventional superdirective arrays. There is still an impedance matching problem due to the radiation resistance. Best et al. (2008) detail a two-element

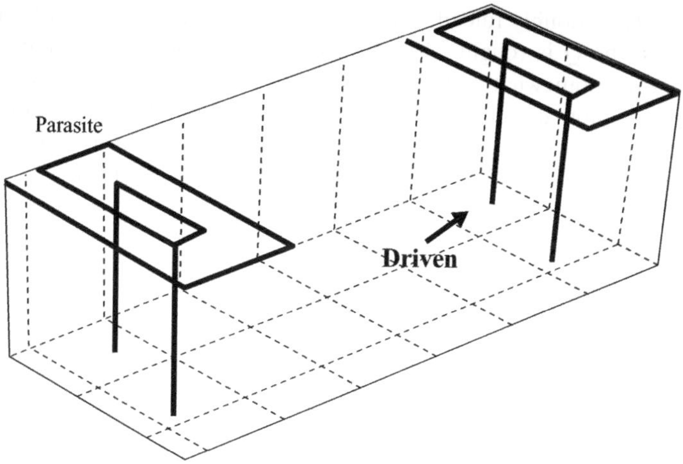

FIGURE 6.7 Top-loaded, folded, and bent antennas. Courtesy of Yaghjian et al. (2008).

superdirective array using folded monopoles. This array is electrically small only in width. More advanced antennas are described by Yaghjian et al. (2008). Their studies have shown that in two-element superdirective arrays, one driven and one parasitic element are almost as effective, as the spacing becomes small. Figure 6.7 shows top-loaded and folded monopoles in a two-element ESA superdirective array. Peak gain of 6.9 and 8 occurs for an element spacing of 0.15λ. Another two-element ESA parasitic array is shown in Figure 6.8; only one element is shown,

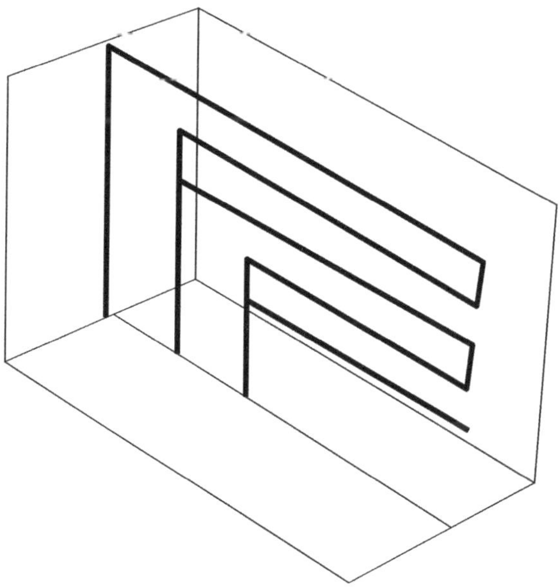

FIGURE 6.8 Doubly folded and bent element. Courtesy of Best et al. (2008).

and it is bent and doubly folded. Peak gain is 20 dB at a spacing of 0.1λ. For both these arrays, the Q is roughly 40; the bandwidth is narrow. Also the tolerance problem, while somewhat better, is still serious. Note that the gain values have been quoted; not directivity. Clearly, the use of resonant elements has significantly improved the efficiency. A related development uses quarter-wave monopoles in an endfire array (Altshuler et al., 2005), but these of course are not ESA.

6.3 CONSTRAINED SUPERDIRECTIVITY

To avoid impractical values of bandwidth, tolerances, and efficiency, it is useful to constrain the optimization process. Constraints can involve SDR, sidelobe level, Q, tolerance, or efficiency. Probably the simplest, and most investigated, is a sidelobe constraint.

6.3.1 Dolph–Chebyshev Superdirectivity

The principles and design equations for arrays with equal-level sidelobes, Dolph–Chebyshev arrays, were covered in detail in Hansen (1998). However, Dolph's derivation (Dolph, 1946) and the formulas of Stegen are limited to $d \geq \lambda/2$. Riblet (1947) showed that this restriction could be removed, but only for N odd. For spacing below half-wave, the space factor is formed by starting at a point near the end of the Chebyshev ± 1 region,[1] tracing the oscillatory region to the other end, and then retracing back to the start end and up the monotonic portion to form the main beam half. Because the Mth-order Chebyshev has $M-1$ oscillations, which are traced twice, and the trace from 0 to 1 and back forms the center sidelobe (in between the trace out and back), the space factor always has an odd number of sidelobes on each side, or into a Chebyshev array for $d < \lambda/2$. The pattern is given by

$$T_M(a \cos \Psi + b) \tag{6.20}$$

$$a = \frac{z_0 + 1}{1 - \cos kd}$$
$$b = \frac{z_0 \cos kd + 1}{\cos kd - 1} \tag{6.21}$$

where $\Psi = kd \sin \theta$. The value of z_0 is

$$z_0 = \cosh \frac{\text{arc cosh SLR}}{M} \tag{6.22}$$

[1] The exact starting point depends on N and kd.

The sidelobe ratio is

$$\mathrm{SLR} = T_M(z_0) \tag{6.23}$$

Formulas have been developed by DuHamel (1953), Brown (1957, 1962), Salzer (1975), and Drane (1963, 1964). Those of Drane will be used here, as they are suitable for computer calculation of superdirective arrays. The array amplitudes are

$$A_n = \frac{\varepsilon_n}{4M} \sum_{m=0}^{M_1} \varepsilon_m \varepsilon_{M_2-m} T_m(x_n)[T_M(ax_n + b) + (-1)^n T_M(b - ax_n)] \tag{6.24}$$

where $\varepsilon_i = 1$ for $i = 0$ and is equal to 2 for $i > 0$; $x_n = \cos n\pi/M$. The integers M_1 and M_2 are, respectively, the integer parts of $M/2$ and $(M+1)/2$. This result is valid for $d \le \lambda/2$. Small spacings (highly superdirective arrays) may require multiple precision because of the subtraction of terms. Many arrays are half-wave spaced; for these the a and b reduce to

$$a = \frac{1}{2}(z_0 + 1), \quad b = \frac{1}{2}(z_0 - 1) \tag{6.25}$$

For half-wave spacing, this approach and that of Dolph give identical results! In fact, owing to the properties of the Chebyshev polynomial, the two space factors, in precursor form, are equal:

$$T_{N-1}\left(z_0^{N-1} \cos\frac{\Psi}{2}\right) \equiv T_M\left(\frac{z_0^M(\cos\Psi + 1) + \cos\Psi - 1}{2}\right), \quad N - 1 = 2M \tag{6.26}$$

where the superscripts on z_0 indicate that each must be chosen for the proper form. Because many computers have no inverse hyperbolic functions, it is convenient to rewrite the z_0 as

$$z_0 = \frac{1}{2}\left[\mathrm{SLR} + \sqrt{\mathrm{SLR}^2 - 1}\right]^{1/M} + \frac{1}{2}\left[\mathrm{SLR} - \sqrt{\mathrm{SLR}^2 - 1}\right]^{1/M} \tag{6.27}$$

Directivity is given by

$$D = \frac{\left[\sum_{n=1}^N A_n\right]^2}{\sum_{n=1}^N \sum_{m=1}^N A_n A_m \,\mathrm{SINC}[(n-m)kd]} \tag{6.28}$$

where the A_n coefficients are given by Equation 6.18. Directivity for arrays of three, five, seven, and nine elements has been calculated, for sidelobe ratios of 10 and 20 dB. The superdirectivity can be seen in Figures 6.9 and 6.10 for spacing below 0.5λ, as the ordinary directivity (using the Chebyshev coefficients that are independent of spacing) goes smoothly to 0 dB at zero spacing. The figures display these calculated directivities versus element spacing. Thus, a three-element array offers roughly 3 dB directivity for small spacings and a five-element array offers roughly 5 dB.

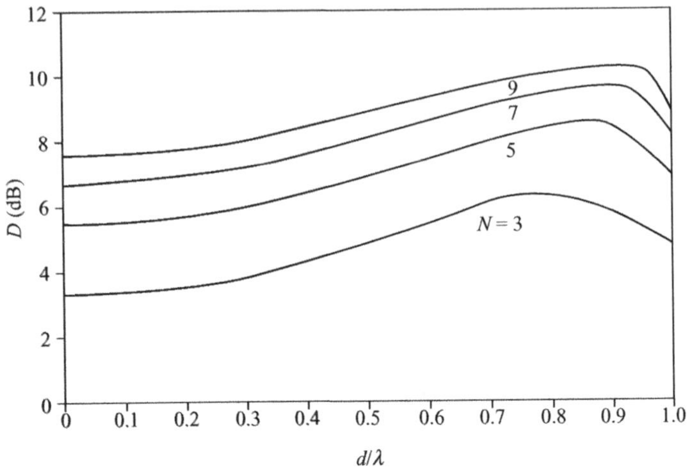

FIGURE 6.9 Chebyshev array directivity, SLR = 10 dB.

The Q is given by

$$Q = \frac{120 \sum A_n A_m^*}{\sum_n \sum_m A_n A_m^* R_{nm}} \tag{6.29}$$

R_{nm} is the mutual resistance between the n and m elements. Normally, the latter would, for thin wire dipoles, be computed by the efficient algorithm using sine and

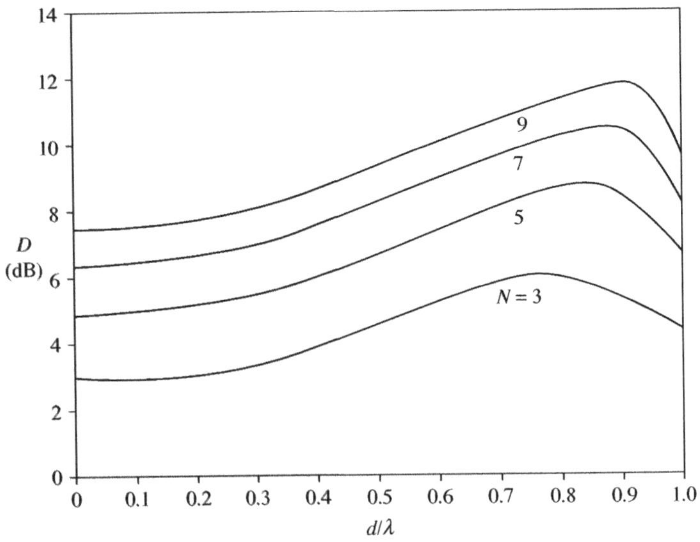

FIGURE 6.10 Chebyshev array directivity, SLR = 20 dB.

cosine integrals developed by Hansen (1972). But superdirectivity typically involves subtracting large numbers, especially for Q. It is necessary to employ double precision in the calculation of the mutual resistances, and hence in the sine and cosine integrals. Because such a subroutine could not be found, a Chebyshev economized series expansion, developed by Luke (1975), was used by this author to construct a double-precision sine and cosine integral subroutine. The results check all digits given in Abramowitz and Stegun (1970, Tables 5.1–5.3). Note the similarity between the D and Q formulas, Equations 2.21 and 2.22.

Riblet (1947) also showed that the Chebyshev polynomials T_n alone, of all real and complex polynomials, have the optimum properties that if the first null angle is specified the sidelobe level is minimized, and that if the sidelobe level is specified the first null angle is minimized. DuHamel (1953) modified the Dolph–Chebyshev design equations so that they would apply to a broadside or an endfire array of odd number of elements. He compared four endfire designs, for a seven-element array with $d/\lambda = 0.25$. These were uniform amplitude, normal phase (U); Hansen–Woodyard (HW); Schelkunoff, where the array polynomial zeroes are equally spaced on the unit circle (S); and an optimum design with the same SLR as the Schelkunoff design (O). Table 6.2 gives beamwidth, SLR, and ratio of largest to smallest current. Computing facilities circa 1953 did not allow calculation of either directivity or input resistances. Note that the element spacing is too large to allow significant superdirectivity, as borne out by the current ratios.

Another case of modest superdirectivity was given by Sanzgiri and Butler (1971). Stepwise sidelobe constraints were employed, and the optimum directivity was formulated as the ratio of two Hermitian quadratic forms, as previously described. Lagrangian methods were used to solve for max D. The array was broadside with nine elements at $d/\lambda = 0.6$. Several sidelobe envelopes were used; the case with constant SLR = 20 dB was typical. Directivity was 14.83, with an SDR of 1.55. This very modest value was due to the large element spacing; significant SDR for a broadside array requires d/λ much less than 0.5.

Multiple power pattern constraints were used by Kurth (1974) with directivity optimization. Constraints on both main beam and sidelobe were used, leading to the common ratio of Hermitian quadratic forms solved by Lagrangian multipliers. A circular array of dishes was used as an example. Cox et al. (1986) obtained a modest superdirectivity for an acoustic endfire array for various angular distributions of white noise. He also discussed "oversteering" past endfire to increase directivity. Apparently, the acoustics community was not familiar with Hansen–Woodyard!

TABLE 6.2 Comparison of Four Endfire Array Designs

	θ_3	SLR	I_4/I_1
U	99	13.3	1
HW	57	7.4	1
S	53	28	9.19
O	52	31	8.78

Dawoud and Anderson (1978) used Chebyshev polynomials to optimize the ratio of beam peak value for a superdirective array to beam peak value for a uniformly excited cophasal array. As the beamwidth narrows, this ratio rapidly decreases. However, there seems to be no simple relationship between this ratio and SDR. Another polynomial approach, by Dawoud and Hassan (1989), used Legendre polynomials instead of Chebyshev polynomials. The former yields slightly greater directivity for a broadside array with small spacing in wavelengths. The calculated directivities (SDR = 6.2) seem to be much too high for the 3 dB beamwidth shown.

6.3.2 Superdirective Ratio Constraint

By specifying an SDR, the designer can design the array and then evaluate bandwidth, tolerances, and efficiency. Using the definition of SDR provided by Taylor, the aperture or array problem was cast by Fong (1967) into a ratio of two Hermitian quadratic forms, which are then solved by Lagrangian methods. Let the scalar product of coefficients be

$$\langle A|A^*\rangle = |A_n|^2 \tag{6.30}$$

where * indicates the complex conjugate. Call the row vector of complex array excitations J and the column vector of path length phases F:

$$F = \begin{vmatrix} \exp(-jkr_1 \sin\theta) \\ \vdots \\ \exp(-jkr_n \sin\theta) \end{vmatrix} \tag{6.31}$$

where r_n is the distance from the reference point to the nth element. For a uniformly spaced array, $r_n = (n-1)d$. Now define matrices A and B, where A is

$$|A| = |F^*\rangle\langle F| \tag{6.32}$$

and the elements of B are

$$B_{nm} = \frac{1}{4\pi}\int f_1(\theta,\phi)\exp[-jkd(n-m)\sin\theta]d\Omega \tag{6.33}$$

The pattern of the ith element is f_i. Isotropic elements and a uniformly spaced array allow simplifications of A and B. The elements then become

$$A_{nm} = \exp[-jkd(n-m)\sin\theta], \quad B_{nm} = \mathrm{SINC}[(n-m)kd] \tag{6.34}$$

Now the directivity can be written in abbreviated form:

$$D = \frac{\langle J|A|J^*\rangle}{\langle J|B|J^*\rangle} \tag{6.35}$$

The tolerance sensitivity S is defined as the ratio of variance of peak field strength produced by errors of variance σ_T^2 (Uzsoky and Solymar, 1956):

$$S = \frac{\sigma_E^2}{\sigma_j^2} = \frac{(\Delta E)^2/E^2}{\sigma_j^2} = \frac{\sigma_j^2 \langle J|J^*\rangle / \langle J|A|J^*\rangle}{\sigma_j^2} = \frac{\langle J|J^*\rangle}{\langle J|A|J^*\rangle} \tag{6.36}$$

Thus, sensitivity is also written in abbreviated form. Q is conveniently found from $Q = SD$, or

$$Q = \frac{\langle J|J^*\rangle}{\langle J|B|J^*\rangle} \tag{6.37}$$

The directivity is a ratio of two Hermitian quadratic forms, with B positive and A at least positive semidefinite. Thus, all the eigenvalues of the associated equation are zero or positive real. Because A is a single-term dyad, there is one nonzero eigenvalue. The eigenvector (excitation) is given by

$$|J\rangle = |B|^{-1}F\rangle \tag{6.38}$$

The corresponding maximum directivity is given by

$$D_{\max} = \langle F|B|^{-1}F^*\rangle \tag{6.39}$$

In many applications, it is important to maximize D/T, directivity/system noise temperature. This is equivalent to maximizing signal-to-noise ratio S/N. To do this, the element pattern in the integral for B_{nm} is multiplied by the antenna noise temperature $T(\theta, \phi)$. Then the excitation vector that maximizes S/N or D/T is that of Equation 6.32.

Another ratio that can be directly minimized is beam efficiency: the fraction of power contained within the main beam, null-to-null. This is

$$\eta_b = \frac{\langle J|A|J^*\rangle}{\langle J|B|J^*\rangle} \tag{6.40}$$

To avoid the practical difficulties of finding the complex roots of a complex polynomial of high order, Winkler and Schwartz (1972) transformed that problem into one of finding the eigenvalues of a real matrix. This is numerically much faster, and readily available subroutines can be used. Maximum directivity (equivalent to SNR for a uniform noise field) was calculated for a four-element array of isotropes. Their calculations show that for broadside array spacing less than $\lambda/5$ the Q rises very fast while directivity is slowly increasing. Thus, for this broadside array the *tolerance factor*[2] $T = D/Q$ increases very fast as $d \rightarrow 0$. For an endfire array, both Q and D are increasing as $d \rightarrow 0$, but again the tolerance factor increases very fast.

A related constrained D optimization used as a constraint the integral of current squared divided by radiated power (Margetis et al., 1998). This constraint is related to the SDR constraint.

A different approach was taken by Rhodes (1971), who optimized directivity of a line source subject to a fixed SDR. The pattern and aperture distribution are expanded in a series of prolate spheroidal functions. The SDR of Taylor was generalized to include edge effects (element pattern). When the SDR approaches a large value, the aperture distribution approaches a delta function spike at each end. For a given SLR, the series of spheroidal functions provides the maximum directivity that can be achieved.

6.3.3 Bandwidth or Q Constraint

Bandwidth is probably the most useful constraint. Constraining Q is equivalent, because for narrowband antennas (and all superdirective antennas are narrowband), the 3 dB bandwidth is simply $1/Q$. Again the problem is formulated as the ratio of two Hermitian quadratic forms, by Lo et al. (1966) and Cheng (1971). Using a scalar p proportional to the Lagrangian multiplier, the problem can be reduced to a complex polynomial of $2(N-1)$ degrees of freedom in p, for an N-element array. Unfortunately, solving such an equation for large N is extremely tedious. See Winkler and Schwartz (1972) for an alternative solution. The paper lists quadratic ratios for directivity and Q, for the uniform case, optimum directivity, and optimum directivity for a prescribed Q. They also compared several endfire cases, as discussed in Section 6.2.2.2. See Section 6.4.1 for additional bandwidth data.

Kovács and Solymár (1956) treated the inverse problem: Given directivity for an array configuration, find the excitation that minimizes Q. They showed that the minimum Q increases as D^2 for a fixed array size.

6.3.4 Phase or Position Adjustment

An iterative perturbation approach was used by Cheng (1971) to provide optimum directivity for an array. This was applied first to a ring array of uniformly spaced isotropic elements. Both amplitudes and phases were adjusted. The directivity increases rapidly with decreasing ring diameter below 2λ. Because amplitudes are difficult (expensive in hardware and losses) to adjust, phase-only optimization was applied to a 12-element ring array; for diameters less than 3λ, or element spacings less than 0.78λ, roughly 2 dB increase in D is obtained.

Phase-only optimization of an endfire array of isotropes was given by Voges and Butler (1972); they used steepest descent to solve for maximum directivity. Their 10-element array had spacing of $d = 0.4\lambda$, so there is some uncertainty about the pattern, due to $d \succ \lambda/4$. The SDR was 1.71. Q was just above 4, a value slightly less than a corresponding Hansen–Woodyard array.

Optimization of D by varying the interelement spacing in a linear array was performed by Butler and Unz (1967). Steepest descent methods were used on quadratic forms. A seven-element array was optimized with uniform amplitude. The maximum D agreed with the maximum calculated by Tai (1964) of 11.5. The SDR was 1.92; the spacings were symmetric, and changed slightly.

6.3.5 Tolerance Constraint

Because tolerances, in excitation amplitude and phase, in element position, and in element orientation, are critical for superdirective arrays, it is logical to optimize superdirectivity subject to a constraint on tolerances. Tseng and Cheng (1967) assumed standard deviations for element amplitude, for phase, and for element position. Directivity was maximized by reducing the Hermitian ratio to a determinantal equation; the matrix equation was solved via standard computer eigenvector routines. An example was calculated: An equally spaced eight-element endfire array of short dipoles; the normal directivity for $d = 0.3\lambda$ was 12.6, while the no-error optimum value was 45, and SDR was 3.57. Modest phase or amplitude errors transformed a directivity curve that was increasing as $d \to 0$ into a curve that turned down around $d \sim 0.35\lambda$ to 0.4λ, and rapidly decreases as $d/\lambda \to 0$. Endfire arrays appear to be more sensitive to errors, perhaps because the sidelobe region occupies more of the pattern volume.

Uzsoky and Solymar (1956) introduce a tolerance factor T that is the ratio of Q to D: $T = Q/D$. It has also been used by Lo et al. (1966), but called the sensitivity factor S. This factor is important as it tells whether Q is increasing faster than D as the SDR increases. They show that for endfire arrays, the uniform case has $T \sim 1$ over a wide range of spacings; the Hansen–Woodyard case has $T \sim 0.3$ over a wide range of spacings; the amplitude-optimized case gives a rapidly rising T for spacings below about 0.2λ; the fully optimized case T rapidly rises for spacings below about 0.45λ. Thus, the Q is rising fast (roughly exponentially) while the directivity is rising slowly.

A different approach was taken by Newman et al. (1978); they defied a sensitivity factor, which was the sum of element current magnitudes squared divided by the magnitude of the sum of currents. Also defined was a safety factor, and of course errors in terms of rms values of amplitude, phase, and position errors. For sidelobe fidelity in the -20, -30, and -40 dB regions, the sensitivity factor should be 10^3, 10^4, and 10^5, respectively. Using a safety factor of 1, the total allowable error (1σ) was roughly 3%, 1%, and 0.3%. Gilbert and Morgan (1955) maximized directivity subject to a fixed "background pattern" (see also Section 6.4.3).

6.4 BANDWIDTH, EFFICIENCY, AND TOLERANCES

6.4.1 Bandwidth

The first of three major difficulties with superdirective arrays is bandwidth, which rapidly becomes a problem as the element spacing decreases below $\lambda/2$ for broadside arrays or below $\lambda/4$ for endfire arrays. Thus, the Q is of concern; for narrowband antennas, half-power bandwidth $\simeq 1/Q$ and the impedance matched bandwidth $\simeq 2/Q$. Note that the VSWR ≤ 2 bandwidth is given by $1/\sqrt{2}Q$. The ratio of stored to dissipated energy, Q, can similarly be written in terms of array coefficients and mutual coupling, which for isotropic elements at broadside is:

$$Q = \frac{\sum_{n=1}^{N} A_n^2}{\sum_{n=1}^{N} \sum_{m=1}^{N} A_n A_m \, \mathrm{SINC}[(n-m)2\pi d/\lambda]} \tag{6.41}$$

Directivity D at broadside is given by

$$D = \frac{\left(\sum_{n=1}^{N} A_n\right)^2}{\sum_{n=1}^{N} \sum_{m=1}^{N} A_n A_m \, \mathrm{SINC}[(n-m)kd]} \tag{6.42}$$

Calculations have shown (Hansen, 1981b) that, for broadside arrays of fixed length, both directivity and Q increase with the number of elements as expected. Figure 6.11 shows Q of Chebyshev arrays of odd numbers of elements versus element spacing. An array with an even number of elements has a slightly higher Q than the array with the next larger odd number of elements, possibly because the odd-element sampling is more efficient. Figures 6.12 and 6.13 show Q versus directivity for Chebyshev arrays two wavelengths long. For all the cases computed, Q varied approximately linearly with directivity. Figure 6.14 shows log Q versus directivity for odd arrays with lengths 1λ, 2λ, and 5λ. The circles represent points calculated in double precision; extended precision is required for arrays of more elements than those shown in the figure. The straight lines are drawn

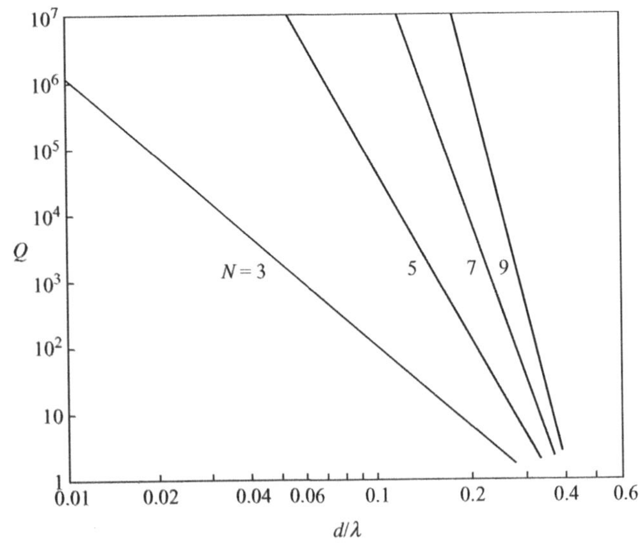

FIGURE 6.11 Q of broadside Chebyshev arrays versus element spacing.

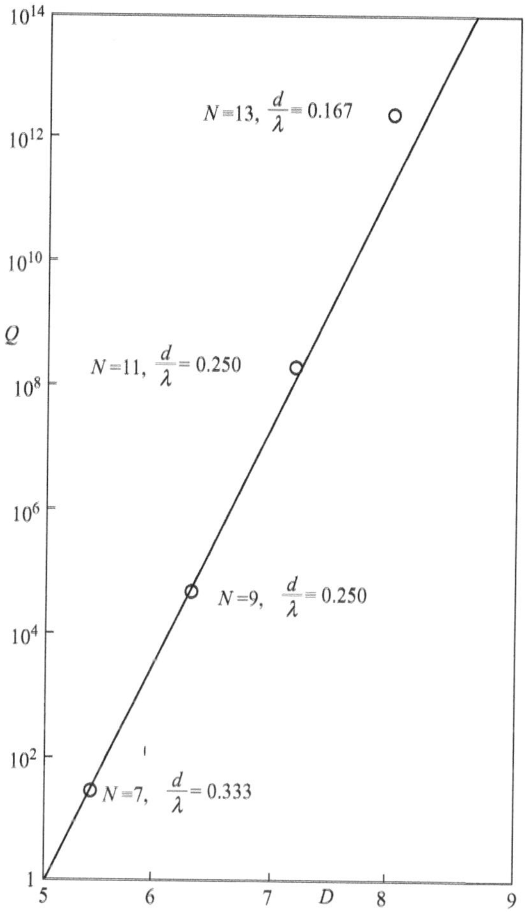

FIGURE 6.12 Q versus directivity, SLR = 10 dB, $L = 2\lambda$.

through the uniform excitation directivity point (equal to D_0) with slope of π, where $D_0 = 1 + 2L/\lambda$. There is at this time no physical significance to using this value of slope, but it is suggested by calculations of Rhodes (1974) on superdirective line sources. In making a best straight-line fit to the set of points for each of the three arrays, the slopes were in fact clustered around the value of 3.14. However, it is difficult to perform this fit with precision because, as pointed out by Rhodes, the curve of log Q versus directivity has an oscillatory behavior for low values of Q. If the assumptions above are true, superdirective broadside array behavior is predicted by the equation

$$\log Q = \pi(D - D_0) \tag{6.43}$$

The 20 dB Chebyshev data of Figure 6.13 are close to this result. Thus, the superdirective array clearly fits into the category of fundamental limitations in

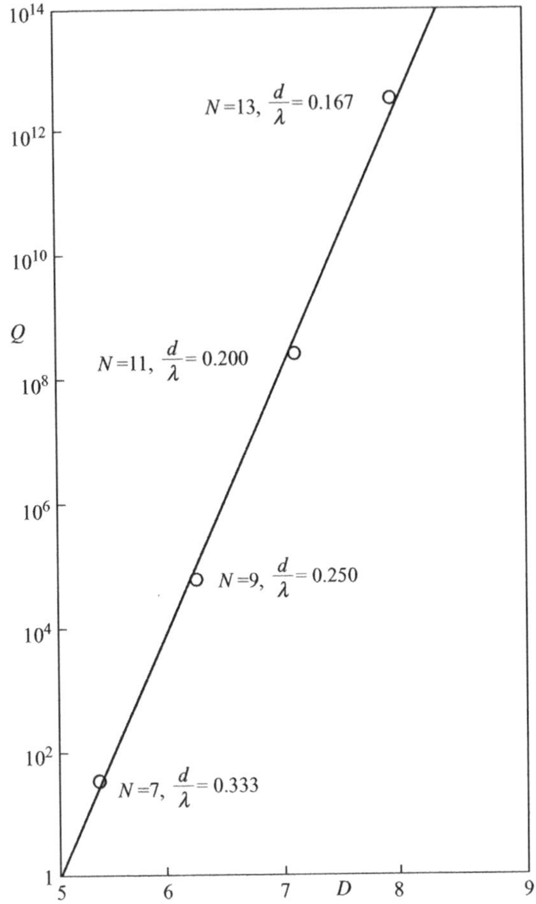

FIGURE 6.13 Q versus directivity, SLR = 20 dB, $L = 2\lambda$.

antennas (Hansen, 1981a). Whether the assumed slope of π can be physically justified remains an interesting problem at this time.

The Q to be expected from an array of isotropes is approximately a function of the number of elements divided by d/λ. Figure 6.15 shows this relationship for many arrays; each array is represented by a circle. Spacings larger than 0.3λ are not included, as the amount of superdirectivity achieved is small for these. log Q is approximately linear with $N\lambda/d$; the straight-line fit in the figure is

$$\log Q \simeq 0.16043\left(\frac{N\lambda}{d}\right) - 1.53476 \tag{6.44}$$

These data allow the array designer to estimate the degree of superdirectivity (Q) for a given array geometry.

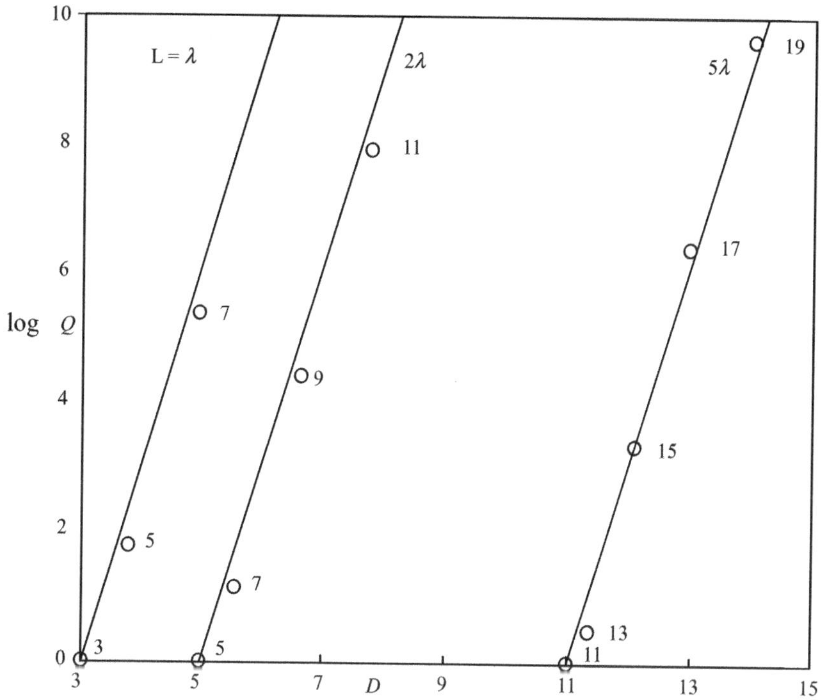

FIGURE 6.14 Q versus maximum directivity, odd arrays of isotropes.

Calculations of performance have been made for superdirective linear arrays of parallel dipoles. Collinear dipoles are not considered, as they would overlap neighbors. Figure 6.16 shows log Q versus directivity for arrays two wavelengths long. Both 0.1λ and 0.5λ dipoles are used. Again, log Q versus G is a straight line. As expected, the half-wave dipole array has higher Q than the corresponding isotropic array, owing to energy storage in the dipole near-field. With short (0.1λ) dipoles, this energy storage is much larger, resulting in much higher Q.

For endfire linear parallel dipole arrays, the variation of log Q with directivity is again linear but the slope changes as the length of the array changes. Figure 6.17 gives data for arrays of lengths 1λ, 2λ, and 5λ; the number of elements in each array is shown in the figure (Hansen, 1998).

6.4.2 Efficiency

A second undesirable feature of superdirective arrays is low efficiency, due to both matching network losses (see Section 6.6) and losses in the antenna elements. Both losses are caused by the low radiation resistances of these arrays. In many cases, but not all, the elements at the array center show the lowest radiation resistance; calculations for many small broadside arrays of an odd number of isotropes show

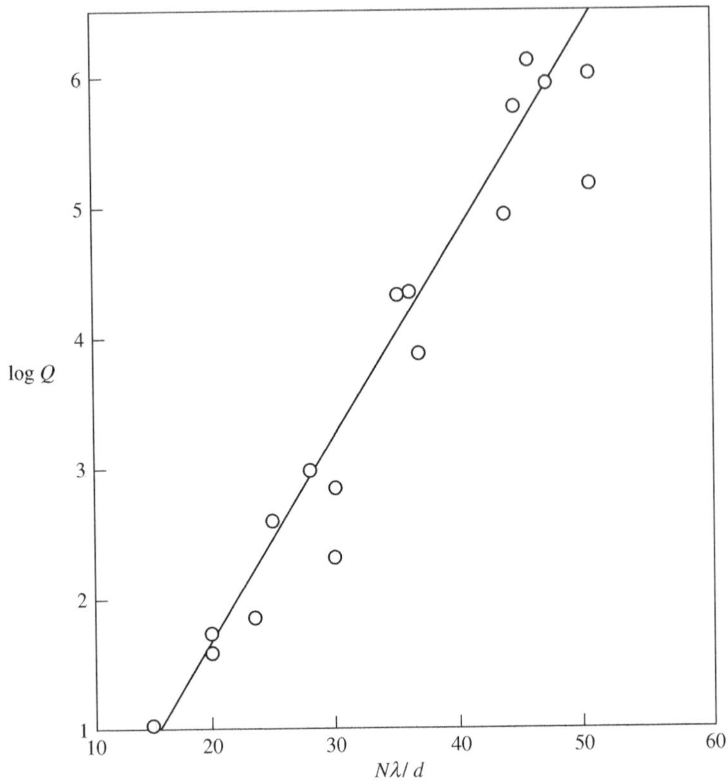

FIGURE 6.15 Q of broadside linear SD arrays of isotropes.

that $R_{\text{rad}} \propto 1/\sqrt{Q}$. Figure 6.18 shows these data, where each circle represents one array. The straight-line log–log fit is

$$R_{\text{rad}} \simeq \frac{0.8058}{\sqrt{Q}} \tag{6.45}$$

Odd arrays were used, as there are more variables per length, allowing better control of superdirectivity.

Loss resistance of cylindrical or strip dipoles is easily computed: A dipole of length l, radius a, and made of material with surface resistance R_s has a loss resistance R_{loss} of

$$R_{\text{loss}} = \frac{R_s l(1 - \text{SINC}\, kl)}{4\pi a \sin^2(1/2)kl} \tag{6.46}$$

For half-wave or resonant dipoles, $R_{\text{loss}} = R_s l/4\pi a$. Strip dipoles are equivalent to cylindrical dipoles, with strip width equal to $4a$. Copper wires have an ideal surface resistivity of

$$R_s \simeq 0.000261\sqrt{f\,\text{MHz}} \tag{6.47}$$

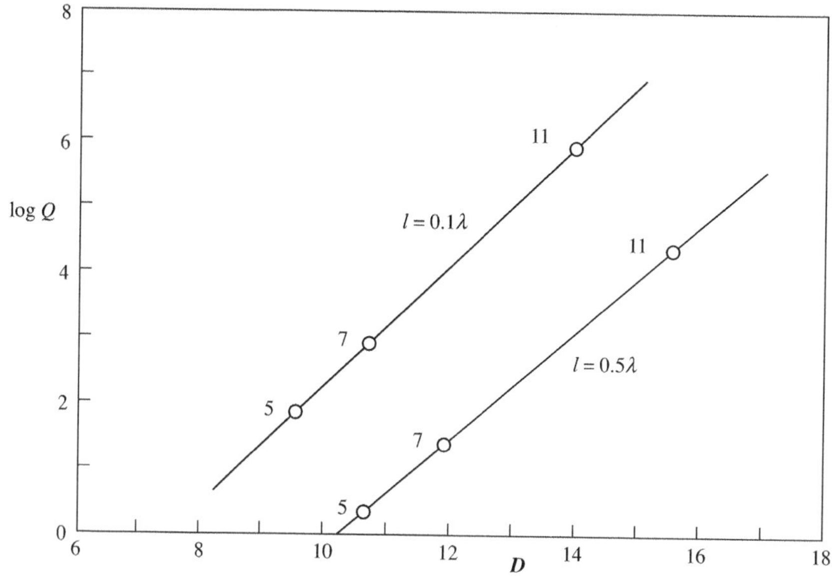

FIGURE 6.16 Q of 2λ, broad side arrays of parallel dipoles.

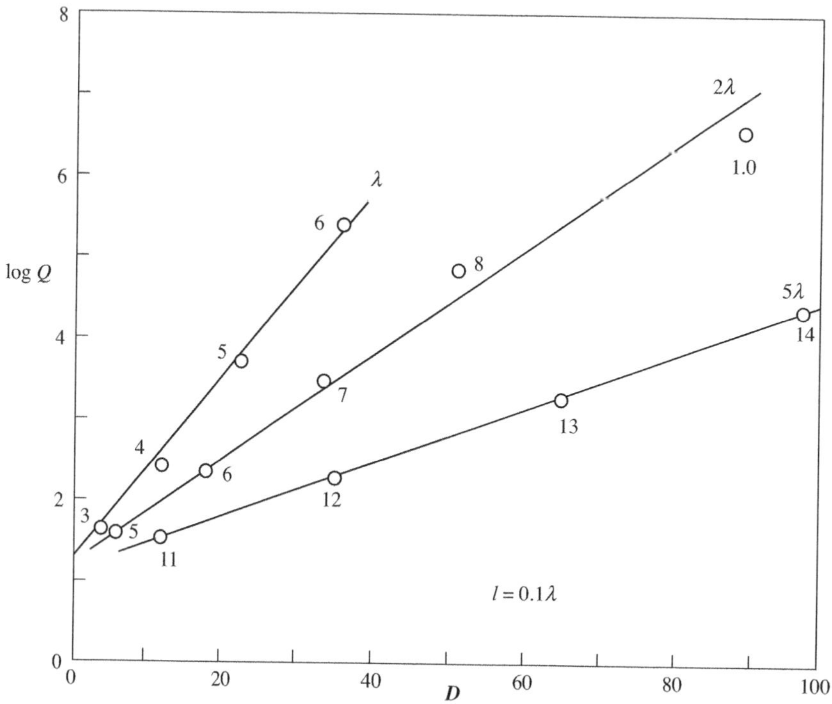

FIGURE 6.17 Q of endfire arrays of parallel dipoles.

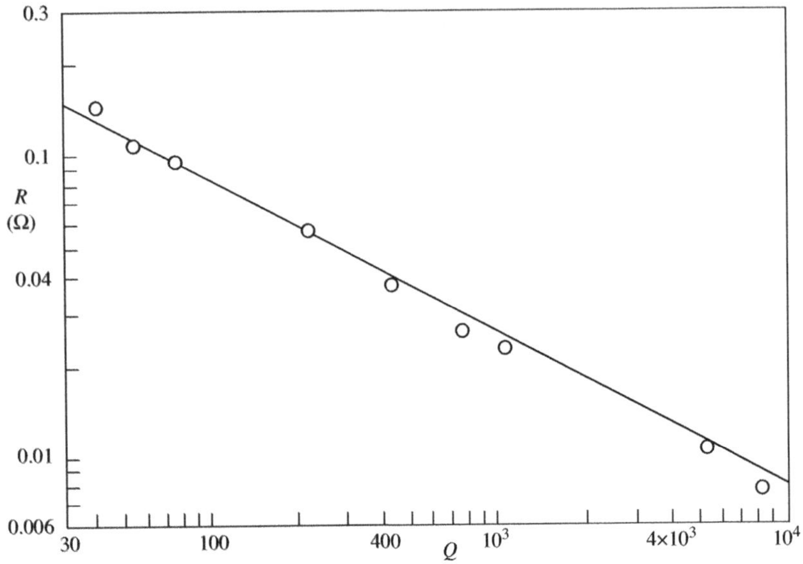

FIGURE 6.18 *Q* versus *R*; broadside arrays.

Over the range of 10–1000 MHz, R_s varies from 0.000825 to 0.00825 Ω/□. Using a value of $l/a = 25$, a moderately fat dipole, the loss resistance varies from 0.0066 Ω at 10 MHz to 0.066 Ω at 1 GHz. When these typical loss resistance numbers are compared with the radiation resistance values of Figure 6.18, it is clear that superdirective array efficiency may be a severely limiting consideration. Use of high-temperature superconductor (HTS) materials in the array and feed network can produce high efficiencies, but now the Q is that from Figures 6.16 and 6.17.

6.4.3 Tolerances

The third significant problem with superdirective arrays is tight tolerances. Because superdirectivity involves a partial cancellation of the element contributions at the main beam peak, with more cancellation for more superdirectivity, the tolerance of each element coefficient (excitation) becomes smaller (tighter) with more superdirectivity (Uzsoky and Solymar, 1956). A simple calculation has been made of these effects for Dolph–Chebyshev arrays by perturbing the center element of an odd array, finding the tolerance to reduce the directivity by 0.5 dB. This is not expected to be sensitive to sidelobe ratio, and a value of 20 dB was used. Calculations were made for $N = 3$ and 5 as a function of spacing, with the results shown in Figure 6.19. It was noted that the percentage error versus d/λ curve is linear up to spacings of roughly 0.1λ. For $N = 3$ the slope is 2:1, and for $N = 5$ the slope is 4:1. Thus, for $N = 3$, if the element spacing is halved, the tolerances must be four times tighter. The percentage tolerance for the center

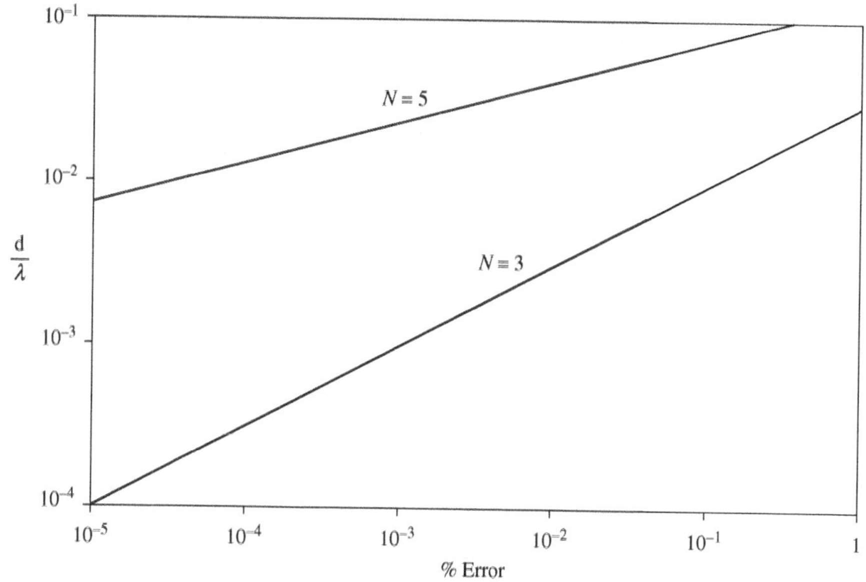

FIGURE 6.19 Center element tolerance versus element spacing; 0.5 dB D drop.

element, to maintain close (0.5 dB) to the expected directivity, is approximately given by

$$\frac{100}{\sqrt{Q}} \tag{6.48}$$

but the constant varies with the number of elements.

Thus, with $Q = 1000$, for example, the tolerance is 3.2%. Bandwidth and radiation resistance are more serious limitations of superdirective arrays.

An extreme example is the array of Yaru described above. An error in the center element excitation of only one part in 10^9 produces a directivity drop of roughly 3 dB. In general, the best way to evaluate tolerances for a given array design is to calculate the currents and then calculate the directivity change for a small change in one of the currents. Typically, an amplitude error of σ_a (standard deviation) is equivalent to a phase error of σ_ϕ in radians.

6.5 MISCELLANEOUS SUPERDIRECTIVITY

Solymar (1958) found the maximum directivity of a line source whose distribution was a Fourier series. Superdirectivity can occur when the number of harmonics N exceeds the source L/λ.

Analogous situations between acoustics and electromagnetic waves were described by Kock (1959). These include waveguides, and other guiding structures, and wave diffraction. Of interest here is his demonstration of superdirectivity using five small loudspeakers with rheostats and reversing switches (normal excitation for comparison). Superdirectivity of about 4 × was easily achieved!

An arc, or partial circle, of open loops or slotted cylinders with only one driven element can be tuned for resonance by adjusting the gaps (King, 1989). At resonance, it may be possible to produce a very narrowband peak of directivity (Veremy and Shestopalov, 1991). What degree of superdirectivity, if any, can be realized is not yet apparent. Measurements of an array of seven open loops on an arc with only one driven element (Bokhari et al., 1992) are promising but not yet conclusive. This area is not yet well understood. Closely related is a circular (ring) array of short, parallel, dipoles, with only one excited element. This ring can support a surface (slow) wave, and the closed ring can produce extremely sharp resonances (King et al., 2002; Fikioris et al., 1990; Fikioris, 1998). The small input resistance, narrow bandwidth, and narrow beams are characteristic of superdirectivity. It was shown by Janning and Munk (2002) that a planar array of short closely spaced dipoles could support a surface wave.

An analogy between superdirectivity and data processing to provide superresolution was provided by Buck and Gustincic (1967). Superresolution provides main beamwidth less than the Rayleigh limit λ/L (see Hansen, 1981a). Processing of the array outputs, for example, is by maximum entropy spectral analysis (Burg) or by maximum likelihood (Capon). The paper shows how noise limits the superresolution, just as errors limit superdirectivity.

6.6 SUPERDIRECTIVE ANTENNA SUMMARY

An aperture can have any superdirectivity; an array can have limited superdirectivity. Both will suffer from low radiation resistance, narrow bandwidth, low efficiency, and tight tolerances. A modest superdirective ratio is usable. Small dipoles and loops are superdirective and share many of these problems. Q is of interest only if it is large, so that bandwidth $= 1/Q$. Antennas with large VSWR need a matching network with very low intrinsic loss. Here HTS can offer a significant advantage. We can have superdirectivity, but only a small supergain is useful. Use of self-resonant ESA allows a better efficiency but bandwidth and tolerances are still limiting.

REFERENCES

Abramowitz, M. and Stegun, I. A. (Eds.), *Handbook of Mathematical Functions*, National Bureau of Standards, 1970.

Altshuler, E. E. et al. A Monopole Superdirective Array. *Trans. IEEE*, Vol. AP-53, August 2005, pp. 2653–2661.

Bach, H. Directivity of Basic Linear Arrays. *Trans. IEEE*, Vol. AP-18, January 1970, pp. 107–110.

BACON, J. M. AND MEDHURST, R. G. Superdirective Aerial Array Containing Only One Fed Element. *Proc. IEE,* Vol. 116, March 1969, pp. 365–372.

BEST, S. R. ET AL. An Impedance-Matched 2-Element Superdirective Array. *IEEE Antenn. Wireless Propag. Lett.*, Vol. 7, 2008, pp. 302–305.

BLOCH, A., MEDHURST, R. G., AND POOL, S. D. A New Approach to the Design of Super-Directive Aerial Arrays. *Proc. IEE,* Vol. 100, September 1953, pp. 303–314.

BLOCH, A., MEDHURST, R. G., AND POOL, S. D. Superdirectivity. *Proc. IRE,* Vol. 48, Part III, 1960, p. 1164.

BOKHARI, S. A. ET AL. Superdirective Antenna Array of Printed Parasitic Elements. *Electron. Lett.*, Vol. 28, July 1992, pp. 1332–1334.

BOUWKAMP, C. J. AND DE BRUIJN N. G. The Problems of Optimum Antenna Current Distribution. *Philips Res. Rep.*, Vol. 1, 1945 /1946, pp. 135–158.

BROWN, J. L. A Simplified Derivation of the Fourier Coefficients for Chebyshev Patterns. *Proc. IEE,* Vol. 105C, November 1957, pp. 167–168.

BROWN, J. L. On the Determination of Excitation Coefficients for a Tchebycheff Pattern. *Trans. IEEE,* Vol. AP-10, March 1962, pp. 215–216.

BUCK, G. J. AND GUSTINCIC, J. J. Resolution Limitations of a Finite Aperture. *Trans. IEEE,* Vol. AP-15, May 1967, pp. 376–381.

BURINGTON, R. S. *Handbook of Mathematical Tables and Formulas,* 5th ed., McGraw-Hill, 1973, Section 365.12.

BUTLER, J. K. AND UNZ, H. Beam Efficiency and Gain Optimization of Antenna Arrays with Nonuniform Spacings. *Radio Sci.*, Vol. 2, July 1967, pp. 711–720.

CHENG, D. K. Optimization Techniques for Antenna Arrays. *Proc. IEEE,* Vol. 58, December 1971, pp. 1664–1674.

CHENG, D. K. AND TSENG, F. I. Gain Optimization for Arbitrary Antenna Arrays. *Trans. IEEE,* Vol. AP-13, November 1965, pp. 973–974.

COX, H., ZESKIND, R. M., AND KOOIJ, T. Practical Supergain. *Trans. IEEE,* Vol. ASSP-34, June 1986, pp. 393–398.

DAWOUD, M. M. AND ANDERSON, A. P. Design of Superdirective Arrays with High Radiation Efficiency. *Trans. IEEE,* Vol. AP-26, November 1978, pp. 819–823.

DAWOUD, M. M. AND HASSAN, M. A. Design of Superdirective Endfire Antenna Arrays. *Trans. IEEE,* Vol. AP-37, June 1989, pp. 796–800.

DOLPH, C. L. A Current Distribution for Broadside Arrays Which Optimizes the Relationship Between Beam Width and Side-Lobe Level. *Proc. IRE,* Vol. 34, 1946, pp. 335–348.

DRANE, C. J. Derivation of Excitation Coefficients for Chebyshev Arrays. *Proc. IEE,* Vol. 110, October 1963, pp. 1755–1758.

DRANE, C. J. Dolph–Chebyshev Excitation Coefficient Approximation. *Trans. IEEE,* Vol. AP-12, November 1964, pp. 781–782.

DUHAMEL, R. H. Optimum Patterns for Endfire Arrays. *Proc. IRE,* Vol. 41, May 1953, pp. 652–659.

FIKIORIS, G. Experimental Study of Novel Resonant Circular Arrays. *IEE Proc. Microwave Antenn. Propag.*, Vol. 145, February 1998, pp. 92–98.

FIKIORIS, G., KING, R. W. P., AND WU, T. T. The Resonant Circular Array of Electrically Short Elements. *J. Appl. Phys.*, Vol. 68, 1990, pp. 431–439.

Fletcher, B., *Practical Methods of Optimization*, Second Edition, 2000, John Wiley & Sons.

Fong, T. S. On the Problem of Optimum Antenna Aperture Distribution. *J. Franklin Inst.*, Vol. 283, March 1967, pp. 235–249.

Franz, K. Remarks on the Absorption Surfaces of Directive Aerials. *Hochfrequenztechnik Elektroakustik*, Vol. 61, 1943, p. 51.

Gilbert, E. N. and Morgan, S. P. Optimum Design of Directive Antenna Arrays Subject to Random Variations. *Bell Syst. Tech. J.*, Vol. 34, May 1955, pp. 637–663.

Goward, F. K. An Improvement in End-Fire Arrays. *Proc. IEE*, Vol. 94, Part III, November 1947, pp. 415–418.

Gröbner, W. and Hofreiter, N., Integraftel, Zweitertail, Springer, Verlag, 1958.

Hansen, R. C. Formulation of Echelon Dipole Mutual Impedance for Computer. *Trans. IEEE*, Vol. AP-20, November 1972, pp. 780–781.

Hansen, R. C. Fundamental Limitations in Antennas. *Proc. IEEE*, Vol. 69, February 1981a, pp. 170–182.

Hansen, R. C. Some New Calculations on Antenna Superdirectivity. *Proc. IEEE*, Vol. 69, October 1981b, pp. 1361–1366.

Hansen, R. C. Linear Arrays. In *Handbook of Antenna Design*, Rudge, A. W. et al. Eds., IEE/Peregrinus, 1983.

Hansen, R. C. Hansen–Woodyard Arrays with Few Elements. *Microwave Opt. Technol. Lett.*, Vol. 5, January 1992, pp. 44–46.

Hansen, R. C. *Phased Array Antennas*, Wiley, 1998.

Hansen, R. C. Antennas. *In Reference Data for Engineers*, 9th ed., Newnes, 2002, Chapter 32.

Hansen, W. W. and Woodyard, J. R. A New Principle in Directional Antenna Design. *Proc. IRE*, Vol. 26, March 1938, pp. 333–345.

Harrington, R. F. Antenna Excitation for Maximum Gain. *Trans. IEEE*, Vol. AP-13, November 1965, pp. 896–903.

Janning, D. S. and Munk, B. A. Effects of Surface Waves on the Current of Truncated Periodic Arrays. *Trans. IEEE*, Vol. AP-50, September 2002, pp. 1254–1265.

Jahnke, E. and Emde, F., Tables of Functions, Fourth Edition, Dover Publications, 1945.

Jordan, E. C. and Balmain, K. G. *Electromagnetic Waves and Radiating Systems*, 2nd ed., Prentice-Hall, 1968.

King, R. W. P. Supergain Antennas and the Yagi and Circular Arrays. *Trans. IEEE*, Vol. AP-37, February 1989, pp. 178–186.

King, R. W. P., Fikioris, G. and Mack, R. B. *Cylindrical Antennas and Arrays*, Cambridge University Press, 2002.

Kock, W. E. Related Experiments with Sound Waves and Electromagnetic Waves. *Proc. IRE*, Vol. 47, July 1959, pp. 1192–1201.

Kovács, R. and Solymár, L. Theory of Aperture Aerials Based on the Properties of Entire Functions of the Exponential Type. *Acta Phys. Hung.*, Vol. 6, 1956, pp. 161–184.

Kurth, R. R. Optimization of Array Performance Subject to Multiple Power Pattern Constraints. *Trans. IEEE*, Vol. AP-22, January 1974, pp. 103–105.

Kyle, R. F. Super-Gain Aerial Beam. *Electron. Radio Eng.*, Vol. 36, September 1959, pp. 338–340.

La Paz, L. and Miller, G. A. Optimum Current Distributions on Vertical Antennas. *Proc. IRE*, Vol. 31, 1943, pp. 214–232.

Lo, Y. T. Array Theory. In *Antenna Handbook*, Lo, Y. T. and Lee, S. W.Eds., Van Nostrand, 1988, Chapter 11.

Lo, Y. T. et al. Optimization of Directivity and Signal-to-Noise Ratio of an Arbitrary Antenna Array. *Proc. IEEE*, Vol. 54, August 1966, pp. 1033–1045.

Luke, Y. L. *Mathematical Functions and Their Approximations*, Academic Press, 1975, Table 4.4.

MA, M. T. Theory and Application of Antenna Arrays, Johan Wiley & Sons, 1974.

Margetis, D. et al. Highly Directive Current Distributions: General Theory. *Phys. Rev. E*, Vol. 58, August 1998, pp. 2531–2547.

Newman, E. H. et al. Superdirective Receiving Arrays. *Trans. IEEE*, Vol. AP-26, September 1978, pp. 629–635.

Oseen, C. W. Die Einsteinsche Nadelstichstrahlung und die Maxwellshen Gleichungen. *Ann. Phys.*, Vol. 69, 1922, p. 202.

Pritchard, R. L. Optimum Directivity Patterns for Linear Point Arrays. *J. Acoust. Soc. Am.*, Vol. 25, September 1953, pp. 879–891.

Reid, D. G. The Gain of an Idealized Yagi Array. *J. IEE*, Vol. 93, Part IIIA, 1946, pp. 564–566.

Rhodes, D. R. On an Optimum Line Source for Maximum Directivity. *Trans. IEEE*, Vol. AP-19, July 1971, pp. 485–492.

Rhodes, D. R. *Synthesis of Planar Antenna Sources*, Clarendon Press, 1974.

Riblet, H. J. Discussion on a Current Distribution for Broadside Arrays Which Optimizes the Relationship Between Beam Width and Side-Lobe Level. *Proc. IRE*, Vol. 35, 1947, pp. 489–492.

Riblet, H. J. Note on the Maximum Directivity of an Antenna. *Proc. IRE*, Vol. 36, May 1948, pp. 620–623.

Salzer, H. E. Calculating Fourier Coefficients for Chebyshev Patterns. *Proc. IEEE*, Vol. 63, 1975, pp. 195–197.

Sanzgiri, S. M. and Butler, J. K. Constrained Optimization of the Performance Indices of Arbitrary Array Antennas. *Trans. IEEE*, Vol. AP-19, July 1971, pp. 493–498.

Schelkunoff, S. A. A Mathematical Theory of Linear Arrays. *Bell Syst. Tech. J.*, Vol. 22, 1943, pp. 80–107.

Seeley, E. W. Two- and Three-Loop Superdirective Receiving Antennas. *J. Res. NBS*, Vol. 67D, March–April 1963a, pp. 215–235.

Seeley, E. W. VLF Superdirective Loop Arrays. *J. Res. NBS*, Vol. 67D, September–October 1963b, pp. 563–565.

Sokolnikoff, I. S. and Sokolnikoff, E. S. *Higher Mathematics for Engineers and Physicists*, 2nd ed., McGraw-Hill, 1941.

Solymar, L. Maximum Gain of a Line Source Antenna if the Distribution Function Is a Finite Fourier Series. *Trans. IRE*, Vol. AP-6, July 1958, pp. 215–219.

Stearns, C. O. *Computer Performance of Moderate Size, Super-Gain Antennas*, NBS Report 6797, September 5, 1961.

Stutzman, W. L. and Thiele, G. A. *Antenna Theory and Design*, 2nd ed., Wiley, 1998.

Tai, C. T. The Optimum Directivity of Uniformly Spaced Broadside Arrays of Dipoles. *Trans. IEEE*, Vol. AP-12, July 1964, pp. 447–454.

Taylor, T. T. A Discussion of the Maximum Directivity of an Antenna. *Proc. IRE*, Vol. 36, 1948, p. 1135.

TAYLOR, T. T. Design of Line-Source Antennas for Narrow Beamwidth and Low Sidelobes. *Trans. IRE*, Vol. AP-3, January 1955, pp. 16–28.

TSENG, F.-I. AND CHENG, D. K. Gain Optimization for Arbitrary Antenna Arrays Subject to Random Fluctuations. *Trans. IEEE*, Vol. AP-15, May 1967, pp. 356–366.

UZKOV, A. I. An Approach to the Problem of Optimum Directive Antennae Design. *C. R. Acad. Sci. URSS*, Vol. 53, 1946, pp. 35–38.

UZSOKY, M. AND SOLYMAR, L. Theory of Super-Directive Linear Arrays. *Acta Phys. Hung.*, Vol. 6, 1956, pp. 185–205.

VEREMY, V. V. AND SHESTOPALOV, V. P. Superdirective Antenna with Passive Resonant Reflector. *Radio Sci.*, Vol. 26, March–April 1991, pp. 631–636.

VOGES, R. C. AND BUTLER, J. K. Phase Optimization on Antenna Array Gain with Constrained Amplitude Excitation. *Trans. IEEE*, Vol. AP-20, July 1972, pp. 432–436.

WINKLER, L. P. AND SCHWARTZ, M. A. Fast Numerical Method for Determining the Optimum SNR of an Array Subject to a *Q* Factor Constraint. *Trans. IEEE*, Vol. AP-20, July 1972, pp. 503–505.

YAGHJIAN, A. D. ET AL. Electrically Small Supergain Endfire Arrays. *Radio Sci.*, Vol. 43, 2008, pp. 1–10.

YARU, N. A. Note on Supergain Antenna Arrays. *Proc. IRE*, Vol. 39, September 1951, pp. 1081–1085.

CHAPTER 7

SUPERCONDUCTING ANTENNAS

7.1 INTRODUCTION

The advent of high-T_c superconducting materials has presented new opportunities for improving antenna performance. Many antenna workers are not familiar with the current findings on superconductors. Considered here are electrically small antennas, and matching of them, and millimeter-wavelength antennas. Not included are SQUID devices that incorporate a loop or other antenna. These are primarily of interest as very low-frequency magnetic field sensors.

7.2 SUPERCONDUCTIVITY CONCEPTS FOR ANTENNA ENGINEERS

There are many excellent books on superconductivity; only a basic overview is given here. An excellent treatise on all aspects of high- and low-temperature superconductivity, including a thorough mathematical treatment, is provided by Tinkham (1996). A physical understanding, with a minimum of mathematics, is provided by Ginzburg and Andryushin (2004). And there are many others. Lancaster (1997) has written a useful book on HTS applications including antennas. Table 7.1 shows a brief chronology up through the fullerenes. It is interesting to note the first step toward HTS: in 1975, Sleight, Gibson, and Bierstedt discovered superconductivity in a barium lead bismuth metallic oxide at 13K. This significant increase in T_c, for a ceramic material, was a forerunner of HTS. In 1986, Bednorz and Muller discovered a material that had $T_c = 35$K, a major advance. This was quickly followed by Chu, who developed the first compound of the HTS family: YBCO, with $T_c = 92$K. The most recent development was the discovery by Hosono in 2008 of iron-based superconductors, initially with $T_c = 26$K, but quickly evolving to 55K. Although

Small Antenna Handbook, Robert C. Hansen and Robert E. Collin.

TABLE 7.1 Brief History of Superconductivity

1911	Discovery by Onnes @ 4K; H_g DC resistance = 0
1933	Meissner effect: H expulsion on cooling
1934	London: H penetration depth, two-fluid model of normal and SC electrons, flux quantization
1950	Ginzburg and Landau: microscopic quantum theory
1956	Cooper: paired electrons
1957	BCS microscopic theory: H_c, Y_c relationship, and so on
1957	Abrikosov: current vortices in type II materials, each containing one flux quantum
1961	Kunzler: high-field, high-current superconductors
1962	Josephson: tunneling effects
1965	Silver et al.: squid, SC loop with JJ
1975	Sleight et al.: BaPbBiO, 13K
1984	Yagubskii et al.: organic SC
1986	Bednorz and Muller: high-H_c materials, 35K
1987	Chu: YBCO materials, 92K
1991	Hebard et al.: doped fullerenes
2008	Hosono: LaOFeAs, 26K
2009	Chen and others: SmOFeAs, 55K

YBCO has a higher T_c, the iron-based materials are offering new insights into the mechanisms of superconductivity (Day, 2009).

Superconductors have two remarkable properties: zero DC resistance and magnetic flux expulsion except for a thin shell of λ thickness (Meissner effect). The superconducting state occurs when the temperature of certain materials drops below a critical temperature T_c; see Table 7.2. Then the material is in the lowest energy state, and conduction electrons form coupled pairs, called Cooper pairs, through phonon (lattice) interactions, for LTS (low-temperature superconductors). These Cooper pair electrons flow without loss; they are coupled over the coherence length. For example, for LTS such as niobium it is roughly 44 nm and for aluminum it is 1600 nm. For HTS, the coherence length for YBCO is 140 nm. As temperature increases from 0K, the thermal lattice vibrations excite some electrons out of the ground state, breaking up some Cooper pairs. See Table 7.3 for penetration depth[1] λ and coherence length ξ. Crystal defects and impurities can change the coherence length, so purity is always a goal. The normal current flow produces loss. However, the increase in resistivity is small until T_c is approached. As frequency increases from zero, the electron pair kinetic energy produces a delay that creates an electric field along the surface. An applied magnetic field can also force all electrons to normal; this is the critical field, H_c.

The conventional definition of fields in metals is not suitable; a discussion of definitions for superconductors has been given by Campbell (2007).

[1] In most of this book, λ is the wavelength; in this chapter, the SC terminology is used, except in Section 7.4.

TABLE 7.2 Some Superconductor Critical Temperatures

Elements		
Al	1.2K	
Sn	3.7K	
Hg	4.2K	
Pb	7.2K	
Nb	9.3K	
Alloys		
NbTi	10.5K	
NbN	16K	
Compounds		
Nb_3Sn	18.4K	Widely used for magnets because of strong pinning forces
Nb_3Ge	23.2K	
MgB_2	40K	
Ceramics		
$La_2Ba_2Cu_1O_4$	35K	Bednorz and Muller (April 1986)
$Y_1Ba_2Cu_3O_7$	92K	Chu et al. (February 1987)
$Bi_2Sr_2Ca_2Cu_2O_{10}$	110K	Asano and Hermann (1988)
$Th_2Ba_2Ca_2Cu_3O_{10}$	125K	Beyers and Parkin (1988)
SmFeAsO	55K	Chen and others (2009)

It is useful to compare a superconductor with a perfect electric conductor (PEC). As seen in the sketch of Figure 7.1, the PEC at any temperature and any applied magnetic field contains an internal magnetic field; current can flow without loss. For the superconductor, when a magnetic field is applied and the temperature lowered below T_c, the magnetic field is expelled, except at the surface. Again, current can flow without loss.

Bardeen–Cooper–Schrieffer theory makes a quantum mechanical formulation for the paired electrons. It relates critical temperature, critical magnetic field H_c, coherence length, and so on. It represents LTS well, but only partially for HTS. For both LTS and HTS, there are two types of materials: type I and type II. Type I

TABLE 7.3 Penetration Depth and Coherence Length

		λ (nm)	ξ (nm)
Type I	Pb	39	87
	Nb	44	38
	Sn	51	230
Type II	Nb_3Sn	65	3
	PbBi	200	20
	YBaCuO	140	0.2–0.6
	LaSrCuO	200	0.6

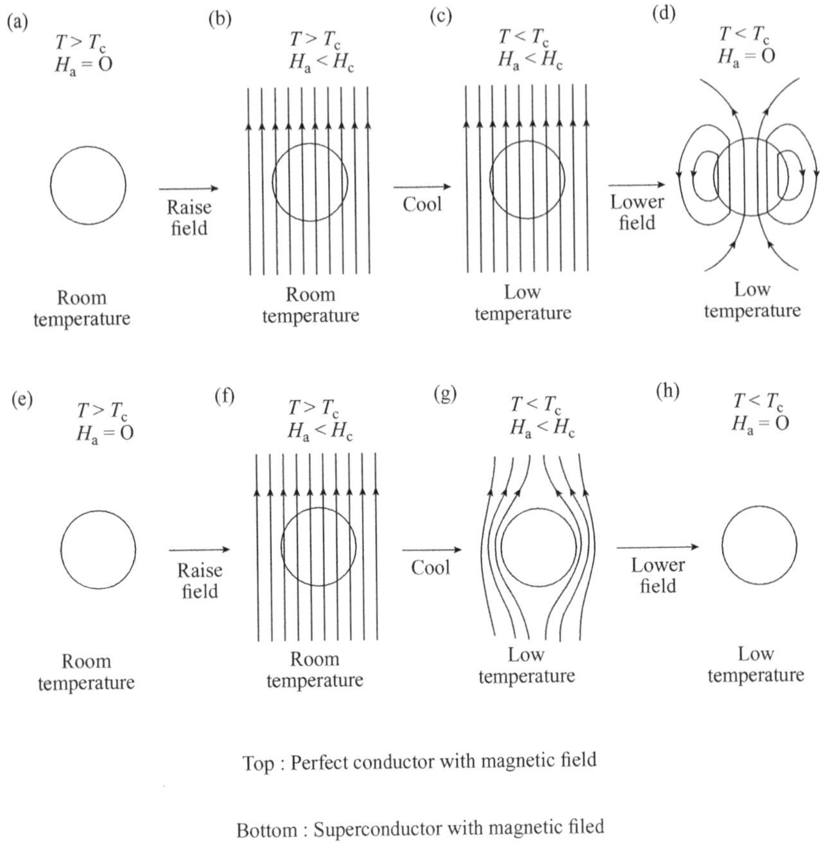

FIGURE 7.1 PEC versus SC.

materials are typically soft metallic elements, exhibiting the Meissner effect up to H_c. Coherence length is roughly comparable to penetration depth. The London penetration depth is the depth of current induced by a magnetic field. It is analogous to skin depth, which is the depth of current in a normal conductor. Type II materials have two critical magnetic fields, H_{c1} and H_{c2}. These have short coherence length and long penetration depth. The Meissner effect exists for fields up to H_{c1}; at H_{c2}, the conductor is normal. Between H_{c1} and H_{c2} flux cores or vortices exist, with the spacing between them decreasing as H approaches H_{c2}. Each vortex contains one flux quantum; a supercurrent circulates around the core. Vortices are typically cylindrical, with radius of one or two coherence lengths. Table 7.4 gives critical fields for representative materials.

Superconductivity is destroyed not only by high current density, by high magnetic fields, or by high temperature, but also by high frequency, in particular by frequency greater than the energy gap frequency. For LTS, the energy gap frequency is roughly 1 THz, whereas for HTS it is roughly 10 THz. Thus, HTS are more suitable for submillimeter applications.

TABLE 7.4 Some Superconductor Critical Fields

Material	H_{c1} (at/cm)	H_{c2} (at/cm)
Nb	1600	2400
NbTi	2800	88,000
Nb_3Sn	2800	185,000

The HTS crystal structure is perovskite; electrically active planes of copper and oxygen are sandwiched between other layers that are reservoirs of charge. Figure 7.2 is a sketch of YBCO. Such a crystal structure is highly anisotropic, with poor conductivity along the copper axis. The HTS parents are insulators, and the symmetry of order parameter is *d*-wave. Although a superconductor below T_c, the material, like most ceramics, is a poor conductor at room temperature. The SC current (Cooper paired electrons) flows along the CuO planes. Because of the crystal structure, the electron pairing is more complex than that of the BCS theory. The two-fluid model is useful for HTS: a mixture of normal and SC currents flow. For the normal currents, $J = \sigma_1 E$. For the SC currents, the London equations apply:

$$\nabla \times J_{sc} = -H/\lambda^2, \qquad \frac{dJ_{sc}}{dt} = \frac{E}{\mu_0 \lambda^2} \tag{7.1}$$

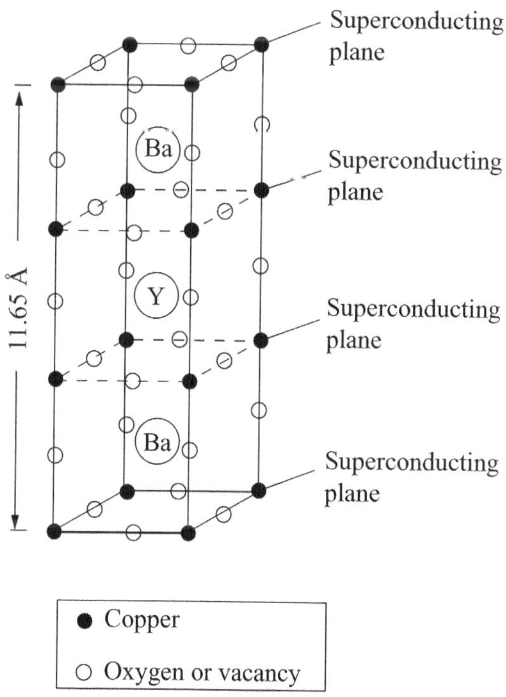

FIGURE 7.2 YBCO lattice.

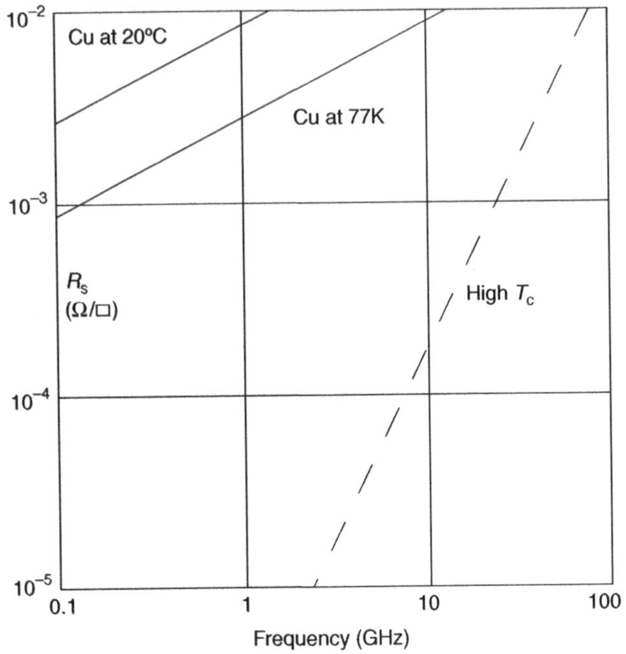

FIGURE 7.3 Surface resistivity trends.

Typical data on surface resistivity versus frequency are shown in Figure 7.3, for copper at room temperature and for copper and HTS at liquid nitrogen temperature. Data on the frequency where HTS resistivity is equal to that of copper at 77K were given by Alford et al. (1991). Data on LTS niobium were given by Piel and Muller (1991); measurements and BCS theory agree very well. The surface resistance, from measurements and theory, is given by

$$R_s \simeq \frac{1}{2}\omega'^2 \mu_0^2 \lambda^3 \sigma_1 \tag{7.2}$$

where σ_1 is the normal state conductivity and ω is angular frequency. Note that R_s varies as frequency squared. In contrast, for a normal conductor the surface resistance varies as $f^{1/2}$. Surface reactance is simply

$$X_s = \omega \mu_0 \lambda \tag{7.3}$$

In both formulas for R_x and X_s, the penetration depth λ is

$$\lambda = \frac{\lambda_0}{\sqrt{1-(T/T_c)^4}} \tag{7.4}$$

where λ_0 is the penetration depth at 0K. Most antenna and circuit work now utilizes SC thin films, with the CuO plane parallel to the substrate. With films, it is easier to control lattice defects and impurities in order to reduce RF losses. Substrates ideally should have a lattice match, have low loss and low ε, and be nonreactive with HTS at all processing temperatures.

In contrast to the ceramic oxides of copper, such as YBCO, the parents of the new iron-based materials are semimetals, and the symmetry of order parameter is s-wave. Both types have alternating layers of atoms.

An indication of how superconductors might affect antennas can be gleaned from considerations of external and internal fields. The performance of almost all antennas is governed by size and shape in free space wavelengths, that is, external fields. Examples are dipoles, slots, spirals, log-periodics, Yagi–Uda, horns, and reflectors. Superconductivity generally has a small effect on external fields, so the size/shape of most antennas will not be reduced. Internal fields exhibit only loss, thereby reducing efficiency. Superconductors can improve efficiency, often at the expense of bandwidth. Most important, superconductors can make a significant increase in the efficiency of an impedance matching network.

It is useful to reflect on the progress of superconductivity. There have been three periods of spectacular progress, roughly 40 years apart. The period 1905–1911 saw liquefaction of helium, helium cryostats, discovery of mercury superconductivity, and discovery of tin and lead superconductivity. The period 1952–1962 saw microscopic quantum theory, Cooper pairs, BCS theory, current vortices, Josephson effect, high-field superconductors, and superconducting magnets. The period 1986–1991 saw the discovery of HTS, the YBCO family, the lanthanum family, fullerenes, and organics. And 2008 brought a new class of HTS, the iron-based family. Will we see room-temperature superconductors around 2030?

7.3 DIPOLE, LOOP, AND PATCH ANTENNAS

Immediately after the discovery of high-temperature superconductors, antenna people looked at how these new materials might improve the performance of antennas. One company actually proposed building the surface of a dish antenna out of HTS in the vain hope that the typical dish aperture efficiency of 65% would be increased to close to 100%. Of course, the conduction loss in a dish antenna is usually less than 1% of the total loss, and the 65% is due to aperture taper and blockage losses. In the following, only antennas that have been built and tested are discussed. Antenna configurations incorporating Josephson junctions are outside the scope of this book.

With the discovery of HTS, LaBaCuO in 1986 by Bednorz and Muller and YBaCuO in 1987 by Chu, it was inevitable that the new technology would be applied to antennas. Intensive activity was reported between 1988 and 1995. This work is reported by type of antenna. No measured data are reported here on ESA; it will appear later that none of the ESA needs HTS, only the matching network for short dipoles should employ HTS (Lancaster et al., 1992b).

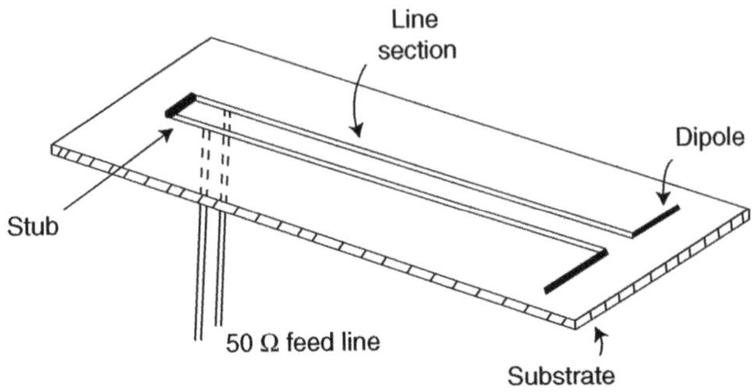

FIGURE 7.4 Short dipole with twin line match. Courtesy of Lancaster et al. (1992a).

7.3.1 Loop and Dipole Antennas

Khamas et al. (1988), Wu et al. (1989), and Lancaster et al. (1992a) measured a short dipole and matching twin line, all made of YBCO, and compared results to a similar system made of copper (see Figure 7.4). The HTS antenna gain was 12 dB higher. However, essentially all of the 12 dB improvement was due to the HTS matching (Hansen, 1990). Section 2.4 discusses network losses. The reduction in gain of this configuration as the input power increased was noted by Gough et al. (1989); see also Portis et al. (1991). Dinger and colleagues (Dinger and White, 1990; Dinger et al., 1991) calculated the efficiency for such a twin line fed dipole and showed that dielectric loss tangents should be less than 10^{-4}. Additional work on HTS dipoles was done by He et al. (1990, 1991). A YBCO thick film loop and matching network was developed and measured by Lancaster et al. (1993) (Figure 7.5).

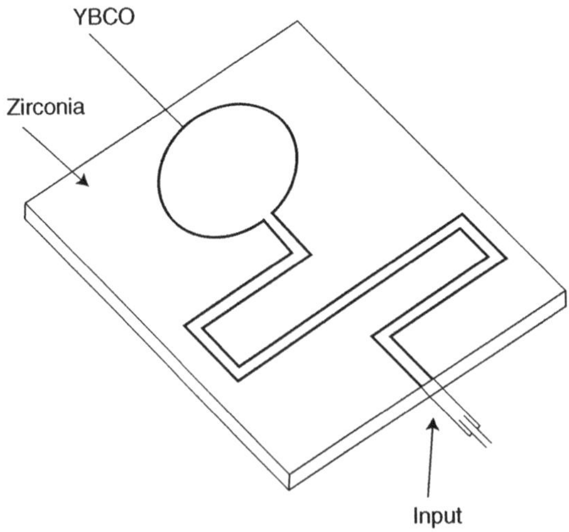

FIGURE 7.5 Loop and matching circuit. Courtesy of Lancaster et al. (1993).

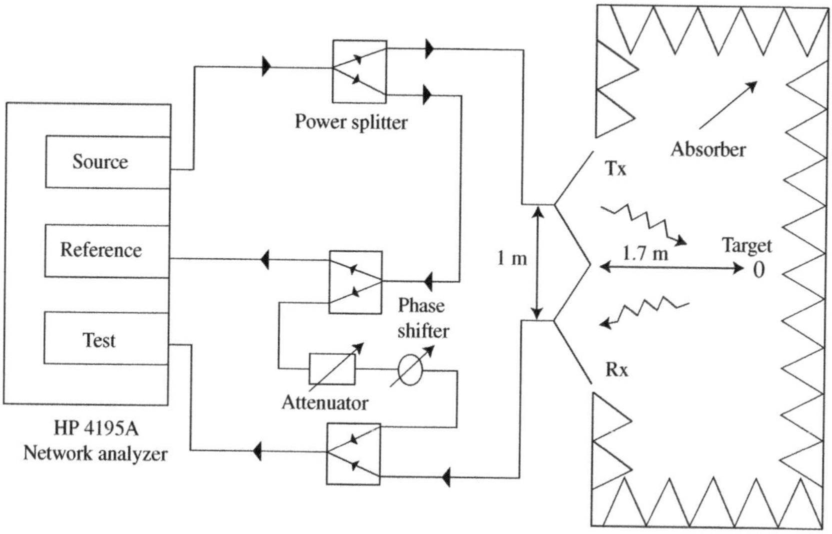

FIGURE 7.6 Backscatter measurement using nulling. Courtesy of Khamas et al. (1993).

Determination of loop-alone performance was made using backscatter measurements (Khamas et al., 1993; see Figure 7.6). Multiturn loops made of BSCCO were tested by Itoh et al. (1993). Matching sections made of copper reduced the overall gain, due to matching loss. In addition, the bandwidth was narrow as the loop radiation resistance was small compared to the reactance. Several ways of feeding small loops were compared by Ivrissimtzis et al. (1994a) (Figure 7.7). Effects of impurities on bandwidth were predicted with a two fluid model by Cook et al. (1992). A related study evaluated the efficiency of an HTS ESA over a lossy ground plane, using Sommerfeld integrals (Cook et al., 1994, 1995). Environmental perturbations were investigated by James and Andrasic (1994). A novel technique is the control of loop or dipole radar cross section by temperature (Cook and Khamas, 1993) or by applied magnetic field (Cook and Khamas, 1994). In both cases, the RCS decreases as the surface resistance increases.

7.3.2 Microstrip Antennas

Work on arrays of microstrip patches is reported below; single patch efforts have been sparse. Richard et al. (1993) compared edge feeding and gap feeding. H-shaped patches have been investigated by Chaloupka et al. (1991) and by Lancaster et al. (1998); see Figure 7.8. A meanderline patch fed by coax (Figure 7.9) was constructed by Chaloupka (1992); Wang and Lancaster (1999) fed a meanderline patch by an H-shaped coupling aperture (Figure 7.10). Resonant size is reduced to roughly $\lambda/10$. A meanderline antenna using EuBaCuO was described by Suzuki et al. (1992). Current density on patches has been measured using the kinetic inductance photoresponse (Newman and Culbertson, 1993). The real part of the

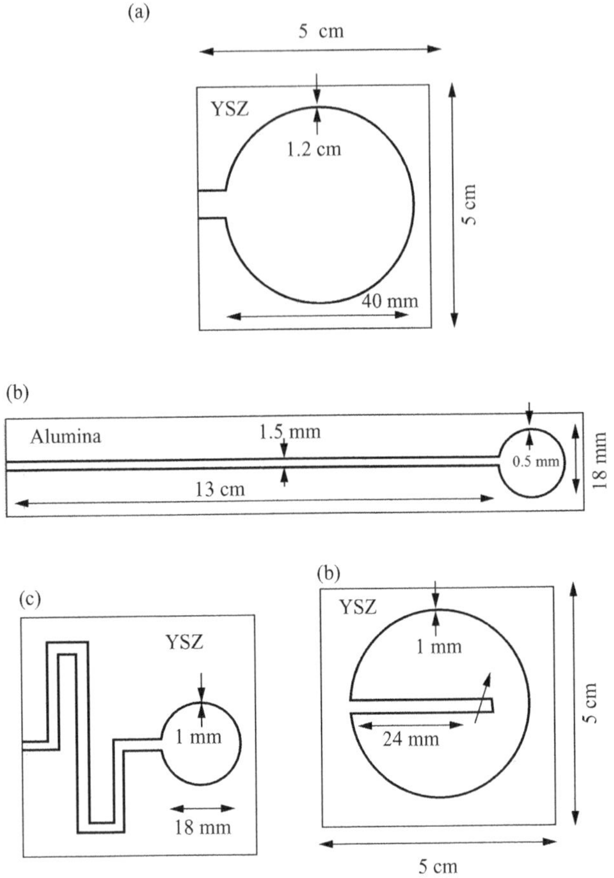

FIGURE 7.7 Loop feeding and matching: (a) resonant loop; (b) coplanar strip and capacitor; (c) coplanar strip meander; (d) coplanar strip inside. Courtesy of Ivrissimtzis et al. (1994a).

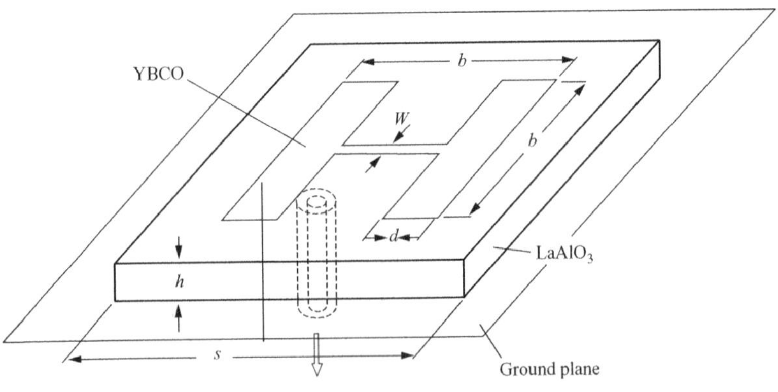

FIGURE 7.8 H-patch. Courtesy of Chaloupka et al. (1991).

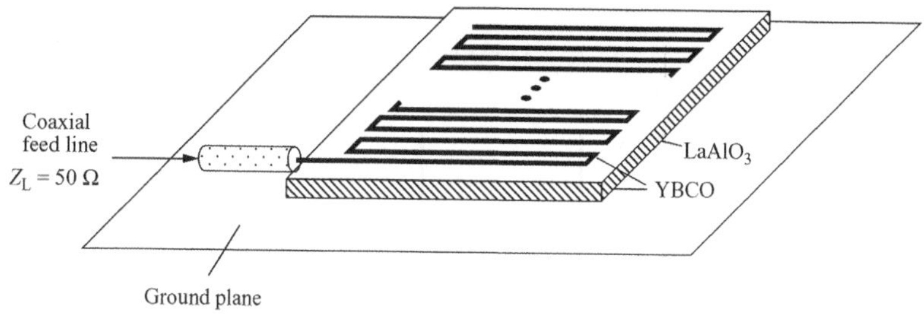

FIGURE 7.9 Meander patch. Courtesy of Chaloupka (1992).

surface impedance affects the patch input impedance, whereas the imaginary part alters the resonant frequency (Ali et al., 1999a). Yoshida et al. (2001) and Tsutsumi et al. (2005) proposed a slot fed by coplanar waveguide, with a multisection matching circuit. The HTS matching section was critical, as the radiation resistance was less than 1 Ω. A bowtie patch with HTS matching circuit was tested by Chung (2001). Bandwidth was small compared to a bowtie dipole in free space. Richard et al. (1992b) built an annular ring antenna with YBCO, but only a trivial improvement in efficiency over a silver ring resulted. Oda et al. (2007) and Kanaya et al. (2007) developed U-shaped and straight slot antennas with YBCO; the unique feature was use of superconducting matching circuits, see Section 2.4 on matching circuit loss.

7.3.3 Array Antennas

A superdirective array of two parallel dipoles in an endfire mode showed an increase in directivity, an increase in Q (Huang et al., 1991; Lancaster et al., 1992a), and a decrease in efficiency (Altshuler, 2005). A linear array of parallel dipoles was series fed with feed wires crossed between dipoles (Ivrissimtzis et al., 1994b, 1994c; see Figure 7.11). Gain was improved but was less than the potential superdirectivity.

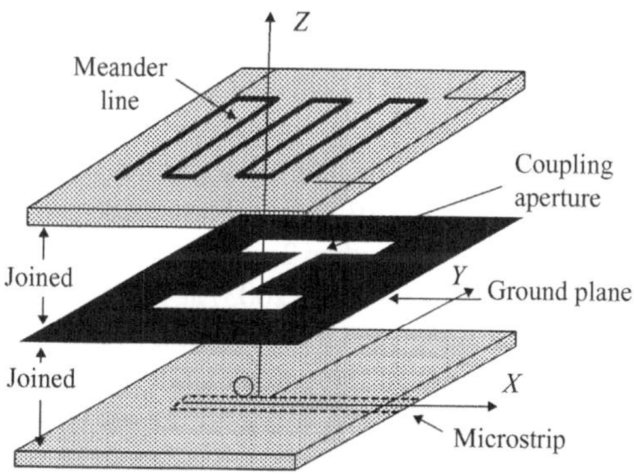

FIGURE 7.10 Aperture coupled meander patch. Courtesy of Lancaster et al. (1998).

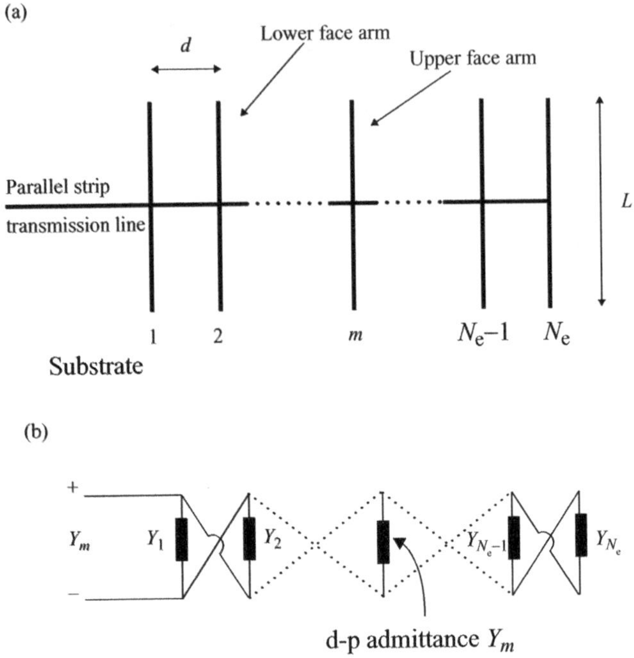

FIGURE 7.11 Series fed array: (a) dipoles; (b) equivalent circuit. Courtesy of Ivrissimtzis et al. (1994b).

Four parallel dipoles were excited in a superdirective endfire manner (Ivrissimtzis et al., 1995a; see Figure 7.12). Gain was good, but bandwidth was narrow. This was extended to a 4 × 4 endfire dipole array (Figure 7.13), with series feeding of each of the four subarrays and a corporate overall feed. Again gain was good, but bandwidth was narrow (Ivrissimtzis et al., 1995b). An endfire array of two helices and matching circuit in BSCCO was investigated by Itoh et al. (1993). Performance was several dB better than a copper array. A 4 × 4 12 GHz patch array with proximity coupled feed network, all HTS, was developed by Herd et al. (1993) (see Figure 7.14). A two-layer 4 × 4 patch array was developed at 20 GHz (Herd et al., 1996); see Figure 7.15. Another two-layer array is developed by Ali et al. (1999b). A broadside corporate fed 4 × 4 patch array at 20 GHz in YBCO was developed by Morrow et al. (1999). At 30K gain was roughly 2 dB better than that of a copper equivalent array, but at 77K the improvement was only about 0.5 dB. A 2 × 2 broadside patch array was built and tested by Richard et al. (1992a). Another 2 × 2 patch array used phase rotation to produce circular polarization (Chung et al., 2003) at 12 GHz. Cryostat and mounting details were also provided (Chung et al., 2005). Gain over a comparable gold array was about 2 dB, probably because of the complexity of the HTS feed and matching network. A departure from YBCO was made by Lewis et al. (1993). They used thallium calcium barium copper oxide in an 8 × 8 patch array, with corporate feed, at 30 GHz (see Figure 7.16). This is probably the most advanced application of HTS to

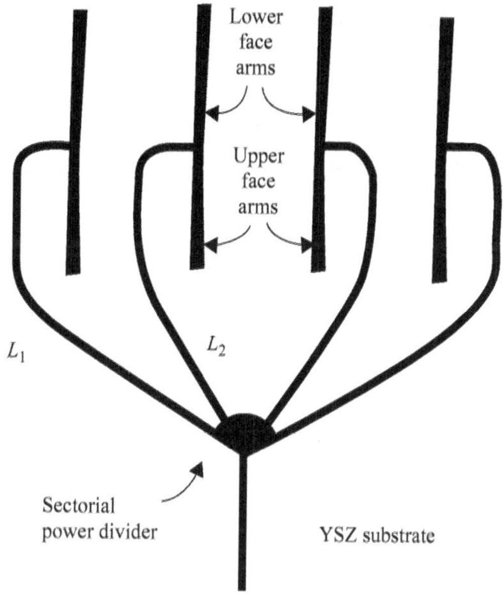

FIGURE 7.12 Endfire dipole array. Courtesy of Ivrissimtzis et al. (1995a).

arrays to date. A YBCO thin film patch array at 12 GHz showed improved gain over a comparison copper array, apparently because of feed loss reduction (Ali et al., 1999b). Superdirective arrays used for adaptive beamforming or for multiport applications, because of the strong mutual coupling, benefit from the insertion of decoupling networks (Chaloupka, 1993, 2001; Chaloupka et al., 2003).

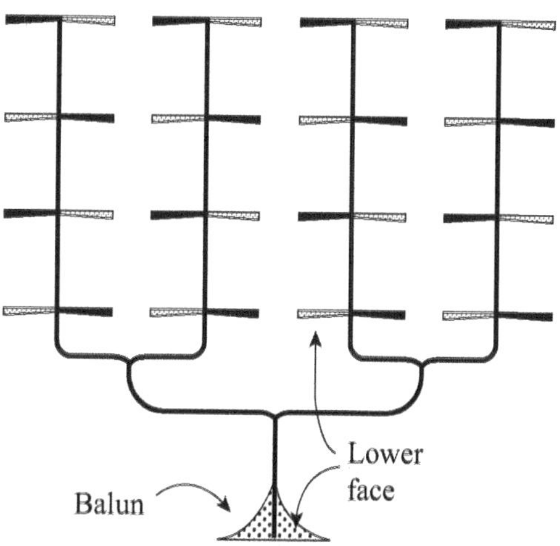

FIGURE 7.13 Endfire array with tapered baluns. Courlesy of Ivrissimtzis et al. (1995b).

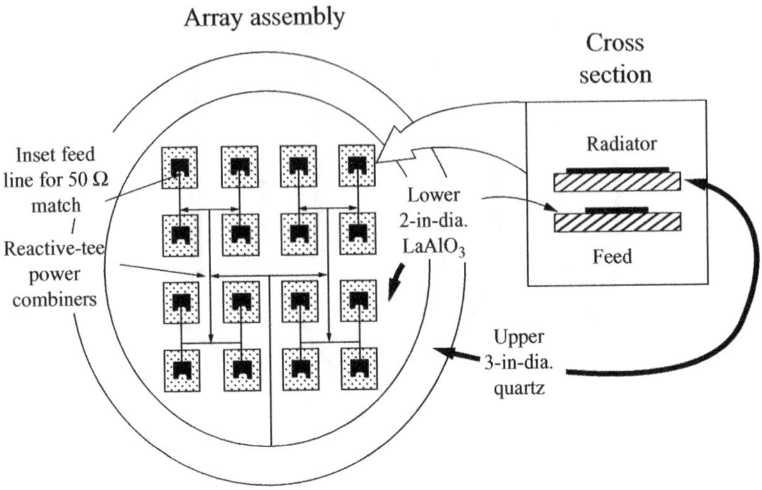

FIGURE 7.14 Proximity coupled patch array. Courtesy of Herd et al. (1993).

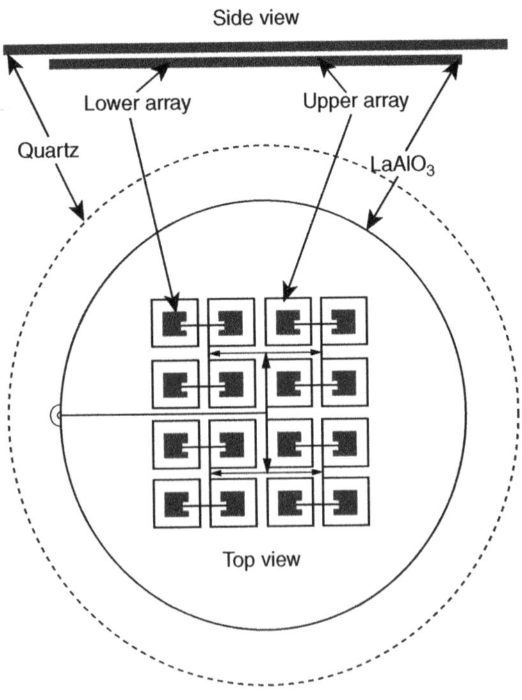

FIGURE 7.15 Multilayer patch array. Courtesy of Herd et al. (1996).

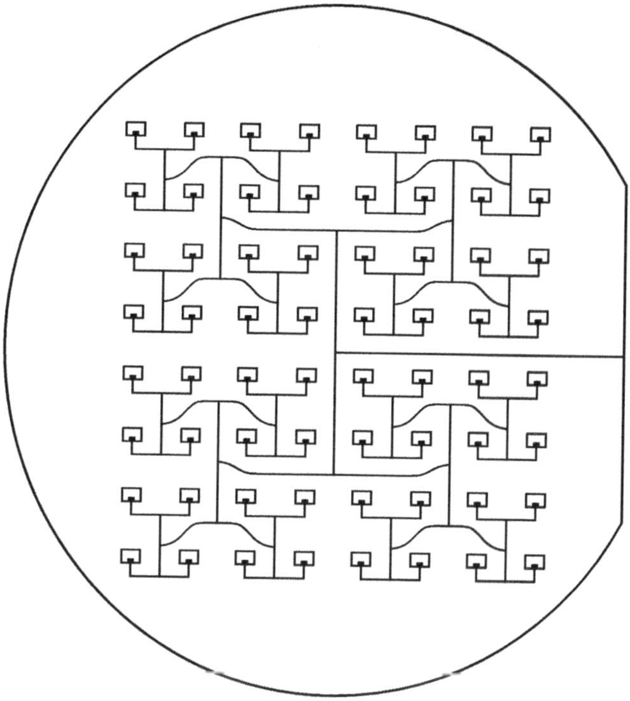

FIGURE 7.16 8 × 8 corporate fed patch array. Courtesy of Lewis et al. (1992).

7.3.4 Millimeter-Wave Antennas

Most reflector and array antennas have low dissipative losses (losses are primarily due to impedance mismatches). However at millimeter wavelengths, transmission line loss, whether waveguide, stripline, or microstrip, is important in determining the feasibility of an array. To illustrate this point, examples are calculated for both waveguide and microstrip planar arrays.

7.3.4.1 Waveguide Flat Plane Array A planar array of waveguide slots is typically constructed of side-by-side waveguide linear slot arrays (sticks), with these fed by another waveguide at right angles, utilizing cross-guide couplers (see Hansen, 2009, Section 10.2). Often the array is divided into quadrants for monopulse operation. Resonant stick array design produces a fixed, broadside beam. For a square array of width L, the feed path length is $2L$. And for small to moderate loss, the array efficiency due to waveguide loss is just

$$\text{efficiency} = 1 - 4\alpha L \tag{7.5}$$

where α is the attenuation coefficient. Formulas for α are widely available and are not repeated here. For an example, the lower portion of a waveguide band is used:

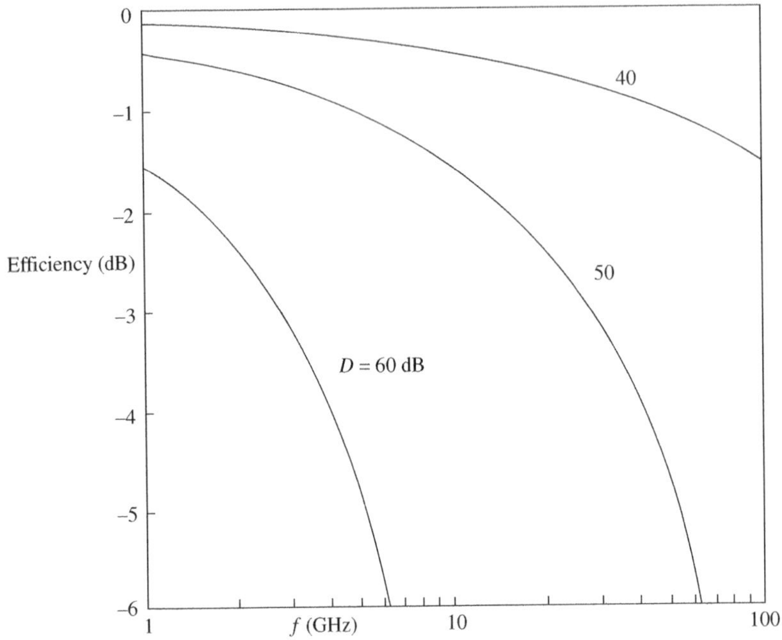

FIGURE 7.17 Waveguide slot flat plane array efficiency.

$\beta/k = 0.5$ and $a = 2b$, using common waveguide notation. Assuming the conductivity of copper, array efficiency due to guide loss is calculated for arrays of directivity 40, 50, and 60 dB (see Figure 7.17). Gain then is the directivity minus the efficiency (in dB). Of course, for low efficiencies the results are only approximate, as the usual low-loss assumptions have been used (tangential magnetic field at the waveguide walls is unchanged). Although the curves show that modest gain (40 dB) may be realizable at 100 GHz, the higher frequencies often require large gains to offset increased path loss.

Traveling wave array sticks have an effective path length of half (Begovich, 1966), or L, which reduces the losses compared with resonant arrays. However, the losses are still appreciable. Use of a corporate feed probably incurs an effective path length greater than $2L$, so its loss will exceed that of the resonant array. In practice, for all arrays, the actual loss will be greater, because of surface roughness, metal imperfections, and so on. Thus, waveguide loss has been a major factor against construction and utilization of high-gain arrays in the 40–100 GHz range.

Use of superconducting waveguides would in principle allow the efficiency to approach 0 dB, and thus high-T_c materials may allow a significant extension of array techniques.

7.3.4.2 Microstrip Planar Array A microstrip array of patch elements is assumed, with loss only in the connecting microstrip lines. Because several approximate results for microstrip loss exist, that used here is given below (Pucel et al., 1968; Wheeler, 1977):

$$\frac{\alpha Z_0 h}{R} = \frac{[((w'/h)+(w'/\pi h))/((w'/2h)+0.94)][1+(h/w')+(h/\pi w')(\ell n(2h/t)-(t/h))]}{[(w'/h)+(2/\pi)\ell n[2\pi e((w'/2h)+0.94)]]^2} \tag{7.6}$$

where

$$\frac{w'}{h} = \frac{120}{Z_0} - \frac{2}{\pi}\left[1+\ell n\left(\frac{120\pi^2}{Z_0}-1\right)\right] \tag{7.7}$$

In this formula, w/h has been written in terms of Z_0, for thin strips. The dielectric is air, to remove dielectric losses. Microstrip conductor width is w, thickness is t, and the spacing is h. A path length of L, giving efficiency equal to $1-2\alpha L$, is used. Parameters used in the calculation are $Z_0=50$, $h/\lambda=0.03$, and $t/h=0.01$. Figure 7.18 shows efficiency for arrays of directivity of 40, 50, and 60 dB, and the results are similar to those for waveguide arrays: Only modest gains are obtainable for 40–100 GHz.

Superconducting microstrip behavior is more complex than that of superconducting waveguide for several reasons. As edge current behavior is different, the strip

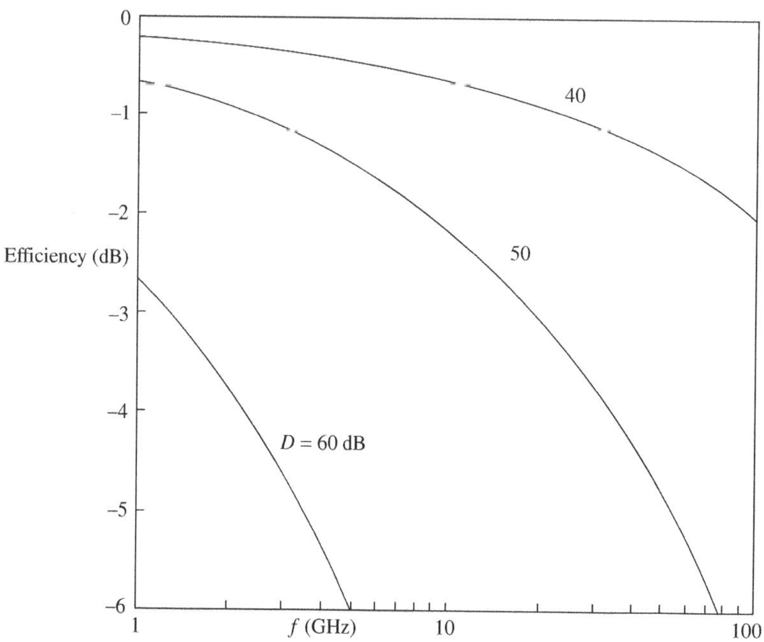

FIGURE 7.18 Microstrip flat plane array efficiency.

conductors may be very thin without the high loss engendered by edge current singularities in normal conductors. And use of narrow strips allows the dielectric thickness to be reduced, thereby reducing dielectric losses (Kautz, 1979). If the dielectric loss needs to be reduced further, low-loss structures such as suspended substrate microstrip or inverted microstrip can be used (Young and Itoh, 1988). The relative value of conductor and dielectric loss will depend on frequency, as the supercurrent component of surface resistance increases with the square of frequency, as predicted for low frequencies by the London two-fluid theory, and for frequencies near the energy gap frequency by the Mattis–Bardeen theory (Kautz, 1978). It is assumed that the residual resistance component is negligible. It appears that microstrip of conventional dimensions could experience a loss reduction of the order of 20 dB at 100 GHz, while thin film microstrip should show another 10 dB of loss reduction. Thus, superconductors will allow high-gain microstrip arrays to be used at millimeter wavelengths.

7.3.5 Submillimeter Antennas

A slot antenna with quasi-optical mixer using an LTS(Nb) trilayer junction operates in the range of 400–810 GHz (Zmuidzinas and Leduc, 1992; Zmuidzinas et al., 1995; Zmuidzinas and Richards, 2004; Gaidis et al., 1996). For these short wavelengths, two parallel slots, with the trilayer SIS chip[2] in between, are all cooled to about 4K by locating them on the cold finger of a helium Dewar. The incoming wave is focused on the double slot by a hemispherical silicon lens. This technology has been used in U.S. space probes. These SIS mixers apparently work below 4.5K, so an HTS version may not be possible.

7.3.6 Low-Temperature Superconducting Antennas

One of the first experiments measured efficiency of a short dipole at 4, 77, and 290K (Moore and Travers, 1966). Significant increases in efficiency were measured as the temperature was lowered. The dipole did not become superconductive; the material was not specified. Low-temperature measurements have been performed on a loop (Walker and Haden, 1969; Walker et al., 1977) and on an endfire array of lead-plated loops in a liquid helium environment (4.2K). Efficiency rose to near 100% over room temperature, but the bandwidth became extremely narrow. Adachi et al. (1976) worked on dipoles, and later on dipole arrays. Russian work using niobium has been reported: Pavlyuk et al. (1978). Bob Hansen measured relative gain and Q for a loop and for a dipole; Krivosheev and Pavlyuk (1979) measured a two-loop endfire array and Vendik et al. (1981) determined the input power a loop could accept while remaining superconductive.

[2] See Tucker and Feldman (1985) for a comprehensive review of SIS technology.

7.4 PHASERS AND DELAY LINES

Phasers, or phase shifters, are used at each element of a phased array to steer the beam. For wideband arrays, time delay (all time delay is "true") is also needed for frequency-independent beam steering. At microwave frequencies, the loss, especially for long room-temperature time delays, may become unacceptable. Superconductors possess an interesting property where the SC Cooper pair currents have kinetic inductance, allowing appreciable delay in a short segment of transmission line. Pond et al. (1987, 1989) measured LTS delay on a NbN transmission line; the dielectric loss was critical. With the kinetic inductance of the Cooper paired electrons goes loss; the SC layer thickness must be carefully chosen to reach an acceptable compromise between low phase velocity (thin film) and low loss (thick film). For an SC, the conductivity is complex, and it varies with the penetration depth and SC layer thickness (Ma and Wolff, 1996). The penetration depth changes with temperature, going to zero at T_c. The real part of conductivity relates to the loss, and the imaginary part relates to the kinetic inductance. Strong currents or strong magnetic fields will affect all parameters. Sheen (1991) has formulated surface resistance, normal inductance, kinetic inductance, and current distribution for stripline, all as a function of penetration depth and film thickness. See also Ma and Wolff (1996) for formulations of complex conductivity and effective dielectric constant.

There was much interest in delay lines in the early 1990s. Liang et al. (1991) used YBCO on a lanthanum aluminum oxide substrate. A reflection configuration with PIN diode switches was used, with 4 bits of phase produced. Each bit utilizes a 3 dB hybrid and two diodes; see Figure 7.19. The difference in line length between the switched and unswitched lines gives the phase bit. Loss at 10 GHz was 1.1 dB, a value considerably less than that for copper phasers. Because a 4-bit phaser provides

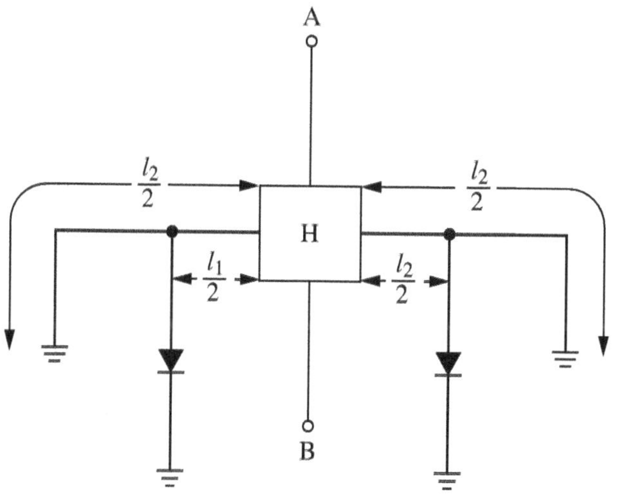

FIGURE 7.19 3dB hybrid PIN diode phaser. Courtesy of Liang et al. (1991).

up to 0.9375λ of delay, the loss per wavelength was 1.2 dB. However, most of this loss was due to substrate loss, diode loss, and mismatch loss. Shen et al. (1991) built a spiral delay line in coplanar line, using ThBaCaCuO on a LaAlO substrate. Delay time was 11 ns, and the loss was 0.25 dB/ns at 8 GHz. Liang et al. (1993) worked on spiral delay lines in YBCO; delays were 27 and 44 ns, with losses of 6 and 16 dB at 6 GHz. These losses are 0.22 and 0.36 dB/ns, and 0.042 and 0.061 dB/λ. A spiral YBCO delay line on sapphire (Liang et al., 1994) produced 9 ns of delay at 77K. Surface resistance was 0.5 mΩ at 10 GHz; insertion loss was 1.5 dB at 6 GHz. This is 0.167 dB/ns at 6 GHz. Also, at 6 GHz the delay line is 54λ long, giving a loss of 0.0278 dB/λ. Another spiral delay line (Hofer and Kratz, 1993) has 3 ns delay but no data on loss. Track et al. (1993) and Martens et al. (1993) built both NbN and YBCO delay lines, using a meanderline pattern. The delay was 8 ns, with a loss of 3 dB at 20 GHz, or 0.375 dB/ns. At 20 GHz, this loss is 0.0125 dB/λ. Other work on meanderline delay lines was done by Hattori et al. (1999) with 2.8 ns delay and 1.1 dB loss at 70K, and at 10 GHz, a loss of 0.39 dB/ns or 0.039 dB/λ; and by Hohenwarter et al. (1993), using NbN and coplanar waveguide, with 45 ns delay and with loss of 0.01 dB/ns and 0.001 dB/λ at 10 GHz. Huang et al. (1993) developed a 3 ns YBCO delay line; loss at 8 GHz was 0.73 dB/ns or 0.092 dB/λ. Talisa et al. (1995) used a closely wrapped spiral of YBCO on LaAlO, in stripline. The line is 1.5 m long! See Figure 7.20. Surface resistance at 10 GHz and 77K is 0.5 mΩ. Delay is 45 ns. Loss is 2 dB at 10 GHz and 6 dB at 20 GHz. At 10 GHz the loss is 0.0044 dB/λ, and at 20 GHz it is 0.0067 dB/λ. Loss is 0.044 dB/ns at 10 GHz and 0.133 dB/ns at 20 GHz. Fenzi et al. (1994) developed a 100 ns delay line using TBCCO on a LaAlO substrate with meandered coplanar waveguide; see Figure 7.21. Loss at 6 GHz was below 0.08 dB/ns, which is 0.013 dB/λ. This is the type of delay that might be suitable for wideband phased array steering. Fabrication details of a YBCO meanderline delayer using epitaxial liftoff were given by Koh and Hohkawa (1999). Important practical implementation considerations such as insulation

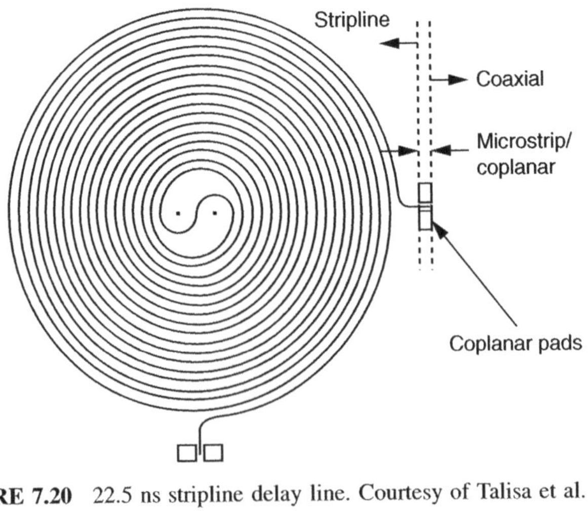

FIGURE 7.20 22.5 ns stripline delay line. Courtesy of Talisa et al. (1995).

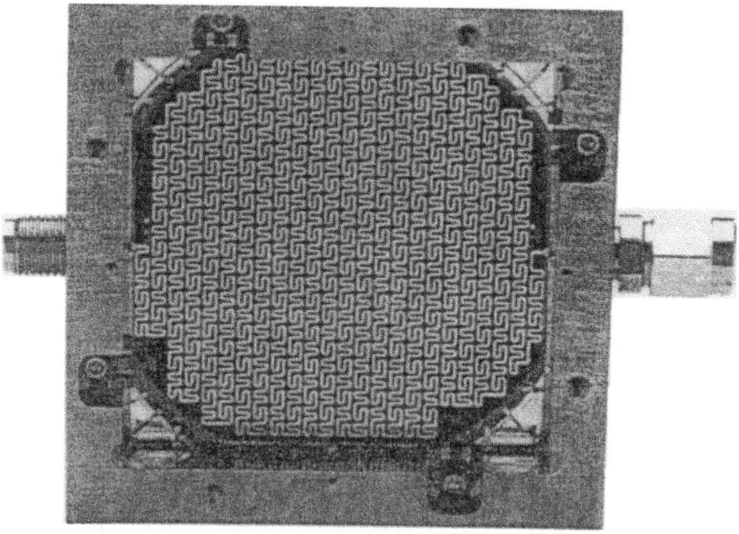

FIGURE 7.21 100 ns delay line. Courtesy of Fenzi et al. (1994).

package, cooler, vacuum package, and electronics package were discussed by Kapolnek et al. (1993). A convenient formula relates line length and delay:

$$L/\lambda = t_{ns} f_{GHz}$$

Su et al. (2004) developed a wideband delay line, with low insertion loss up to 20 GHz. It is a double spiral meander line, in YBCO; see Figure 7.22.

A different approach uses an HTS transmission with many SQUID devices coupled to it. Each SQUID contains one Josephson junction. The first work was

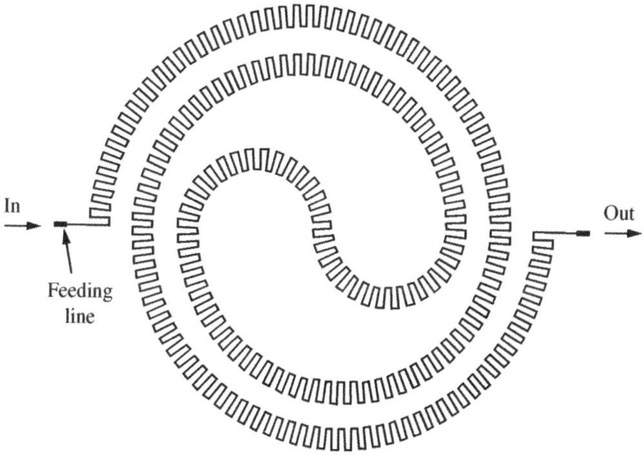

FIGURE 7.22 Double spiral meander delay line. Courtesy of Su et al. (2004).

done on LTS using niobium (Durand et al., 1992). This was followed by an HTS YBCO (Takemoto-Kobayashi et al., 1992) delay line. A variable magnetic field on the SQUIDs changes the delay. The experimental model with 40 SQUIDs produced 60° of phase shift at 10 GHz. Problems of temperature, dynamic range, and complexity appear to make SQUID phasers less attractive than those using single line length.

7.5 SUPERCONDUCTING ANTENNA SUMMARY

The field of superconducting antennas was changed by the emergence of three important principles (Hansen, 1990, 1991; Khamas et al., 1990). *Principle 1*: Dipole-type ESA made of copper or aluminum have radiation resistance much larger than loss resistance; efficiencies are close to 100%. For example, a dipole 0.02λ long has a radiation resistance of 100 mΩ, a value much larger than typical loss resistance. *Principle 2*: Loop-type ESA usually have radiation resistance well below loss resistance, so use of HTS will greatly increase loop efficiency. However, Q is unacceptably high. *Principle 3*: A network matching an ESA to 50 Ω sees a high VSWR, and this greatly increases the intrinsic loss in the matching network.

Thus, it may be concluded that HTS ESA are not useful or cost effective except in special circumstances. The matching circuit can benefit significantly by employing HTS components. HTS delay lines with long delays are promising for steering of wideband phased arrays. Millimeter-wave arrays can benefit from HTS, not necessarily in the antenna elements, but in the feed network. Submillimeter antennas using integrated SIS sensors and antennas are a promising area.

REFERENCES

Adachi, S. et al. An Experiment on Superconducting Antennas. *Trans. IECE Jpn.*, Vol. J59-B, 1976, pp. 299–300.

Alford, McN. N. et al. Surface Resistance of Bulk and Thick Film $YBa_2Cu_2O_7$. *IEEE Trans. Magnet.*, Vol. 27, March 1991, pp. 1510–1518.

Ali, M. I. et al. Effect of Surface Impedance on the Antenna Properties in Superconducting Microstrip Antenna. *Physica C*, Vol. 325, 1999a, pp. 143–152.

Ali, M. I., Ehata, K., and Ohshima, S. Superconducting Patch Array Antenna on Both-Side YBCO Thin Film for Satellite Communication. *Trans. IEEE*, Vol. AS-9, June 1999b, pp. 3077–3080.

Altshuler, E. A. A Method for Matching an Antenna Having a Small Radiation Resistance to a 50-ohm Load. *Trans. IEEE*, Vol. AP-53, September 2005, pp. 3086–3089.

Begovich, N. A. Frequency Scanning. In *Microwave Scanning Antennas*, Vol. 3, Hansen, R. C. Ed., Academic Press, 1966;Peninsula Publishing, 1985.

Campbell, A. M. Maxwell's Equations in Superconductors. *Trans. IEEE*, Vol. AS-17, June 2007, pp. 2531–2536.

CHALOUPKA, H. J. High-Temperature Superconductor Antennas: Utilisation of Low r.f. Losses and of Nonlinear Effects. *J. Supercond.*, Vol. 5, 1992, pp. 405–416.

CHALOUPKA, H. J. Superconducting Multiport Antenna Arrays. *Microwave Opt. Technol. Lett.*, Vol. 6, October 1993, pp. 737–744.

CHALOUPKA, H. J. Antennas. In *Microwave Superconductivity*, Weinstock, H. and Nisenoff, M.,Eds., Kluwer Academic Publishers, 2001, pp. 353–386.

CHALOUPKA, H. J. ET AL. Miniaturised High-Temperature Superconductor Microstrip Patch Antenna. *Trans. IEEE*, Vol. MTT-39, September 1991, pp. 1513–1521.

CHALOUPKA, H. J., WANG, X., AND COETZEE, J. C. A Superdirective 3-Element Array for Adaptive Beamforming. *Microwave Opt. Technol. Lett.*, Vol. 36, March 2003, pp. 425–430.

CHUNG, D.-C. Broadband HTS Microstrip Antennas for Satellite Communication. *Trans. IEEE*, Vol. AS-11, March 2001, pp. 107–100.

CHUNG, D.-C. ET AL. Circularly Polarized HTS Microstrip Antenna Array. *Trans. IEEE*, Vol. AS-13, June 2003, pp. 301–304.

CHUNG, D.-C. ET AL. HTS Microstrip Antenna Array for Circular Polarization with Cryostat. *Trans. IEEE*, Vol. AS-15, June 2005, pp. 1048–1051.

COOK, G. G. AND KHAMAS, S. K. Temperature Control of Radar Cross-Section Using Electrically Small High-T_c Superconducting Antenna Elements for Stealth Applications. *Proc. IEE*, Vol. 140H, August 1993, pp. 326–328.

COOK, G. G. AND KHAMAS, S. K. Control of Radar Cross Sections of Electrically Small High Temperature Superconducting Antenna Elements Using a Magnetic Field. *Trans. IEEE*, Vol. AP-42, June 1994, pp. 888–890.

COOK, G. G. ET AL. Performance Prediction of High-T_c Superconducting Small Antennas Using a Two-Fluid-Moment Method Model. *Appl. Phys. Lett.*, Vol. 60, January 1992, pp. 123–125.

COOK, G. G. ET AL. Predictions of the Efficiencies of Superconducting Small Antennas Connected to Lossy Ground Planes Using a Sommerfeld Integral Technique. *J. Appl. Phys.*, Vol. 76, July 1994, pp. 1266–1268.

COOK, G. G., KHAMAS, S. K., AND BOWLING, D. R. Efficiencies of Superconducting Small Antennas Connected to Lossless Disks Over Lossy Groundplanes. *Trans. IEEE*, Vol. AP-43, June 1995, pp. 631–633.

DAY, C. Iron-Based Superconductors. *Phys. Today*, August 2009, pp. 36–40.

DINGER, R. J. AND WHITE, D. J. Theoretical Increase in Radiation Efficiency of a Small Dipole Antenna Made with a High Temperature Superconductor. *Trans. IEEE*, Vol. AP-38, August 1990, pp. 1313–1316.

DINGER, R. J., BOWLING, D. R., AND MARTIN, A. M. A Survey of Possible Passive Antenna Applications of High-Temperature Superconductors. *Trans. IEEE*, Vol. MTT-39, September 1991, pp. 1498–1507.

DURAND, D. J. ET AL. The Distributed Josephson Inductance Phase Shifter. *Trans. IEEE*, Vol. AS-2, March 1992, pp. 33–38.

FENZI, N. ET AL. Development of High Temperature Superconducting 100 Nano-Second Delay Line. *Proc. SPIE*, Vol. 2156, 1994, pp. 143–151.

GAIDIS, M. C. ET AL. Characterization of Low-Noise Quasi-Optical SIS Mixers for the Submillimeter Band. *Trans. IEEE*, Vol. MTT-44, July 1996, pp. 1130–1139.

GINZBURG, V. L. AND ANDRYUSHIN, E. A. *Superconductivity*, World Scientific, 2004.

GOUGH, C. E. ET AL. Critical Currents in a High-T_c Superconducting Short Dipole Antenna. *Trans. IEEE*, Vol. MAG-25, March 1989, pp. 1313–1314.

HANSEN, R. C. Superconducting Antennas. *Trans. IEEE*, Vol. AES-26, March 1990, pp. 345–355.

HANSEN, R. C. Antenna Applications of Superconductors. *Trans. IEEE*, Vol. MTT-39, September 1991, pp. 1508–1512.

HANSEN, R. C. *Phased Array Antennas*, 2nd ed., Wiley, 2009.

HATTORI, W., TOSHITAKE, T., AND TAHARA, S. Reentrant Delay-Line Memory Using a $YBa_2Cu_3O_7$ Coplanar Delay Line. *Trans. IEEE*, Vol. AS-9, June 1999, pp. 3829–3832.

HE, Y. S. ET AL. Liquid Nitrogen Temperature Superconducting Antennas Made from Metal Oxide Ceramics. *Cryogenics*, Vol. 30, Suppl., 1990, pp. 946–950.

HE, Y. S. ET AL. Progress in High T_c Superconducting Ceramic Antennas. *Supercond. Sci. Technol.*, Vol. 4, 1991, pp. S124–S126.

HERD, J. S. ET AL. Experimental Results on a Scanned Beam Microstrip Antenna Array with a Proximity Coupled YBCO Feed Network. *Trans. IEEE*, Vol. AS-3, March 1993, pp. 2840–2843.

HERD, J. S. ET AL. Twenty-GHz Broadband Microstrip Array with Electromagnetically Coupled High T_c Superconducting Feed Network. *Trans. IEEE*, Vol. MTT-44, July 1996, pp. 1384–1389.

HOFER, G. J. AND KRATZ, H. A. High Temperature Superconductor Coplanar Delay Lines. *Trans. IEEE*, Vol. AS-3, March 1993, pp. 2800–2803.

HOHENWARTER, G. K. G. ET AL. Forty Five Nanoseconds Superconducting Delay Lines. *Trans. IEEE*, Vol. AS-3, March 1993, pp. 2804–2807.

HUANG, F. ET AL. A Superconducting Microwave Linear Phase Delay Line Filter. *Trans. IEEE*, Vol. AS-3, March 1993, pp. 2778–2781.

HUANG, Y. ET AL. A High Temperature Superconductor Superdirectional Antenna Array. *Physica C*, Vol. 180, 1991, pp. 267–271.

ITOH, K. ET AL. High T_c Superconducting Small Antennas. *Trans. IEEE*, Vol. AS-3, March 1993, pp. 2836–2839.

IVRISSIMTZIS, L. P. ET AL. On the Design and Performance of Electrically Small Printed Thick Film $YBa_2Cu_3O_7$ Antennas. *Trans. IEEE*, Vol. AS-4, 1994a, pp. 33–40.

IVRISSIMTZIS, L. P. ET AL. High Gain Printed Dipole Array Made of Thick Film High-T_c Superconducting Material. *Electron. Lett.*, Vol. 30, January 1994b, pp. 92–93.

IVRISSIMTZIS, L. P. ET AL. High-Gain Series Fed Printed Dipole Arrays Made of High-T_c Superconductors. *Trans. IEEE*, Vol. AP-42, October 1994c, pp. 1419–1429.

IVRISSIMTZIS, L. P., LANCASTER, M. J., AND ALFORD, N. MCN. Supergain Printed Arrays of Closely Spaced Dipoles Made of Thick Film High-T_c Superconductors. *IEE Proc. Microwaves Antenn. Propag.*, Vol. 142, February 1995a, pp. 26–34.

IVRISSIMTZIS, L. P., LANCASTER, M. J., AND ALFORD, N. MCN. A High Gain YBCO Antenna Array with Integrated Feed and Balun. *Trans. IEEE*, Vol. AS-5, June 1995b, pp. 3199–3202.

JAMES, J. R. AND ANDRASIC, G. Environmental Coupling Loss Effects in Superconducting HF Loop Antenna Design. *Proc. IEE Microwave Antenn. Propag.*, Vol. 141, April 1994, pp. 94–100.

KANAYA, H. ET AL. Development of a HTS Slot Antenna with Multi-Bandpass Filters. *Trans. IEEE*, Vol. AS-17, June 2007, pp. 882–885.

KAPOLNEK, D. J. ET AL. Integral FMCW Radar Incorporating an HTSC Delay Line with User-Transparent Cryogenic Cooling and Packaging. *Trans. IEEE*, Vol. AS-3, March 1993, pp. 2820–2823.

KAUTZ, R. L. Picosecond Pulses on Superconducting Striplines. *J. Appl. Phys.*, Vol. 49, January 1978, pp. 308–314.

KAUTZ, R. L. Miniaturization of Normal-State and Superconducting Striplines. *J. Res. NBS*, Vol. 84, May–June 1979, pp. 247–259.

KHAMAS, S. K. ET AL. High-T_c Superconducting Short Dipole Antenna. *Electron. Lett.*, Vol. 24, April 1988, pp. 460–461.

KHAMAS, S. K. ET AL. Significance of Matching Networks in Enhanced Performance of Small Antennas When Supercooled. *Electron. Lett.*, Vol. 26, May 1990, pp. 654–655.

KHAMAS, S. K. ET AL. Investigation of the Enhanced Efficiencies of Small Superconducting Loop Antenna Elements. *J. Appl. Phys.*, Vol. 74, August 1993, pp. 2914–2918.

KOH, K. AND HOHKAWA, K. Fabrication of Superconducting Delay Line with GaAs Schottky Diode. *Trans. IEEE*, Vol. AS-9, June 1999, pp. 3224–3227.

KRIVOSHEEV, E. F. AND PAVLYUK, V. A. High-Gain Cryogenic Antenna Array. *Sov. Tech. Phys. Lett.*, Vol. 5, August 1979, pp. 384–385.

LANCASTER, M. J. *Passive Microwave Device Applications of High-Temperature Superconductors*, Cambridge University Press, 1997.

LANCASTER, M. J. ET AL. Supercooled and Superconducting Small-Loop and Dipole Antennas. *Proc. IEE*, Vol. 139, Part H, June 1992a, pp. 264–270.

LANCASTER, M. J. ET AL. Superconducting Antennas. *Supercond. Sci. Technol.*, Vol. 5, 1992b, pp. 277–279.

LANCASTER, M. J. ET AL. YBCO Thick Film Loop Antenna and Matching Network. *Trans. IEEE*, Vol. AS-3, March 1993, pp. 2903–2905.

LANCASTER, M. J., WANG, H. Y., AND HONG, J.-S. Thin-Film HTS Planar Antennas. *Trans. IEEE*, Vol. AS-8, December 1998, pp. 168–177.

LEWIS, L. L. ET AL. Performance of TlCaBaCuO 30 GHz 64 Element Antenna Array. *Trans. IEEE*, Vol. AS-3, 1993, pp. 2844–2847.

LIANG, G. C. ET AL. High-Temperature Superconductor Resonators and Phase Shifters. *Trans. IEEE*, Vol. AS-1, March 1991, pp. 58–66.

LIANG, G. C. ET AL. High-Temperature Superconducting Delay Lines and Filters on Sapphire and Thinned $LaAlO_3$ Substrates. *Trans. IEEE*, Vol. AS-3, September 1993, pp. 3037–3042.

LIANG G. C. ET AL. High-Temperature Superconductive Devices on Sapphire. *Trans. IEEE*, Vol. MTT-42, January 1994, pp. 34–40.

MA, J.-G. AND WOLFF, I. Electromagnetics in High-T_c Superconductors. *Trans. IEEE*, Vol. MTT-44, April 1996, pp. 537–542.

MARTENS, J. S. ET AL. HTS-Based Switched Filter Banks and Delay Lines. *Trans. IEEE*, Vol. AS-3, March 1993, pp. 2824–2827.

MOORE, J. D. AND TRAVERS, D. N. Radiation Efficiency of a Short Cryogenic Antenna. *Trans. IEEE*, Vol. AP-14, March 1966, p. 246.

MORROW, J. D. ET AL. Circularly Polarized 20-GHz High-Temperature Superconducting Microstrip Antenna Array. *Trans. IEEE*, Vol. AS-9, December 1999, pp. 4725–4731.

NEWMAN, H. S. AND CULBERTSON, J. C. Measurement of the Current-Density Distribution in High-Temperature Superconducting Microstrip by Means of Kinetic-Inductance Photoresponse. *Microwave Opt. Technol. Lett.*, Vol. 6, October 1993, pp. 725–728.

ODA, S. ET AL. Electrically Small Superconducting Antennas with Bandpass Filters. *Trans. IEEE*, Vol. AS-17, June 2007, pp. 878–881.

PAVLYUK, V. A. ET AL. Superconducting Antenna. *Sov. Tech. Phys. Lett.*, Vol. 4, February 1978, p. 80.

PIEL, H. AND MULLER, G. The Microwave Surface Impedance of High Temperature Superconductors. *Trans. IEEE Magnet.*, Vol. 27, March 1991, pp. 854–862.

POND, J. M., CLAASSEN, J. H., AND CARTER, W. L. Measurements and Modeling of Kinetic Inductance Microstrip Delay Lines. *Trans. IEEE*, Vol. MTT-35, December 1987, pp. 1256–1262.

POND, J. M. KROWNE, C. M., AND CARTER, W. L. On the Application of Complex Resistive Boundary Conditions to Model Transmission Lines Consisting of Very Thin Superconductors. *Trans. IEEE*, Vol. MTT-37, January 1989, pp. 181–190.

PORTIS, A. M. ET AL. Power-Induced Switching of an HTS Microstrip Patch Antenna. *Supercond. Sci. Technol.*, Vol. 4, 1991, pp. 436–438.

PUCEL, R. A., MASSE, D. J., AND HARTWIG, C. P. Losses in Microstrip. *Trans. IEEE*, Vol. MTT-16, June 1968, pp. 342–350; see also MTT-16, December 1968, p. 1064.

RICHARD, M. A. ET AL. Performance of a Four-Element Ka-Band High-Temperature Superconducting Microstrip Antenna. *IEEE Microwave Guided Wave Lett.*, Vol. 2, April 1992a, pp. 143–145.

RICHARD, M. A. ET AL. Performance of a K-Band Superconducting Annular Ring Antenna. *Microwave Opt. Technol. Lett.*, Vol. 5, June 1992b, pp. 257–259.

RICHARD, M. A., BHASIN, K. B., AND CLASPY, P. C. Superconducting Microstrip Antennas: An Experimental Comparison of Two Feeding Methods. *Trans. IEEE*, Vol. AP-41, July 1993, pp. 967–974;correction AP-52, March 2004, pp. 911–912.

SHEEN, D. M. Current Distribution, Resistance, and Inductance for Superconducting Strip Transmission Lines. *Trans. IEEE*, Vol. AS-1, June 1991, pp. 108–115.

SHEN, Z.-Y. ET AL. High T_c Superconducting Coplanar Delay Line with Long Delay and Low Insertion Loss. *IEEE MTT Symp. Dig.*, 1991, pp. 1235–1238.

SU, H. T. ET AL. Wide-Band Superconducting Microstrip Delay Line. *Trans. IEEE*, Vol. MTT-52, November 2004, pp. 2482–2487.

SUZUKI, M. ET AL. Superconductive Small Antennas Made of Eu-Ba-CuO Thin Films. In *Advances in Superconductivity*, Vol. V, Bando, Y. and Yamauchi, H., Eds., Springer, 1992, pp. 1127–1130.

TAKEMOTO-KOBAYASHI, J. H. ET AL. High-T_c Superconducting Monolithic Phase Shifter. *Trans. IEEE*, Vol. AS-2, March 1992, pp. 39–44.

TALISA, S. H. ET AL. High-Temperature Superconducting Wide Band Delay Lines. *Trans. IEEE*, Vol. AS-5, 1995, pp. 2291–2294.

TINKHAM, M. *Introduction to Superconductivity*, McGraw-Hill, 1996.

TRACK, E. K., DRAKE, R. E., AND HOHENWARTER, G. K. G. Optically Modulated Superconducting Delay Lines. *Trans. IEEE*, Vol. AS-3, March 1993, pp. 2899–2902.

TSUTSUMI, Y., KANAYA, H., AND YOSHIDA, K. Design and Performance of an Electrically Small Slot Loop Antenna with a Miniaturized Superconduction Matching Circuit. *Trans. IEEE*, Vol. AS-15, June 2005, pp. 1020–1023.

TUCKER, J. R. AND FELDMAN, M. J. Quantum Detection at Millimeter Wavelengths. *Rev. Mod. Phys.*, Vol. 57, 1985, pp. 1055–1113.

VENDIK, O. G., KOZYREV, A. B., AND MOROZIK, V. P. Maximum Power of a Miniature Superconducting Radiator. *Sov. Phys. Tech. Phys.*, Vol. 26, July 1981, pp. 888–890.

WALKER, G. B. AND HADEN, C. R. Superconducting Antennas. *J. Appl. Phys.*, Vol. 40, 1969, pp. 2035–2039.

WALKER, G. B., HADEN, C. R., AND RAMER, O. B. Superconducting Superdirectional Antenna Arrays. *Trans. IEEE*, Vol. AP-25, November 1977, pp. 885–887.

WANG, H. Y. AND LANCASTER, M. J. Aperture-Coupled Thin-Film Superconducting Meander Antennas. *Trans. IEEE*, Vol. AP-47, May 1999, pp. 829–836.

WHEELER, H. A. Transmission-Line Properties of a Strip on a Dielectric Sheet on a Plane. *Trans. IEEE*, Vol. MTT-25, August 1977, pp. 631–647.

WU, Z. ET AL. High T_c Superconducting Small Loop Antenna. *Physica*, Vol. C162-4, 1989, pp. 385–386.

YOSHIDA, K. ET AL. Superconducting Slot Antenna with Broadband Impedance Matching Circuit. *Trans. IEEE*, Vol. AS-11, March 2001, pp. 103–106.

YOUNG, B. AND ITOH, T. Loss Reduction in Superconducting Microstrip-Like Transmission Lines. *IEEE Microwave Theory Tech. Symp. Dig.*, 1988, pp. 453–456.

ZMUIDZINAS, J. AND LEDUC, H. G. Quasi-Optical Slot Antenna SIS Mixers. *Trans. IEEE*, Vol. MTT-40, September 1992, pp. 1797–1804.

ZMUIDZINAS, J. AND RICHARDS, P. L. Superconducting Detectors and Mixers for Millimeter and Submillimeter Astrophysics. *Proc. IEEE*, Vol. 92, October 2004, pp. 1597–1616.

ZMUIDZINAS, J. ET AL. Low-Noise Slot Antenna SIS Mixers. *Trans. IEEE*, Vol. AS-5, June 1995, pp. 3053–3056.

APPENDIX A

A WORLD HISTORY OF ELECTRICALLY SMALL ANTENNAS

In the beginning, all antennas were electrically small, as Marconi in 1908 recognized that long wavelengths were needed for long-distance transmission. The first commercial transatlantic service was initiated in 1910 at 82 kHz! An extensive network of VLF stations was built around the world before World War I. There were stations at Kootwijk, Malabar, and Java (Dutch); at Paris, Lyon, and Saigon (French); at Carnarvon (UK); at Naven (Germany); and at Sayville and Tuckerton (USA). Between World War I and World War II, VLF stations were added at Balboa, Cavite, Criggion, Leafield, Rigby, and Kalbe. Watt (1967) lists just short of 100 VLF stations. Finally, the most advanced and most powerful stations were built at Jim Creek (Washington) and Cutler (Maine). The U.S. Navy VLF communications system was started after World War I and continues till date. Frequencies ranged from roughly 10 to 100 kHz, with wavelength from 30 to 3 km. Tall towers were used, many hundreds of meters high. Antenna engineering consisted primarily of tower design, structural engineering, properties of cables, and properties of high-voltage insulators. Umbrella loading, as sketched in Figure A.1, was used starting in 1913. An analysis was given by Smith and Johnson (1947). It was recognized almost immediately that top loading would significantly improve radiation resistance, and thereby the critical parameter efficiency (Carter and Beverage, 1962). Figure A.2 shows an idealized top-loaded fan, while Figure A.3 shows the Cutler Maine VLF antenna with its extensive top-loaded structure. In addition to the worldwide VLF communications network, the OMEGA navigation system started in 1968 with nine

Small Antenna Handbook, Robert C. Hansen and Robert E. Collin.

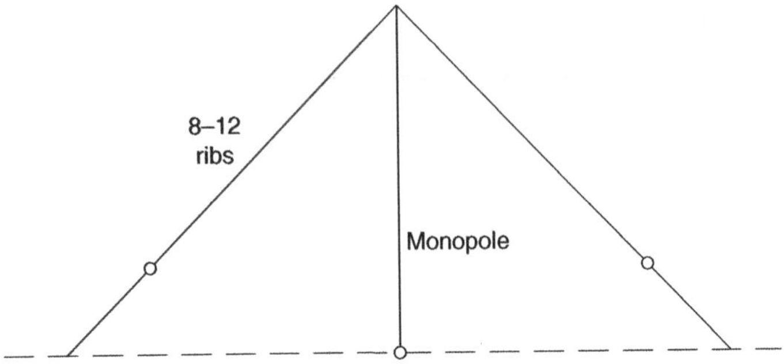

FIGURE A.1 VLF umbrella antenna.

transmitters around the world, operating at 10–14 kHz. A few years later, this system was superseded by the LORAN navigation system, which operated at 90–110 kHz. The shorter wavelength allowed a significant increase in location accuracy. The efficacy of these simple loading schemes for improving the radiation resistance of a monopole is shown by their use today in various configurations.

Interestingly, a recent survey of ESA (Sylusar, 2007) references only U.S. authors, all of which are discussed later in this appendix.

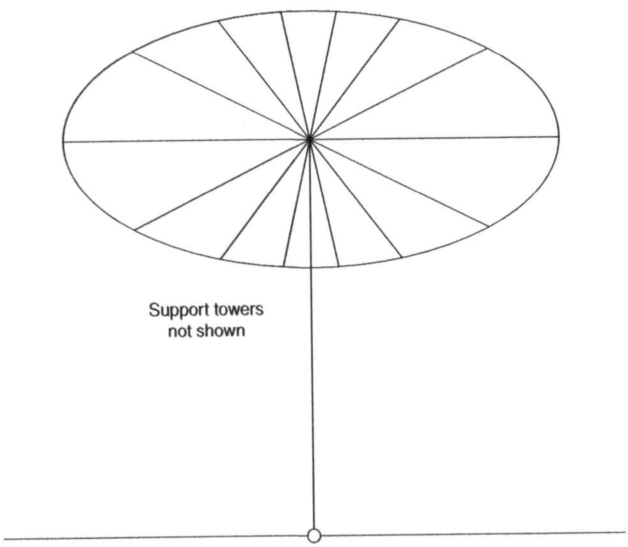

FIGURE A.2 VLF top-loaded monopole.

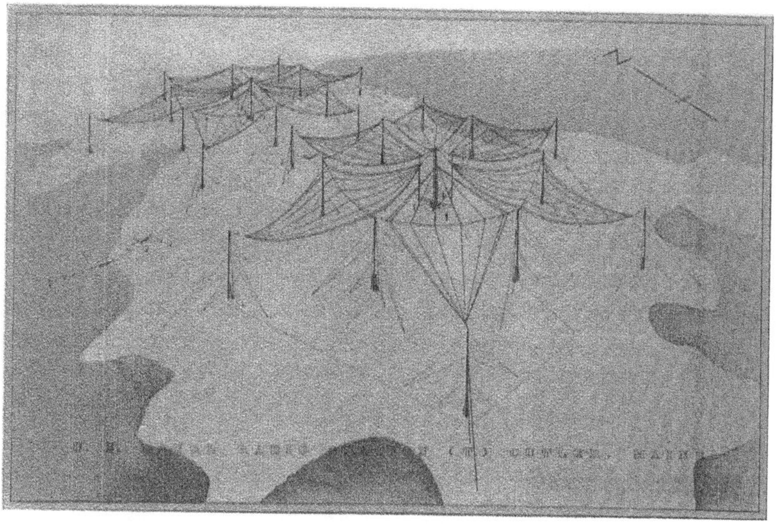

FIGURE A.3 Cutler maine VLF antennas. Courtesy of Watt (1967).

A.1 ELECTRICALLY SMALL ANTENNA CHRONOLOGY

1889 loop; Hertz
1901 long wire; Marconi
1936 inductive loading; Brown
1939 ferrite loopstick; Polydoroff
1947 fundamental limitations; Chu and Wheeler
1948 dielectric loading; Wheeler
1950 partial sleeve, PIFA; Jordan and Nash
1952 cavity antenna; Master
1953 patch; Deschamps
1960 active antenna; Frost and Meinke
1964 resonant wire; Fenwick and Best
1971 non-Foster; Poggio and Mayes
1976 dual mode; Goubau
1997 $Q = 0$ antenna; Grimes and Grimes
1998 fractal; Puente and Baliarda
2003 antennas with metamaterials
2009 double resonance; Stuart
2010 magnetic loading; Kim, Breinberg, and Yaghjian

A.2 LOOPS

Loops are usually electrically small, as a single-turn loop approaches resonance as the diameter (circular loop) approaches 0.1λ. Larger loops experience large impedance resonances and pattern breakup. Thus, the rule of thumb is that the total wire length for a multiturn loop should be no greater than $\lambda/10$ (Storer, 1956). Segmented loops can be larger and still maintain constant current but they are outside the scope of this book. Often a loop is tuned by a series capacitor; this arrangement significantly increases the output but the bandwidth becomes very narrow. Because the radiation resistance of a small loop varies as the diameter in wavelengths to the fourth power, typical radiation resistances are in the tens or hundreds of milliohms. Because of this, typical loop efficiencies can be very small.

A novel coupling scheme was invented by Dunlavy (1971); this is the loop-coupled loop. It consists of a small loop coupled to the source or receiver and a much larger usually coplanar enveloping loop. This larger parasitic loop is usually tuned with a capacitor. See Figure A.4. Another implementation uses a copper

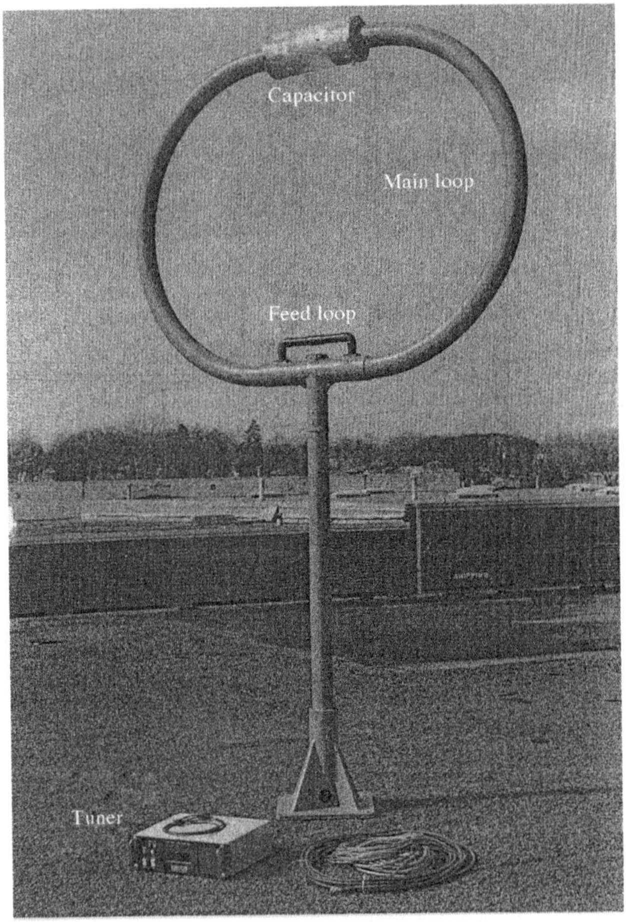

FIGURE A.4 Loop-coupled loop. Courtesy of Dunlavy (1971).

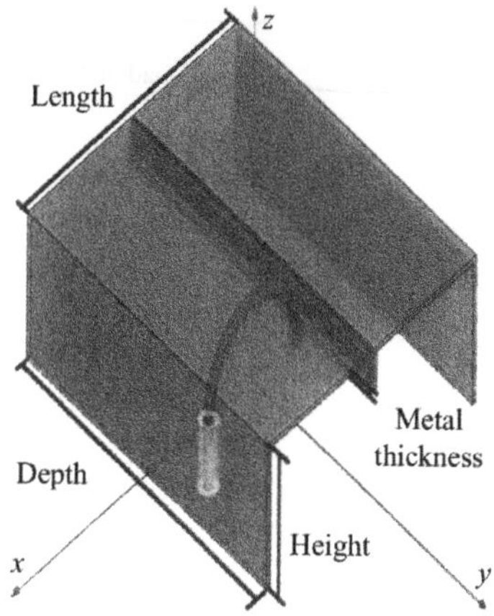

FIGURE A.5 Copper sheet loop-coupled loop. Courtesy of Erentok and Ziolkowski (2007).

sheet half-loop over a ground plane with parallel plate capacitor and a small coupling loop. See Figure A.5. Analyses of the loop-coupled loop (Hansen, 2006, Section 1.5.3; Belrose, 2007) show that the efficiency is very low. Loop-coupled loop antennas were manufactured by ARA for a while but eventually the customers discovered the above facts. A better antenna uses only the larger tuned loop, with the feed across the capacitor.

A.3 MONOPOLES WITH INDUCTIVE LOADING

Probably the first work on inductive loading of an electrically short monopole antenna was by George Brown of RCA (Brown, 1936). No documented work on series inductive loading of dipoles had been found for the period during World War II and the years immediately after. Brown's work was extended by Belrose (1953); he was primarily concerned with antennas located on vehicles near the ground. At about the same time, Bulgerin and Walters (1954) were working at the Naval Ordnance Lab, Corona, on monopoles with series inductive loading. They calculated input resistance and reactance versus height of the loading coil as well as losses in the coil. These results were experimentally obtained and laid the foundation for more detailed computer analyses later on. A decade later, Harrison (1963) used King–Middleton theory to analyze the inductively loaded dipole using a superposition of asymmetrically excited dipoles. However, for the asymmetrically excited dipole, only zero-order solutions are available. As the feed point (loading point) moves toward the end, the results become less accurate. The U.S. Army Electronics Lab, Fort Monmouth, was also working on inductive loading, with a brief summary given by Czerwinski (1965). Several years later, the faculty at Michigan State University was investigating series inductive loading of dipoles (Lin et al., 1970). Their limited results were based on approximate solutions to the integral equations. The design of inductively loaded

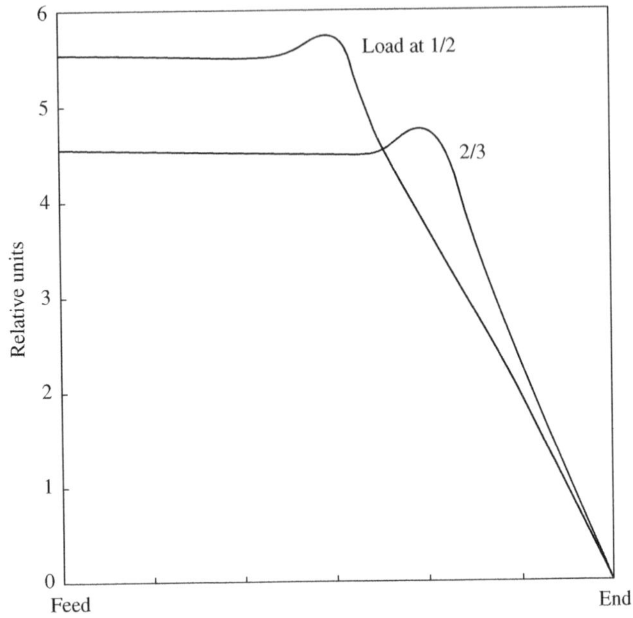

FIGURE A.6 Current distribution. Courtesy of Hansen (1972).

dipole (or monopole) antennas was put on a sound basis by Hansen (1973, 1975a, 1975b). Modeling the antenna with the piecewise sinusoidal moment method allowed accurate determination of efficiency, input resistance, and ratio of loading reactance to antenna reactance. Extensive data of input resistance versus loading point (where the load is placed) for monopole lengths over wavelength from 0.05 to 0.15, radiation resistance improvement factors, resonant loading reactance, efficiency versus loading point, and ratio of loading reactance to antenna reactance were given. These data allow design trades to be readily performed. Examples are Figure A.6, which shows the current distribution peaking at the load point, and Figure A.7, which shows the radiation resistance improvement factor versus load point for three dipole lengths ($l = 2h$).

Experimental measurements of inductively loaded short monopoles were made by Fournier and Pomerleau (1978). Their results largely confirm the moment method results just mentioned. Resistive loading, either discrete or continuous, can be considered. However, the efficiency rapidly decreases as the loading resistance is increased.

The art of calculating and measuring inductively loaded antennas has not changed significantly since the papers cited above. What has changed is the wider availability of ferrite materials, covering a wider range of frequencies and permeabilities. In general, losses have been reduced somewhat in recent years. A new material, which is suitable for frequencies above which ferrites are not suitable, has appeared. This is metaferrite from Metamaterials Limited, Austin, TX. This

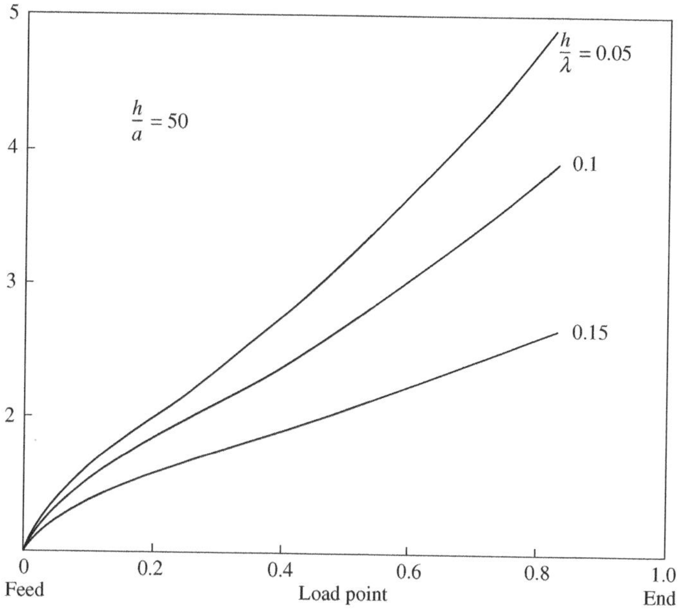

FIGURE A.7 Radiation resistance improvement factor. Courtesy of Hansen (1975a).

material consists of a stack of many laminates where each layer is a thin plastic sheet with a deposited film of magnetic compounds chosen for each frequency band, and patterned so as to minimize eddy current loss. This material promises to allow inductively loaded monopoles to be designed at higher frequencies than with ferrites. Loading of dipoles via capacitance, either using series capacitors or dielectric sleeves, is covered in Section A.6.

A.4 FERRITE LOOPSTICK

During World War II, the loopstick antenna was developed in Germany. It consisted of a cylindrical plastic tube filled with powered iron, which served as a magnetic core, with a coil wrapped around the core near the center of the core. It may have been based on the work of Polydoroff (1941); he patented a loop with a hollow ferromagnetic core, with the core length roughly twice the coil diameter. See Figure A.8. One of the German antennas was evaluated for WFAFB by University of Illinois—Urbana Antenna Lab. The results were published in the *1952 Allerton USAF Antenna Research and Development Symposium* (Jordan and Hansen, 1952). A ferrite loopstick was also tested, and as expected the losses were considerably lower and the affective permeability higher. Ferroxcube (Van Suchtelen, 1954) was used. A short ferrite loopstick antenna for aircraft application is shown in Figure A.9. Note that it uses a hollow core. This antenna lab ferrite loopstick was apparently the first. Of course, they are now widely used in portable electronic devices. A typical

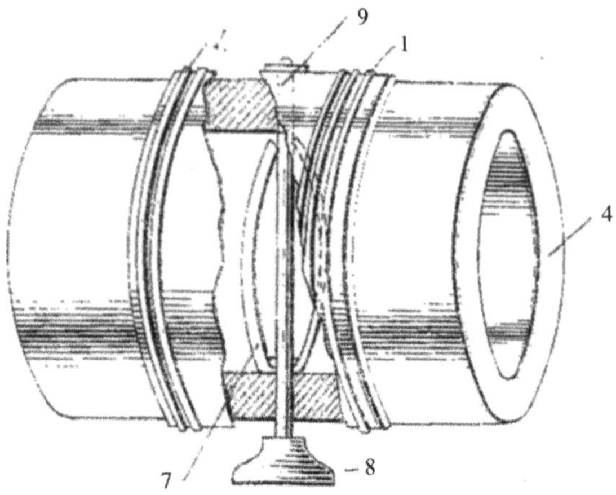

FIGURE A.8 Polydoroff loopstick antenna. Courtesy of Polydoroff (1941).

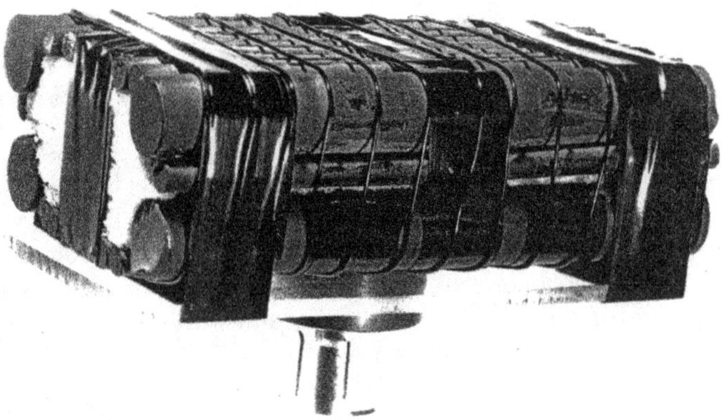

FIGURE A.9 Experimental loopstick. Courtesy of Jordan and Hansen (1952).

loopstick antenna with a distributed winding is shown in Figure A.10. The winding can also be a short multilayer concentrated at the core center.

The effective permeability of a loopstick core was calculated by Jim Wait in a pair of classic papers published in 1953. He replaced the cylindrical core with a prolate

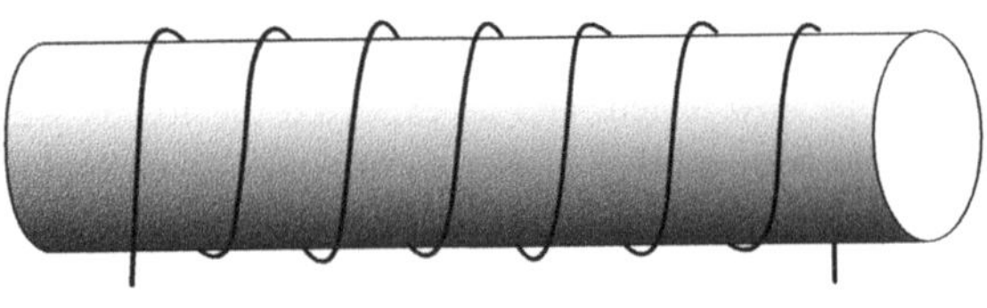

FIGURE A.10 Loopstick antenna.

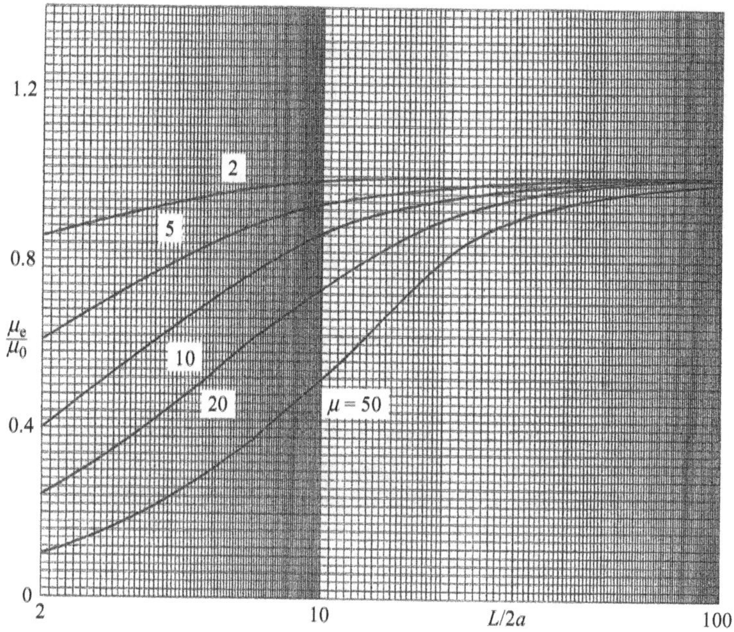

FIGURE A.11 Effective permeability versus length/diameter. Courtesy of Wait (1953).

spheroidal core, where the two lengths are the same and the central diameter of the spheroid is the same as the cylinder diameter. The effective permeability is equal to the intrinsic permeability times a factor, which involves spheroidal Q_1 wave functions and a parameter that depends upon the core length-to-diameter ratio. It was discovered that cores with higher permeability needed to be longer to realize the intrinsic permeability. See Figure A.11 for the effective permeability of a solid core. Also, most of the flux density is near the surface of the core so that hollow cores could be used. This problem was also solved by Wait again using prolate spheroidal geometry. The hollow inside of the core is represented by a second prolate spheroid. Again the effective permeability is obtained from an equation involving many spheroidal wave functions, and again a parameter depending on length-to-diameter ratio. Data for both solid cores and hollow cores, and the appropriate equations, are conveniently provided in a recent book (Hansen, 2006, Section 1.3.3.3).

If the graphs in the reference are not adequate, the effective permeability is readily calculated because the prolate spheroidal wave functions are tabulated in Abramowitz and Stegen (1970).

A.5 FUNDAMENTAL LIMITATIONS

Harold Wheeler, along with Lan Chu, was a founder of the basic understanding of electrically small antennas (ESA). In a series of three papers (Wheeler, 1947, 1958, and 1959), he established several widely used and important concepts. The radian sphere of radius a encompasses electrically small antennas; now it is commonly taken to be $ka \leq 0.5$, where $k = 2\pi/\lambda$. This is a practical limit for ESA. The second concept was radiation power factor. This was a hangover from the power community, and is equal to the reciprocal of Q. Wheeler called his limits on ESA "fundamental," but it is more realistic to call them practical limitations. For example, for dipole-type antennas he considered a cylindrical volume, which included any top hat loading; he derived a practical limitation on Q for this geometry, based on electrostatics. See Section A.6. Another limitation involved spherical coils, with or without magnetic cores. Again the limitation was in terms of practical coil and materials. These formulas have proved useful, but are not rigorous. In contrast, the fundamental limitations derived by Chu utilize spherical modes, both TE and TM.

Chu (1948) derived the fundamental limitation for an ESA, which could be contained within a hypothetical sphere, the radian sphere of Wheeler. The normalized radial wave impedance was expressed as a continued fraction, equivalent to a ladder network. For TM spherical modes, the network R, L, and C were calculated, along with the stored energy external to the sphere and the radiated power. From these, the minimum Q was obtained for each mode. For the lowest mode, TM_1, the well-known result is

$$Q = \frac{1}{ka} + \frac{1}{k^3a^3}$$

Here a is the radius of the sphere. It has been pointed out by many people that the formula for minimum Q should include energy stored within the enveloping sphere. This was done by Thal (2006), using the same ladder network theory of Chu. Subsequently, Hansen and Collin (2009) provided exact formulas for minimum Q for both TE and TM modes. These formulas are in terms of spherical Bessel and Hankel functions, and the results agree with the numerical calculations of Thal.

Because the original Chu Q formulation is expressed in spherical Bessel functions, which have closed forms, the exact Chu TE and TM values for the first three modes are (Collin and Rothschild, 1964)

$$Q_1 = \frac{1}{ka} + \frac{1}{(ka)^3}$$

$$Q_2 = \frac{3}{ka} + \frac{6}{(ka)^3} + \frac{18}{(ka)^5}$$

$$Q_3 = \frac{6}{ka} + \frac{21}{(ka)^3} + \frac{135}{(ka)^5} + \frac{675}{(ka)^7}$$

As mentioned by Chu, Q becomes large when $ka \ll n$.

When the energy inside the radian sphere is included, the Q formula has an infinite number of terms, because now the formula has spherical Hankel functions in the denominator. A best fit for the TM_1 mode was made by Hansen and Collin and is the following:

$$Q = \frac{1}{\sqrt{2}ka} + \frac{1.5}{k^3a^3}$$

Inclusion of permeable material inside the radian sphere for a TE mode antenna, as discussed later, can reduce the TE factor of 3 to close to the original Chu factor of 1.

A.6 DIELECTRIC LOADING

Probably the first to consider loading a dipole or a monopole antenna with dielectric material was Wheeler (1947). A dielectric cylinder was assumed to surround the dipole, with the dipole on the cylinder axis. Wheeler derived Q (actually radiation power factor, which is the inverse of Q) as

$$Q = \frac{6}{k_a k^3 a^2 l}$$

In this equation, the radius and length of the cylinder are a and l, respectively. As usual, $k = 2\pi/\lambda$. The factor k_a is Wheeler's shape factor and it was evaluated only for limiting cases. For tall cylinders, where length is much greater than radius, k_a approaches $4l/\pi a$. At the other extreme, for short cylinders, k_a approaches 1. These factors were apparently based on electrostatics. Thus, the nominal Q for short cylinders is $6/k^3a^2l$, while for long cylinders it is $3\pi/2k^3al^2$. At least two groups of researchers are working on extending Chu theory to a cylindrical enclosing volume, but no results are yet available. Raymond and Webb (1948) investigated a monopole loaded with a dielectric cylinder; they assumed a sinusoidal current distribution and calculated the resulting radiation resistance for several cases. They also performed measurements using dielectric material consisting of polystyrene mixed with varying amounts of titanium dioxide to give dielectric constants of 1.5–22. Their measurements roughly corroborate their theoretical calculations. The dielectric loading, as one might expect, reduces the radiation resistance and also reduces the bandwidth.

Polk (1959) evaluated biconical antennas, which were loaded with either dielectrics or ferromagnetic materials. In both cases, the material was in the form of a sphere, which included the biconical antenna. For electrically small antennas, the result was a reduced bandwidth. A suitable combination of antenna length and dielectric constant could produce resonance and a further increase in either could induce a condition of superdirectivity, with all its limitations. Work on a short dipole in a dielectric sphere was continued by Galejs (1962). Again it was found that the Q increased by a factor of $(\varepsilon+2)^2/9\varepsilon$. The Q is unchanged for $\varepsilon = 1$ and 4, and there is a small diminution for ε between these two values. Larger values of ε can significantly raise the Q as expected. Galejs pointed out that the cylindrical results of Wheeler do not apply, as the dielectric surface here is not tangential to the electric field, as it was in Wheeler's case. Careful measurement of a dipole with a cylindrical sleeve was made by Lamensdorf (1967). He measured admittance and current distributions for various antenna lengths, antenna diameters, and dielectric constants. Unfortunately, no bandwidth or Q for a dipole measurements were made. In general, the principal effect of the dielectric sleeve was to increase the electrical length of the dipole. James and Burrows (1973) used powdered dielectric to form a cylindrical cover. Both radiation resistance and bandwidth were reduced compared to the monopole alone. Dielectric constants of 18 and 90 were used.

The previous theoretical investigations approximately solved integral equations for assumed current distribution. The more accurate moment method approach for dielectric sleeve coded dipoles was developed by Richmond and Newman (1976). Their moment method approach used piecewise sinusoidal expansion and test functions. The dielectric sleeve was accounted for by modifying the matrix equation. The thin dielectric sleeve was replaced by an equivalent source with electric current density and from this a correction was calculated for the impedance matrix. A similar approach was used by Popovic (1982, 1991) except that the current expansion function was an entire domain polynomial. This is an interesting alternative to subsectional expansion functions, but it is not clear that this is the best approach for many problems.

Another investigation concerned insulated long wire antennas in lossy media (Chen and Warne, 1992). A special case was the insulted wire in free space. However, the long wire results are not useful for short antennas. Finally, the most recent paper

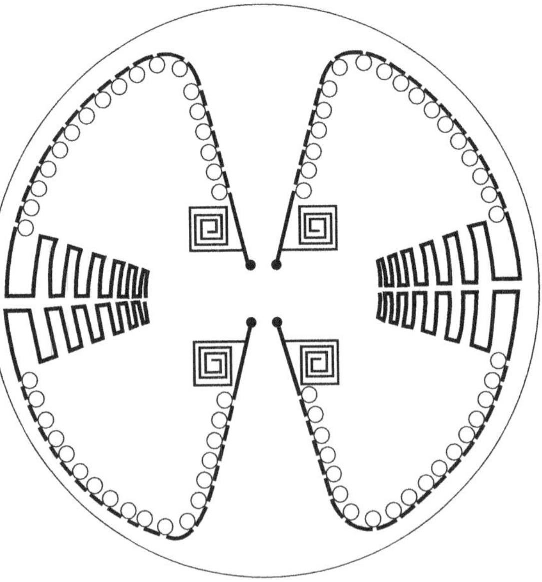

FIGURE A.12 Two loaded dipoles. Courtesy of Turner (1977).

(Jaisson, 2008) derived closed-form results for a monopole with a dielectric sleeve. A sinusoidal current distribution is assumed and the quasi-static effect of the sleeve used by Popovic and Richmond and Newman was used. Admittance was calculated for dielectric constants of 3.2, 9, and 15 but only for one length. Thus, his results are less useful than some work previously cited.

In the nut case category is an electrically small dipole antenna consisting of two adjacent dipoles, with each bent into a large U such that the four dipole arms fit inside a circle (Turner, 1977). The dipole arms have many series capacitor loads and each dipole end has a zigzag inductor. As mentioned elsewhere, the series capacity loading degrades the performance, and the inductive loads are located at the end instead of near the center of each dipole arm. See Figure A.12.

To summarize the effects of dielectric loading on a monopole or dipole, the radiation resistance is increased but the bandwidth is decreased.

A.7 PIFA, TRANSMISSION LINE, AND PATCH ESA

An early ESA was the partial sleeve antenna developed at the University of Illinois—Urbana Antenna Lab in the early 1950s. It was described in detail in a report (Jordan, 1953) (since declassified). Figure A.13 shows the partial sleeve, which is a curved plate, shorted to the correspondingly curved fuselage skin, tuned at one end and fed roughly near the center. When the sleeve is located over a flat ground plane, it takes the form shown in Figure A.14 and this may be recognized as the planar inverted-F antenna (PIFA), widely used in mobile communications and computer devices of all types. The capacitive tuning allows the PIFA to be electrically short. PIFA design details are given by Hirasawa and Haneishi (1991). For multiple frequencies, PIFA often have fingers, cutouts, additional posts, and so on. When the planar strip reduces to a wire, it is the transmission line antenna (Hansen, 2006); see Figure A.15.

The patch antenna was invented by Deschamps and Sichak (1953). The patch antenna is normally not considered an ESA. A high dielectric constant substrate

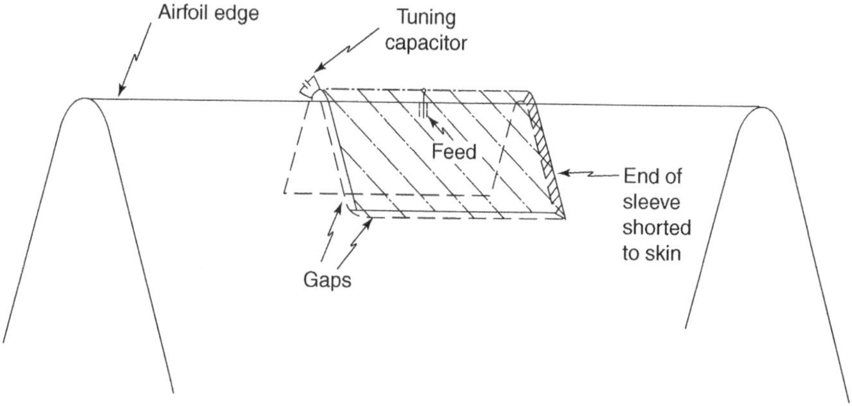

FIGURE A.13 Partial sleeve antenna. Courtesy of Jordan (1953).

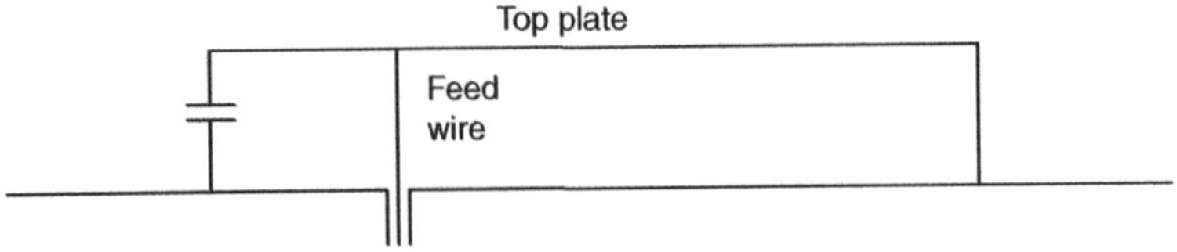

FIGURE A.14 Basic planar inverted-F antenna.

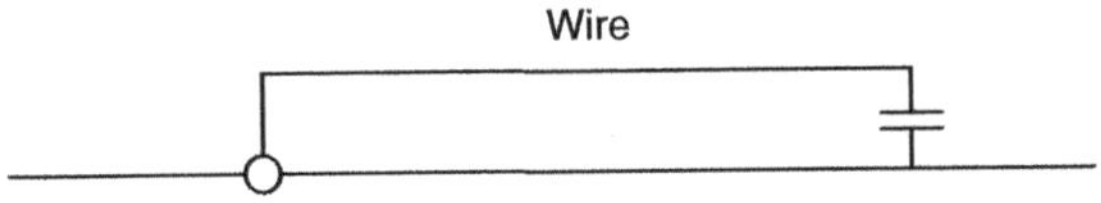

FIGURE A.15 Transmission line antenna.

reduces not only the resonant size, but also the bandwidth. However, a low-loss magnetic substrate also reduces the resonant size, but the bandwidth is increased (Hansen and Burke, 2000). One such material is metaferrite (from Metamaterials Limited, Austin, TX).

A.8 ELECTRICALLY SMALL CAVITY ANTENNAS

The electrically small cavity antenna consists of rectangular cavity with a feed post across the width of the cavity; the feed post is interrupted at its center by two perpendicular capacitor plates, which act to tune the antenna. A feed connection across the gap couples energy in or out (Masters, 1952). Another effort on reducing cavity size used a ferrite powder loaded cavity with the slot across the face; closely related is the loading of a cavity backed slot antenna with a ferrite powder (Adams et al., 1962). Further work on a ferrite-loaded cavity slot antenna used bars of ferrite material (Adams and Lyon, 1963). A major advance uses low-loss metaferrite material in the cavity as it allows a shallow cavity. The material is a many-layer laminate of thin dielectric sheets, each with a patterned deposition of magnetic alloys (Metamaterials Limited, Austin TX).

A.9 ACTIVE ANTENNAS

Antennas are typically connected to an amplifier, either for receiving or for transmitting. These are sometimes called "active antennas," but this nomenclature is deprecated. Herein active antenna means that an electronic device such as a transistor is inserted into the antenna itself, not just at the terminals.

The first active antenna seems to be the "parametric amplifier antenna" by Frost (1960). In this, a symmetric coaxial line cavity is terminated at each end by a tunnel diode. Energy is coupled out of the cavity by a small loop. See Figure A.16.

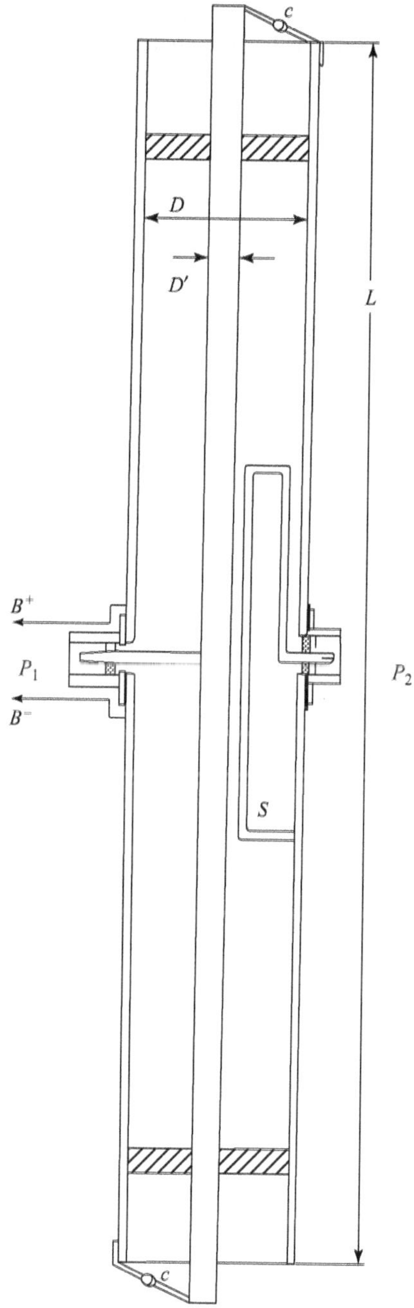

FIGURE A.16 Parametric amplifier antenna. Courtesy of Pedinoff (1961).

The capacity of the tunnel diode allows a shorter coaxial cavity to achieve resonance. An advantage parametric amplifiers have is a potentially low noise figure. This work was followed by a negative-conductance slot amplifier (Pedinoff, 1961). Here a slot is cut into the end plate of a waveguide and a tunnel diode is placed across the slot.

Due to the many problems associated with tunnel diodes, such as nonlinearity, the next step was to replace the diode with a transistor. This was initiated by Copeland et al. (1964), who called their antenna the "antennafier." The antenna is used as a circuit element of the transistor amplifier. The transistor is connected between the midpoint of the dipole and a gamma match rod. Although the dipole was half-wave, the notion of integrating antenna and transistor was important.

The next step was to insert the transistor into each arm of a dipole, with a parasitic conductor providing the third terminal. See Figure A.17. This concept was promoted by Ed Turner of the WPAFB, in particular with contracts at the Technische Hochschule in Munich (Turner, 1970). Papers were published by Meinke (1966), Meinke and Landstorfer (1968), Landstorfer and Meinke (1971), and Flachenecker et al. (1972).

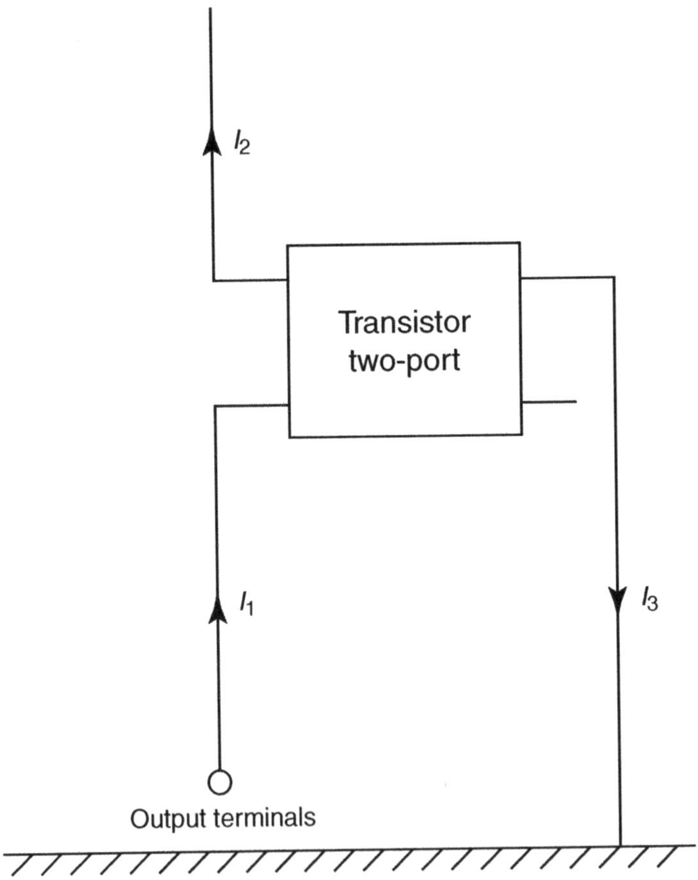

FIGURE A.17 Dipole with transistor and shunt feed.

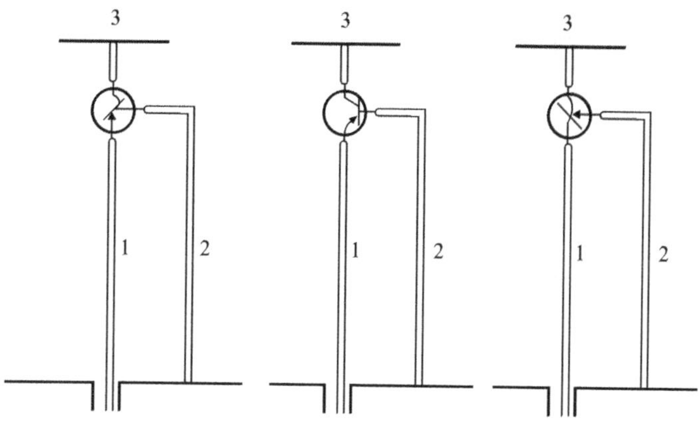

Q3 **FIGURE A.18** Three possible connections of a transistor. Courtesy of Meinke (1966).

Figure A.18 shows the three possible connections. These papers were primarily concerned with handling the noise problem. The major difficulty with these circuits is that there is no noise bandwidth filtering except through the impedance of the antenna. Nor is it possible to adjust transistor circuit parameters to minimize the noise. As a result of these difficulties, active antennas quickly fell into disuse. Maclean and Ramsdale (1975) and Maclean and Morris (1975) analyzed system noise versus transistor position in the dipole or monopole. Another active antenna is a folded dipole with a transistor in the center of the arm opposite to the feed (Daniel et al., 1975).

In 1971, a new paradigm appeared. Poggio and Mayes (1971) proposed and analyzed a loaded dipole, where negative inductances were placed at several points along each dipole arm. This landmark paper ushered in the era of non-Foster circuits applied to antennas. They have application primarily to the matching of electrically small antennas, as opposed to the loading technique investigated by Poggio and Mayes. See Section A.11.

A.10 RESONANT ESA

Increasing wire length in an ESA to improve performance via fractal geometry is not satisfactory. However, there are three ways of increasing wire length that need to be examined. The first is a meander dipole, sketched in Figure A.19. Here each arm of the dipole (or monopole) is continued back and forth parallel to and closely spaced to

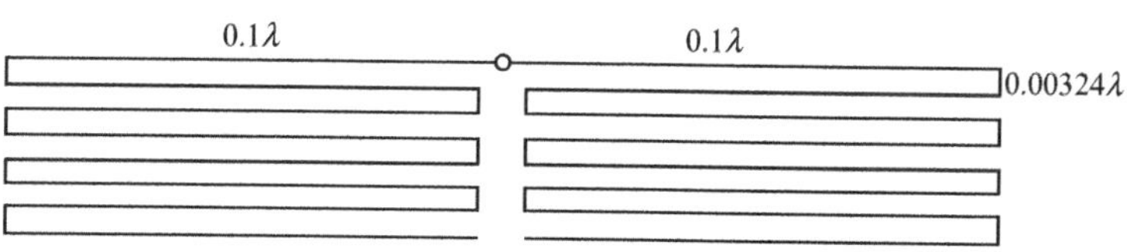

FIGURE A.19 Eight-arm meander antenna.

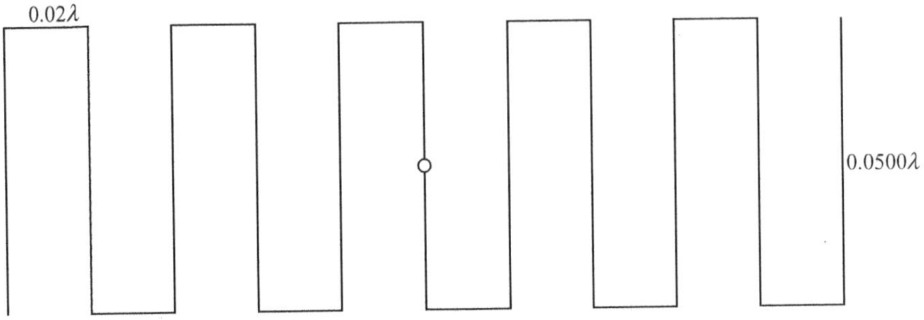

FIGURE A.20 Ten-section zigzag antenna.

the dipole arm. The second configuration called the zigzag antenna is sketched in Figure A.20. Now the back and forth wire excursions are perpendicular to the dipole axis. The third scheme to be discussed later is three dimensional and winds a monopole wire in the form of a spiral helix. Moment method calculations have shown that the zigzag construction has better bandwidth and higher radiation resistance than the meander dipole, as there is less cancellation of currents in the zigzag antenna. There is some confusion in nomenclature as some authors call our zigzag antenna a meanderline antenna (Best, 2004a, 2004b).

The spherical helix, sketched in Figure A.21, uses multiple turns of wire to get performance close to resonance. A radiation resistance of 50 Ω can be achieved. Both spherical helix monopole and a folded version can be used (Best, 2005).

A less effective way of using wire length involves a flat spiral winding parallel to the ground plane with the center of the spiral connected to the ground plane. The edge of the spiral is excited by a monopole. See Figure A.22. This scheme

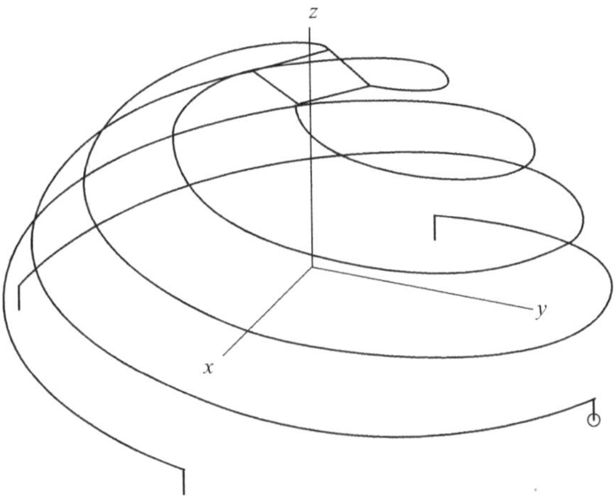

FIGURE A.21 Four-arm folded spherical helix monopole. Courtesy of Best (2005).

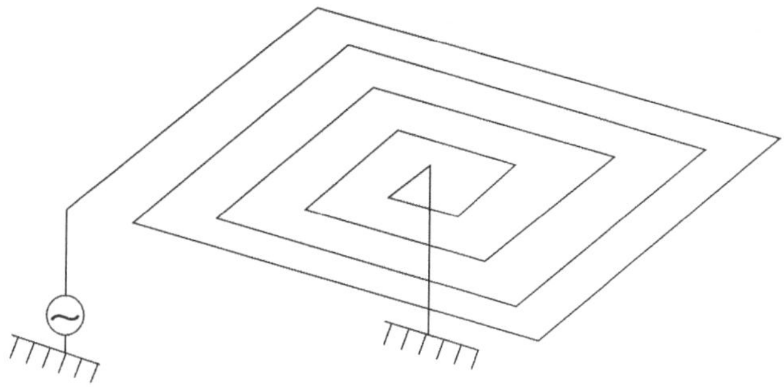

FIGURE A.22 Monopole with spiral load. Courtesy of Fenwick (1964).

(Fenwick, 1964) is less effective as the radiating monopole and its parasite monopole are both very short.

Another scheme for using multiple turns of wire to improve performance uses four monopoles each made of a tightly coiled wire, with the four monopoles following the Chu hypothetical hemisphere (Adams and Bernhard, 2008); see Figure A.23. The four curved monopoles are approximating the ideal sin θ current distribution over the hemisphere. Sufficient wire can be used to provide resonance; however, the radiation resistance of $16\,\Omega$ is divided by 4, yielding $4\,\Omega$, because the four monopoles are fed in parallel. Thus, this arrangement is less efficient than the spherical helix.

The intelligent use of wire length in an electrically small volume to improve performance is one of the significant recent contributions to electrically small antenna understanding and practice.

FIGURE A.23 A photograph of the constructed antenna. Courtesy of Adams and Bernhard (2008).

A.11 NON-FOSTER ANTENNAS

Foster's reactance theorem states that any passive, lossless, one-port network has a reactance or susceptance slope always increasing with increasing frequency (Foster, 1924). Antennas are not non-Foster as the radiation resistance makes them lossy even without conduction loss. However, an equivalent for antennas is that the impedance locus on a Smith chart always moves clockwise with increasing frequency.

In principle, non-Foster circuits can provide negative resistance, negative inductance, or negative capacitance. It is also possible to transform the resistance by a frequency-squared factor. The first application of non-Foster circuits was not to antennas but to telephony. In the early 1930s, Bell Laboratories developed a long-line amplifier that employed a modest amount of negative resistance to partially offset the long-line resistance (Crisson, 1931; Merrill et al., 1954). These long-line repeaters enabled transcontinental telephony and were used until the microwave links became available after World War II.

A major advance occurred when transistors were applied to negative-impedance converter circuits by Linvill (1953). This bypassed the significant limitations of tubes for non-Foster circuits. There have been many papers on various types of circuits for non-Foster performance using transistors.

The earliest application to antennas appears to be a patent filed in 1966 and granted in 1970 by Albee for a loop antenna with a non-Foster circuit (Albee, 1966). Next was the application of non-Foster to a monopole by Harris and Myers (1968). The first formal publication of non-Foster applied to antennas was by Poggio and Mayes (1971), where in theory several non-Foster circuits were inserted along each of the arms of a dipole. A closely related effort by Quirin (1970) investigated dipole antennas with a non-Foster match at the feed point. The first experimental results appear to be those of Bahr (1977). He used a current inverting NIC and a short monopole. More extensive work was reported by Skahill et al. (1998).

Non-Foster circuits can also be used to produce broadband phased arrays, particularly of the connected type. In a linear or planar array of dipoles, the dipoles in a collinear line can be connected end-to-end by a non-Foster circuit (Hansen, 2003, 2004a). In a conventional array, as the frequency decreases below that where the dipoles are roughly half-wave, the scan reactance increases rapidly. However, connecting the collinear dipole ends by a negative inductance non-Foster circuit allows the entire line of dipoles to act as a single dipole. Thus, the bandwidth of a connected array (Hansen, 2009) is roughly from the frequency where the line of dipoles is $\lambda/4$ long up to where each dipole is $\lambda/2$. Thus, an array of many octaves can be achieved. There is some degradation near half-wave frequency, as there the dipoles would like to be unconnected. Note that connecting the dipole ends with capacitors has exactly the wrong physics. As the frequency decreases, the dipoles become less coupled, and the reactance higher. As the frequency approaches for dipole length of half-wave, the capacitors connect the contiguous dipoles and this degrades the impedance. No non-Foster connected arrays have yet been built although the results are very promising (Hansen, 2004b).

Non-Foster circuits for antenna matching must be evaluated in terms of power dissipated, added noise, impedance matching performance, but most of all stability. Initially, open- and short-circuit stability criteria were used but more recent studies have shown that these are necessary but not sufficient. Non-Foster circuits can have internal loops whose effects do not appear on open and short circuits. A more careful stability design, employing Nyquist stability, is more appropriate and has begun to yield positive results. Of the many types of non-Foster circuits, most have one or more poles in the right half Nyquist plane; a small number of circuits can be made stable but this work is just underway. All in all, non-Foster matching is one of the few options available to ESA.

A.12 GOUBAU ANTENNA

A multimode antenna was developed by Goubau (1976); see Figure A.24. It consists of two monopoles fed in parallel, and two fat parasitic monopoles connected to the ground plane. The four monopoles are in a square shape. Each monopole is top loaded by a petal; those attached to the thin feed posts are each larger than a quarter circle, while those attached to the fat parasitic monopoles are smaller than a quarter circle. Just below the top-loaded plates are wire loops, each of which connects the adjacent plates. Use of both TM_1 and TE_1 modes roughly doubles the bandwidth. The Goubau antenna is electrically small only for the lower part of its octave bandwidth. At the upper frequency, $ka \simeq 1$.

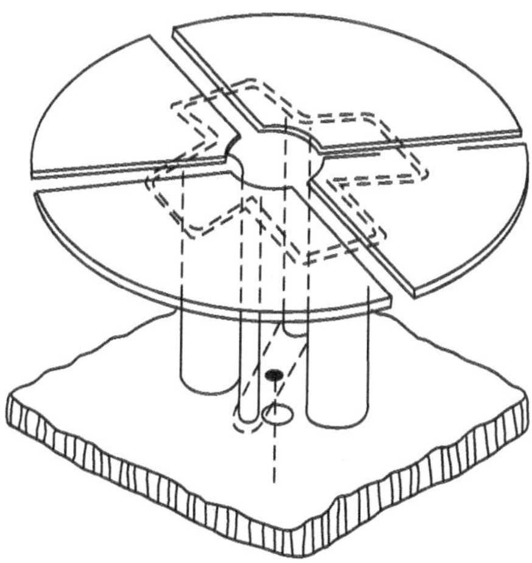

FIGURE A.24 Goubau multimode antenna. Courtesy of Goubau (1976).

A.13 NUT-HOUSE ESA

A.13.1 $Q = 0$ Antenna

In a 1997 paper and in several subsequent publications (1999), Grimes and Grimes announced that an electrically small antenna consisting of several elements properly spaced and phased would produce a Q of zero, that is, infinite bandwidth. No one with either practical or theoretical antenna experience would believe this. Collin (1998) showed that the Grimes attempt to calculate stored energy from the integral of the time-dependent Poynting vector over a spherical surface, without the need to consider the detailed expressions for the energy density in the field, is not possible. Another Grimes approach for calculating energy density used the sum of traveling energy density and standing energy density. They then applied the continuity equation in space and time, and then differentiated with respect to time. But as Collin observed, there is no Lorentz frame corresponding to uniform motion in the radial direction at velocity c. Thus, the Grimes second derivation is also incorrect. The Grimes and Grimes work is now widely discredited.

Nonetheless, Dockon, Inc. (2010) is attempting to piggyback on the $Q = 0$ antenna. They are combining a square single-turn loop with a dipole in the plane of the loop, parallel to one of the loop sides, and centered. See Figure A.25. A feed network presumably provides the feed necessary to optimize the bandwidth. However, basic electromagnetic principles are against the successful operation of this loop and dipole. The total length of conductor in the loop must be roughly less than $\lambda/10$ to avoid pattern breakup and impedance swings. The single-turn square loop will thus have a maximum side length of $\lambda/40$. The resulting radiation resistance will be a maximum of a few

FIGURE A.25 Dockon loop and dipole.

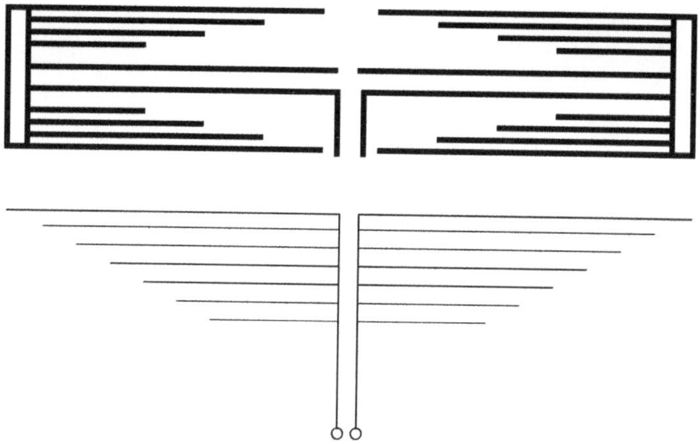

FIGURE A.26 Multi-arm dipoles. Courtesy of Turner and Richard (1968).

ohms. The $\lambda/40$ dipole will have a very high reactance as well as a low radiation resistance. Efficiency will be problem with the loop, and impedance matching will be a problem with the dipole. The best that can be done with any loop–dipole pair is to double the bandwidth. There is also a serious pattern problem that in the plane perpendicular to the dipole the dipole pattern is omnidirectional while the loop pattern is a figure eight. The resulting loop–dipole antenna is narrowband and is not omnidirectional. It is a pity that more antenna people do not read the readily available literature.

A.13.2 Multiarm Dipole

Figure A.26 shows two conceptions of a multiarm dipole (Turner and Richard, 1968). The simple-minded concept is that connecting together dipoles of various lengths gives a broadband dipole. Unfortunately, what results from the multiarm dipole is an equivalent fat single dipole. Increasing the fatness of the dipole does increase the bandwidth by a modest amount over a thin dipole, but this multiarm dipole is far from a broadband antenna.

A.14 FRACTAL MONOPOLES AND LOOPS

Nature, especially plants, is fractal as shown in wonderful detail by Mandelbrot (1977, 1982). For ESA, there are five general categories, each with its initiator and generator. The first stage of all consists of a small number of connected straight-line segments, which may be used as a monopole or fed as a loop. As the fractal stage increases, the segments become shorter and increase greatly in number (Hansen, 2006, Section 1.5.8). Figure A.27 shows a Von Koch monopole, while Figure A.28 shows a Sierpinski monopole. Hilbert and Minkowski monopoles are shown in Figure A.29. A Minkowski loop is shown in Figure A.30; the conductor is open near

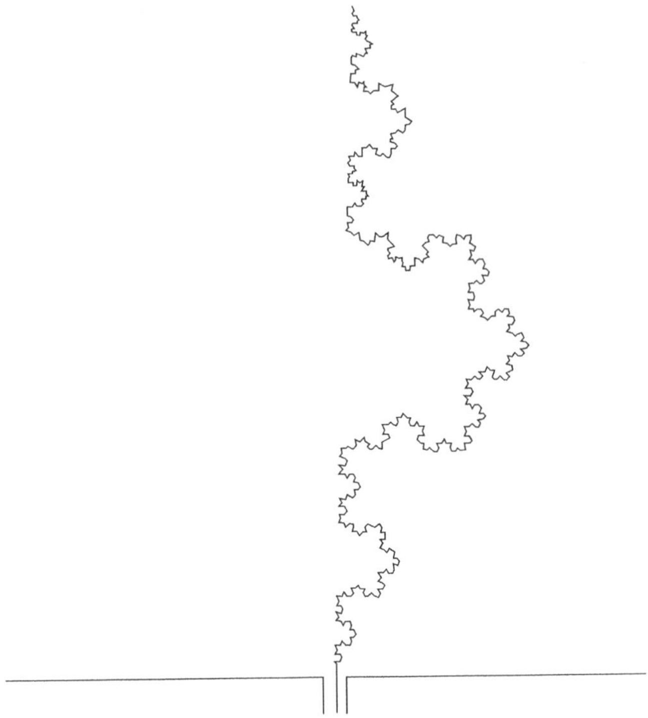

FIGURE A.27 Von Koch monopole.

FIGURE A.28 Sierpinski monopole.

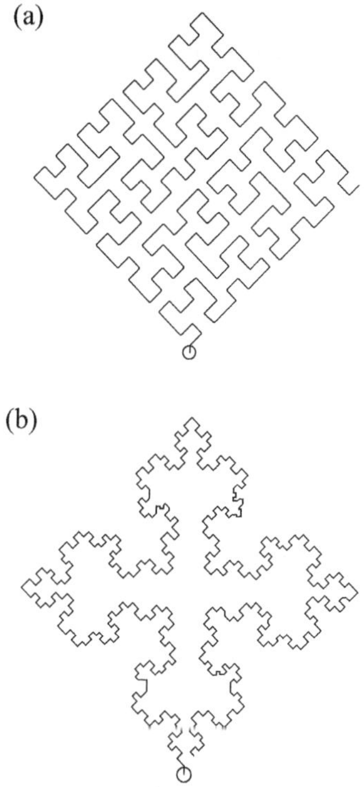

FIGURE A.29 (a) Hilbert and (b) Minkowski monopoles.

FIGURE A.30 Minkowski loop.

the center for the feed. For fractal references, see Hansen (2006, Section 1.5.8). These high-stage fractals, by incorporating a significant length of wire, can become resonant even though the overall width still qualifies as an ESA (Best, 2003). Unfortunately, these configurations store considerable energy in the near-field and have more loss. Thus, the bandwidth and efficiency are both diminished. A closely spaced meanderline monopole (see Section A.10) is better than any fractal monopole.

Alas, Maxwell's equations and the wave equation are not fractal. There are better ways of using longer wire lengths in making an ESA. See Section A.10.

A.15 DOUBLE RESONANCE ESA

Electrically small antennas can be improved by using enough wire to make a resonant ESA, or by exciting the lowest TE and TM modes together (Goubau). Another way was developed by Stuart and Tran (2005, 2007), where several collocated elements are operated together. Typically, one element is excited and one is parasitically coupled. Figure A.31 shows a two-arm antenna where the dipole arms are partially curved to fit inside the hypothetical Chu sphere. Four-arm antennas are shown in

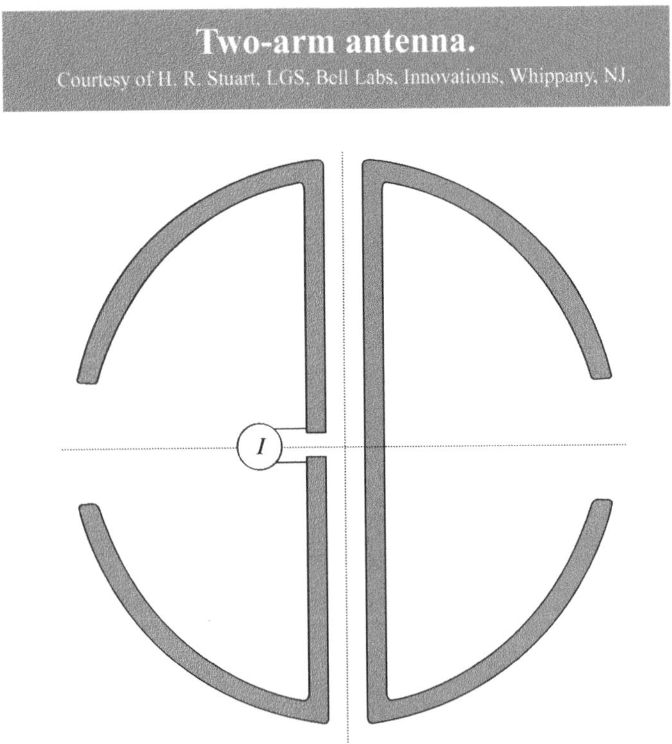

FIGURE A.31 Two-arm antenna. Courtesy of H. R. Stuart, LGS, Bell Labs, Innovations, Whippany, NJ.

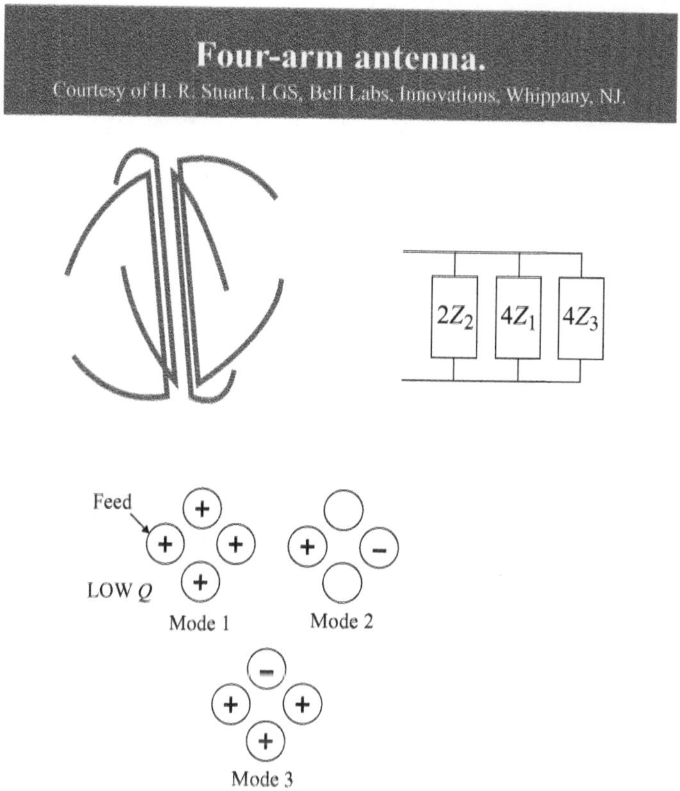

FIGURE A.32 Four-arm antenna. Courtesy of H. R. Stuart, LGS, Bell Labs, Innovations, Whippany, NJ.

Figures A.32 and A.33. The multiple arms allow two or more resonances; it should be noted that these are not simply related to the spherical harmonic modes. Stuart (2008) has done an eigenmode analysis of multielement antennas with one, two, four, and six arms. For example, the four-arm resonator has $ka = 0.67$ and a Q of 9.61. This is exactly twice the value using the basic Chu formula. These resonator antennas all show a double resonant impedance response. As more arms are added, the bandwidths of the antennas improve, for the same size.

It has been well established for single resonance antennas that the bandwidth, for VSWR ≤ 2, is equal to $1/2\sqrt{2}Q$. However, for double resonant ESA, this simple relationship becomes highly inaccurate when the two impedance resonances are closely spaced (Stuart et al., 2007). For a band narrow enough that the resonant peak is well defined, one of the traditional formulas for calculating Q yields a result that will predict the bandwidth. This is (Hansen, 2007)

$$Q = \frac{\omega}{2R_0}\frac{dX}{d\omega} \tag{A.1}$$

FIGURE A.33 Four-arm antenna. Courtesy of H. R. Stuart, LGS, Bell Labs, Innovations, Whippany, NJ.

A.16 MAGNETIC LOADING OF LOOPS

It is well known that dielectric loading of dipoles (TM modes) reduces the available bandwidth. However, magnetic loading of loops (TE modes) can decrease Q as shown by Wheeler (1947). Kim et al. (2010) have extended the work of Thal, including a permeable material inside the Chu sphere. They found that magnetic cores of low loss and large permeability could significantly decrease Q. A value of μ around 100 can reduce the Q to nearly that of the original Chu result, an improvement by nearly a factor of 3. This provides what is probably the ultimate in small loop antennas.

A.17 METAMATERIALS AND ANTENNAS

Metamaterials is a general term, which includes those with negative permeability and/or negative permittivity, as well as other artificial materials. Here the term includes negative parameter materials only. Here the focus is on only one typical application. This is an electrically short dipole surrounded by a thin spherical shell with $-\mu$ and/or $-\varepsilon$; the shell diameter is small in wavelengths.

Ziolkowski and Kipple (2003) used a thin shell of double negative material about a short dipole; see Figure A.34. They assumed a constant drive in voltage or current and calculated the power radiated. This was an unbelievable error, as the power radiated in that case has no relation to gain, directivity, or efficiency. There is also an error in the

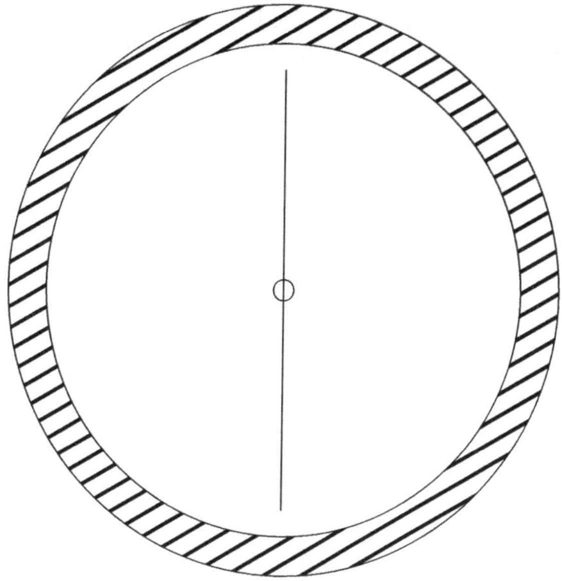

FIGURE A.34 Dipole in metamaterial shell. Courtesy of Ziolkowski and Kipple (2003).

formula for Q for the TM_1 mode. The power radiated problem was pointed out by Kildal (2006). A similar paper (Ziolkowski and Erentok, 2006, 2007) like the earlier paper shows that without dispersion the thin shell can improve the bandwidth of the short dipole. When a dispersion model suitable for optical wavelengths is included, there is a significant reduction in improvement, but still better than the Chu result. Comment will be given later on the proper choice of dispersion model. Part of the improvement in bandwidth is due to the major decrease in efficiency: a 5% efficient antenna has 20 times the bandwidth of 100% efficient antenna. Simulations were performed with HFSS, a finite-element simulation code. However, these simulations, and all others, are of little value unless the detailed fine structure of all the metallic and dielectric pieces that make up the thin shell is simulated.

All evidence to date points to negative parameter properties existing under plane wave excitation. Furthermore, these properties appear to exist only for regular structures large in wavelengths. The mutual coupling in such structure, which is an integral part of their performance, is well represented by Floquet modes in the periodic structure. When there is no unit cell, and the metamaterial structure is small in wavelengths, the mutual coupling among internal components is vastly different. We believe that this in essence removes any negative properties.

Early studies on metamaterial antennas assumed no dispersion. But of course any material exhibiting any negative parameters must be dispersive, and the dispersion can significantly reduce bandwidth, and can sometimes increase loss markedly. Karawas and Collin (2008) use a more appropriate dispersion model and showed that without dispersion the structure can exhibit a non-Foster reactance, and that with dispersion this reactance disappears, and the magical properties of the thin shell also disappear. See Tretyakov et al. (2005). The Erentok paper also uses the incorrect Q formula: it is shown (Hansen and Collin, 2009) that the TM_1 Q formula is a simple two-term formula and it is exact. When the energy inside the sphere is included as by Thal, all of the Chu formulas have an infinite number of terms.

Much effort and computer time have been spent on metamaterial antennas, but there is a dearth of measurements. The only measurement on the dipole with shell that has been published (Holloway et al., 2007; Greegor et al., 2009) was unfortunately inconclusive. The measurements were poorly planned and badly executed. The test antenna was very small compared to the shell, and the feeding structure was unacceptable. Measurements were made in a reverberation chamber. These chambers are excellent for such arcane measurement problems as multipath, and for EMC measurements on devices where cable terminals are unavailable. To have validity, the antenna and shell should have undergone gain measurements on a compact range, in an anechoic chamber, or as last resort on an outdoor range. From the measured gain, one can adduce efficiency. Bandwidth can also be readily measured under such setups. As a result, in spite of much analytical activity on dipoles in shells, no reputable measurements exist.

REFERENCES

Abramowitz, M. and Stegen, I. A. *Handbook of Mathematical Functions*, National Bureau of Standards, 1970, Section 21.

Adams, A. T. and Lyon, J. A. M. Ferrite Loaded Slot and Traveling Wave Antennas. *USAF Antenna Research and Development Symp.*, Allerton, IL, 1963.

Adams, A. T. et al. Pattern and Efficiency Studies of Miniaturized Slot Antennas Utilizing High "Q" Materials. *USAF Antenna Research and Development Symp.*, Allerton, IL, 1962.

Adams, J. J. and Bernhard, J. T. A Class of Electrically Small Spherical Antennas with Near-Minimum Q. *Antenna Applications Symp.*, Allerton, IL, 2008, pp. 86–103.

Albee, T. K. *Submarine Communications Antenna System,* U.S. Patent No. 3,528,014, June 10, 1966.

Bahr, A. J. On the Use of Active Coupling Networks with Electrically Small Receiving Antennas. *Trans. IEEE*, Vol. AP-25, November 1977, pp. 841–845.

Belrose, J. S. Short Antennas for Mobile Operation—Loading the Whip for Low Frequencies. *QST,* September 1953, pp. 30–35, 108.

Belrose, J. S. Performance Analysis of Small Tuned Transmitting Loop Antenna as Evaluated by Experiment and Simulation. *IEEE AP Mag.*, Vol. 49, June 2007, pp. 128–132.

Best, S. R. On the Performance Properties of the Koch Fractal and Other Bent Wire Monopoles. *Trans. IEEE*, Vol. AP-51, June 2003, pp. 1292–1300.

Best, S. R. A Discussion on the Properties of Electrically Small Self-Resonant Wire Antennas. *IEEE AP Mag.*, Vol. 46, December 2004a, pp. 9–22.

Best, S. R. The Radiation Properties of Electrically Small Folded Spherical Helix Antennas. *Trans. IEEE*, Vol. AP-52, April 2004b, pp. 953–960.

Best, S. R. Low Q Electrically Small Linear and Elliptical Polarized Dipole Antennas. *Trans. IEEE*, Vol. AP-53, March 2005, pp. 1047–1053.

Brown, G. H. A Critical Study of the Characteristics of Broadcast Antennas as Affected by Antenna Current Distribution. *Proc. IRE*, Vol. 24, January 1936, pp. 48–81.

Bulgerin, M. A. and Walters, A. B. *Small Antenna Investigation*, NOLC Report 154, 1954, pp. 67–83.

Carter, P. S. and Beverage, H. H. Early History of the Antennas and Propagation Field Until the End of World War I, Part I—Antennas. *Proc. IRE*, Vol. 50, May 1962, pp. 679–682.

Chen, K. C. and Warne, L. K. A Uniformly Valid Loaded Antenna Theory. *Trans. IEEE*, Vol. AP-40, November 1992, pp. 1313–1323.

Chu, L. J. Physical Limitations of Omni-Directional Antennas. *J. Appl. Phys.*, Vol. 19, December 1948, pp. 1163–1175.

Collin, R. E. Minimum *Q* of Small Antennas. *J. Electromagn. Waves Appl.*, Vol. 12, 1998, pp. 1369–1393.

Collin, R. E. and Rothschild, S. Evaluation of Antenna *Q*. *Trans. IEEE*, Vol. AP-12, January 1964, pp. 23–27.

Copeland, J. R., Robertson, W. J., and Verstraete, R. G. Antennafier Arrays. *Trans. IRE*, Vol. AP-12, March 1964, pp. 227–233.

Crisson, G. Negative Impedances and Twin 21-Type Repeater. *Bell Syst. Tech. J.*, Vol. 10, July 1931, pp. 485–513.

Czerwinski, W. P. On Optimizing Efficiency and Bandwidth of Inductively Loaded Antennas. *Trans. IEEE*, Vol. AP-13, September 1965, pp. 811–812.

Daniel, J. P., Dubost, G., and Rospars, J. Transistor-Fed Thick Folded Dipole with Large Bandwidth at Reception. *Electron. Lett.*, Vol. 11, February 1975, pp. 90–92.

Deschamps, G. and Sichak, W. Microstrip Microwave Antenna. *USAF Antenna Research and Development Symp.*, Allerton, IL, 1953.

Dockon, Inc., Reno, NV, *Microwaves and RF*, May 2010.

Dunlavy, J. H. *Wide Range Tunable Transmitting Loop Antenna*, U.S. Patent No. 3,588,905, June 28, 1971.

Erentok, A. and Ziolkowski, R. W. An Efficient Metamaterial-Inspired Electrically-Small Antenna. *Microwave Opt. Technol. Lett.*, Vol. 49, June 2007, pp. 1287–1290.

Fenwick, R. C. A New Class of Electrically Small Antennas. *USAF Antenna Research and Development Symp.*, Allerton, IL, 1964.

Flachenecker, G. et al. Active Receiving Antennas, *Elektronica en Telecommunicatie 6*, June 16, 1972, pp. ET 74–ET 80.

Foster, R. M. A Reactance Theorem. *Bell Syst. Tech. J.*, Vol. 3, April 1924, pp. 259–267.

Fournier, M. and Pomerleau, A. Experimental Study of an Inductively Loaded Short Monopole Antenna. *Trans. IEEE*, Vol. VT-27, February 1978, pp. 1–6.

Frost, A. D. Parametric Amplifier Antenna. *Proc. IRE*, Vol. 48, June 1960, pp. 1163–1164.

Galejs, J. Dielectric Loading of Electric Dipole Antennas. *J. Res. NBS D: Radio Propag.*, Vol. 66D, September–October 1962, pp. 557–562.

Goubau, G. Multi-Element Monopole Antennas. *Proc. Workshop on Electrically Small Antennas*, USARO, Ft. Monmouth, NJ, May 6–7, 1976, pp. 63–67.

Greegor, R. B. et al. Demonstration of Impedance Matching Using a mu-Negative (MNG) Metamaterial. *IEEE AWPL*, Vol. 8, 2009, pp. 92–95.

GRIMES, D. M. AND GRIMES, C. A. Power in Modal Radiation Fields: Limitations of the Complex Poynting Theorem and the Potential for Electrically Small Antennas. *J. Electromagn. Waves Appl.*, Vol. 11, 1997, pp. 1721–1747.

GRIMES, D. M. AND GRIMES, C. A. Radiation *Q* of Dipole-Generated Fields. *Radio Sci.*, Vol. 34, March–April 1999, pp. 281–296.

HANSEN, R. C. Formulation of Echelon Dipole Mutual Impedance for Computer. *Trans. IEEE*, Vol. AP-20, November 1972, pp. 780–781.

HANSEN, R. C. Efficiency Transition Point for Inductively Loaded Monopole. *Electron. Lett.*, Vol. 9, March 1973, pp. 117–118.

HANSEN, R. C. Efficiency and Matching Tradeoffs for Inductively Loaded Short Antennas. *Trans. IEEE*, Vol. COM-23, April 1975a, pp. 430–435.

HANSEN, R. C. Optimum Inductive Loading of Short Whip Antennas. *Trans. IEEE*, Vol. VT-24, May 1975b, pp. 21–29.

HANSEN, R. C. Wideband Dipole Arrays Using Non-Foster Coupling. *Microwave Opt. Technol. Lett.*, Vol. 38, September 25, 2003, pp. 453–455.

HANSEN, R. C. Linear Connected Arrays. *IEEE Antenn. Wireless Propag. Lett.*, Vol. 3, 2004a, pp. 154–156.

HANSEN, R. C. Non-Foster and Connected Planar Arrays. *Radio Sci.*, Vol. 39, 2004b, RS4004.

HANSEN, R. C. *Electrically Small, Superdirective, and Superconducting Antennas*, Wiley, 2006.

HANSEN, R. C. *Q* and Bandwidth of Electrically Small Antennas. *Microwave Opt. Technol. Lett.*, Vol. 49, May 2007, pp. 1170–1171.

HANSEN, R. C. Phased Array Antennas, Second Edition, Wiley, 2009.

HANSEN, R. C. AND BURKE, M. Antennas with Magneto-Dielectrics. *Microwave Opt. Technol. Lett.*, Vol. 26, July 2000, pp. 75–77.

HANSEN, R. C. AND COLLIN, R. E. A New Chu Formula for *Q*. *IEEE AP Mag.*, Vol. 51, October 2009, pp. 38–41.

HARRIS, A. D. AND MYERS, G. A. *An Investigation of Broadband Miniature Antennas,* Naval Postgraduate School Report 52MV8091A, September 1968.

HARRISON, JR., C. W. Monopole with Inductive Loading. *Trans. IEEE*, Vol. AP-11, July 1963, pp. 394–400.

HIRASAWA, K. AND HANEISHI, M. *Analysis, Design, and Measurement of Small and Low-Profile Antennas,* Artech House, Dedham, MA, 1991, Chapter 5.

HOLLOWAY, C. L. ET AL. Metamaterials and Metafilms: Overview and Applications. *IEEE EMC Symp.*, 2007.

JAISSON, D. Simple Model for the Input Impedance of a Wire Monopole Radiator with a Dielectric Coat. *IET Microwaves Antenn. Propag.*, Vol. 2, 2008, pp. 316–323.

JAMES, J. R. AND BURROWS, R. M. Resonance Properties of Dielectric-Loaded Short Unipoles. *Electron. Lett.*, Vol. 9, 1973, p. 300.

JORDAN, E. C. Final Engineering Report, Contract W 33-038-ac-20778(20018), February 1953, pp. 30–71.

JORDAN, E. C. AND HANSEN, R. C. Ferrite Antennas for High Speed Aircraft. *USAF Antenna Research and Development Symp.*, Allerton, IL, 1952.

KARAWAS, G. K. AND COLLIN, R. E. Spherical Shell of ENG Metamaterial Surrounding a Dipole Antenna. *Proc. IEEE MILCOM*, 2008, pp. 1–7.

KILDAL, P.-S. Comments on "Application of Double Negative Materials to Increase the Power Radiated by Electrically Small Antennas". *Trans. IEEE*, Vol. AP-54, February 2006, p. 766.

KIM, O. S., BREINBJERG, O., AND YAGHJIAN, A. D. Electrically Small Magnetic Dipole Antennas with Quality Factors Approaching the Chu Lower Bound. *Trans. IEEE*, Vol. AP-58, June 2010, pp. 1898–1906.

LAMENSDORF, D. An Experimental Investigation of Dielectric-Coated Antennas. *Trans. IEEE*, Vol. AP-15, November 1967, pp. 767–771.

LANDSTORFER, F. M. AND MEINKE, H. H. Transistorized Microwave Antenna with 1 GHz Centre Frequency. *European Microwave Conf.*, Stockholm, August 1971, pp. B7/4: 1–4.

LIN, C. J., NYQUIST, D. P., AND CHEN, K. M. Short Cylindrical Antennas with Enhanced Radiation or High Directivity. *Trans. IEEE*, Vol. AP-18, July 1970, pp. 576–580.

LINVILL, J. G. Transistor Negative-Impedance Converters. *Trans. IRE*, Vol. 41, June 1953, pp. 725–729.

MACLEAN, T. S. M. AND MORRIS, G. Short Range Active Transmitting Antenna with Very Large Height Reduction. *Trans. IEEE*, Vol. AP-23, March 1975, pp. 286–287.

MACLEAN, T. S. M. AND RAMSDALE, P. A. Signal/Noise Ratio for Short Integrated Antennas. *Electron. Lett.*, Vol. 11, February 1975, pp. 62–63.

MANDELBROT, B. B. *Fractals: Form, Chance and Dimension*, W. H. Freeman, 1977.

MANDELBROT, B. B. *The Fractal Geometry of Nature*, W. H. Freeman, 1982.

MASTERS, R. W. Small Tunable Cavity Antenna for Nominally Circular Polarization. *USAF Antenna Research and Development Symp.*, Allerton, IL, 1952.

MEINKE, H. H. Transistors Integrated with Electrically Small Radiators. *USAF Antenna Research and Development Symp.*, Allerton, IL, 1966.

MEINKE, H. H. AND LANDSTORFER, F. M. Noise and Bandwidth Limitations with Transistorized Antennas. *IEEE AP Symp.*, 1968, pp. 245–246.

MERRILL, JR., J. L., ROSE, A. F., AND SMETHURST, J. O. Negative Impedance Telephone Repeaters. *Bell Syst. Tech. J.*, Vol. 33, September 1954, pp. 1055–1092.

PEDINOFF, M. E. The Negative-Conductance Slot Amplifier. *Trans. IRE*, Vol. MTT-9, November 1961, pp. 557–566.

POGGIO, A. J. AND MAYES, P. E. Bandwidth Extension for Dipole Antennas by Conjugate Reactance Loading. *Trans. IEEE*, Vol. AP-19, July 1971, pp. 544–547.

POLK, C. Resonance and Supergain Effects in Small Ferromagnetically or Dielectrically Loaded Biconical Antennas. *Trans. IRE*, Vol. AP-7, December 1959, pp. S414–S423.

POLYDOROFF, W. J. *Antenna System for Wireless Communication*, U.S. Patent No. 2,266,262, December 16, 1941.

POPOVIC, B. D. *CAD of Wire Antennas and Related Radiating Structures*, Research Studies Press/Wiley, 1991.

POPOVIC, B. D., DRAGOVIC, M. B., AND DJORDJEVIC, A. R. *Analysis and Synthesis of Wire Antennas*, Research Studies Press/Wiley, 1982.

QUIRIN, J. D. *A Study of High-Frequency Solid-State Negative-Impedance Converters for Impedance Loading of Dipole Antennas*, MS Thesis, University of Illinois, Urbana, 1971.

RAYMOND, R. C. AND WEBB, W. Radiation Resistances of Loaded Antennas. *J. Appl. Phys.*, Vol. 20, August 1948, pp. 328–330.

RICHMOND, J. H. AND NEWMAN, E. H. Dielectric Coated Wire Antennas. *Radio Sci.*, Vol. 11, January 1976, pp. 13–20.

SKAHILL, G., RUDISH, R. M., AND PIERO, J. A. Electrically Small, Efficient, Wide-Band, Low-Noise Antenna Elements. *Antenna Applications Symp.*, Allerton, IL, 1998, pp. 214–231.

Smith, C. E. and Johnson, E. M. Performance of Short Antennas. *Proc. IRE*, Vol. 35, October 1947, pp. 1026–1038.

Storer, J. E. Impedance of Thin-Wire Loop Antennas. *Trans. AIEE, Part 1: Commun. Electron.*, Vol. 75, November 1956, pp. 606–619.

Stuart, H. R. Eigenmode Analysis of Small Multielement Spherical Antennas. *Trans. IEEE*, Vol. AP-56, September 2008, pp. 2841–2851.

Stuart, H. R. and Tran, C. Subwavelength Microwave Resonators Exhibiting Strong Coupling to Radiation Modes. *Appl. Phys. Lett.*, Vol. 87, 2005, 151108.

Stuart, H. R. and Tran, C. Small Spherical Antennas Using Arrays of Electromagnetically Coupled Planar Elements. *IEEE Antenn. Wireless Propag. Lett.*, Vol. 6, 2007, pp. 7–10.

Stuart, H. R., Best, S. R., and Yaghjian, A. D. Limitations in Relating Quality Factor to Bandwidth in a Double Resonance Small Antenna. *IEEE Antenn. Wireless Propag. Lett.*, Vol. 6, 2007, pp. 460–463.

Sylusar, V. I. 60 Years of Electrically Small Antenna Theory. *Int. Conf. on Antenna Theory and Techniques*, Sevastopol, Ukraine, September 17–21, 2007, pp. 116–118.

Thal, Jr., H. L. New Radiation *Q* Limits for Spherical Wire Antennas. *Trans. IEEE*, Vol. AP-54, October 2006, pp. 2757–2763.

Tretyakov, S. A. et al. The Influence of Complex Material Coverings on the Quality Factor of Simple Radiating Systems. *Trans. IEEE*, Vol. AP-53, March 2005, pp. 965–970.

Turner, E. M. Electrically Small Antennas Made Easy. *USAF Antenna R&D Program*, Allerton, IL, 1970.

Turner, E. M. Broadband Passive Electrically Small Antennas for TV Application. *Antenna Applications Symp.*, Allerton, IL, 1977.

Turner, E. M. and Richard, D. J. Development of an Electrically Small Broadband Antenna. *USAF Antenna Research and Development Symp.*, Allerton, IL, October 1968.

Van Suchtelen, H. Ferroxcube Aerial Rods. *Electron. Appl. Bull.*, Vol. 13, No. 6, 1954, pp. 88–100.

Wait, J. R. Receiving Properties of a Wire Loop with a Spherical Core. *Can. J. Technol.*, Vol. 31, January 1953, pp. 9–14.

Watt, A. D. *VLF Radio Engineering*, Pergamon Press, 1967, Table 2.8.1.

Wheeler, H. A. Fundamental Limitations of Small Antennas—A Helical Antenna for Circular Polarization. *Proc. IRE*, Vol. 35, December 1947, pp. 1479–1488.

Wheeler, H. A. The Spherical Coil as an Inductor, Shield, or Antenna. *Proc. IRE*, Vol. 46, September 1958, pp. 1595–1602.

Wheeler, H. A. The Radiansphere Around a Small Antenna. *Proc. IRE*, Vol. 47, August 1959, pp. 1325–1331.

Ziolkowski, R. W. and Erentok, A. Metamaterial-Based Efficient Electrically Small Antennas. *Trans. IEEE*, Vol. AP-54, July 2006, pp. 2113–2130.

Ziolkowski, R. W. and Erentok, A. At and Below the Chu Limit: Passive and Active Broad Bandwidth Metamaterial-Based Electrically Small Antennas. *IET Microwaves Antenn. Propag.*, Vol. 1, 2007, pp. 116–128.

Ziolkowski, R. W. and Kipple, A. D. Application of Double Negative Materials to Increase the Power Radiated by Electrically Small Antennas. *Trans. IEEE*, Vol. AP-51, October 2003, pp. 2626–2640.

APPENDIX B

DEFINITIONS OF TERMS USEFUL TO ESA

Bandwidth	The frequency range over which the impedance mismatch factor does not exceed a specified value. Common values are half-power (VSWR ≤ 5.828) and 88% power (VSWR ≤ 2). Bandwidth is often expressed as a percent; the frequency range is divided by the center frequency.
Broadband antenna	An antenna whose bandwidth is roughly an octave or more.
Directivity	The ratio of the peak of the radiation intensity to the radiation intensity averaged over all directions.
Efficiency	The ratio of peak radiation intensity to the one that would be obtained if conduction and dielectric losses were absent.
ESA	An electrically small antenna; usually taken as fitting in a sphere of diameter $\lambda/4\pi = 0.0796\lambda$.
Gain	Directivity reduced by conduction and dielectric losses, and by impedance mismatch losses. *Note*: The IEEE definition does not include mismatch losses; industry uses the definition given above.
Radian sphere	A hypothetical sphere of radius $\lambda/2\pi$; the sphere encloses the antenna. *Note*: ESA usually use a radian sphere whose radius is half that above.

Small Antenna Handbook, Robert C. Hansen and Robert E. Collin.

Superdirective antenna	An antenna, usually an array, with directivity higher than that of the same array with all elements uniformly excited.
Supergain antenna	The low radiation resistance of most superdirective antennas produces low efficiency, so that supergain is usually not realized.
TE, TM modes	Electric and magnetic fields that can be expressed as a spherical mode. These utilize spherical Bessel and Hankel functions, associated Legendre functions, and trig functions.

APPENDIX C

SPHERICAL SHELL OF ENG METAMATERIAL SURROUNDING A DIPOLE ANTENNA

GEORG K. KARAWAS
US Army CERDEC S&TCD, Fort Monmouth, NJ, USA

ROBERT E. COLLIN
Case Western Reserve University, Cleveland, OH, USA

C.1 INTRODUCTION

In recent years, considerable interest has arisen with respect to the application of materials with negative permittivity and/or negative permeability to various electromagnetic devices. Ideal lossless and nondispersive materials with negative permittivity and permeability do not exist in nature. Artificial materials with these properties are referred to as metamaterials (Smith et al., 2000).

Recently, several papers have appeared that consider the use of a thin spherical shell of metamaterial surrounding a small dipole antenna for the purpose of improving its electrical performance (Ziolkowski and Kipple, 2003; Ziolkowski and Erentok, 2006; Tretyakov et al., 2005).

In Ziolkowski and Kipple (2003), the thin spherical shell consisted of ideal double negative material with permittivity $\varepsilon = -\varepsilon_0$ and permeability $\mu = -\mu_0$. In Ziolkowski and Erentok (2006), the spherical shell had a negative permittivity (epsilon negative or ENG material) with $\varepsilon = -3\varepsilon_0$; the solution for the electromagnetic field was obtained numerically.

In both of the papers mentioned above, it was found that the presence of a thin spherical shell surrounding the small dipole antenna resulted in a sharp resonance effect when the radial thickness of the shell was varied. It was also found that the frequency response of the dipole antenna showed a resonance effect but the resonance curve half-power bandwidth was not nearly as sharp.

Small Antenna Handbook, Robert C. Hansen and Robert E. Collin.

The effects uncovered were somewhat unexpected and, unfortunately, the numerical solutions did not provide a physical insight into the fundamental causes for the observed resonances. For this reason, we have reconsidered the dipole antenna surrounded by a thin shell of ENG material.

We use the field equivalence principle to construct a constant current source consisting of a current sheet on the surface of a perfect magnetic conducting sphere, and its dual, a constant voltage source consisting of a sheet of magnetic current on the surface of a perfect electric conducting sphere.

We then apply spherical transmission line theory, which leads to a straightforward exact analytical solution for the problem involving only TM_{10} modes. The use of transmission line theory makes it unnecessary to solve a system of simultaneous equations, such as that used in the Special Issue of the *IEEE Transactions on Microwave Theory and Applications* (2005).

Our analytical solution, based on transmission line theory, provides a physical insight into why the resonances occur. Within the spherical ENG shell, the formal expression that usually represents the time-averaged electric field energy in the complex Poynting vector theorem, now in the context of the interpretation of the complex Poynting vector theorem, represents inductive energy. Outside the ENG shell, the reactive energy in the near-zone field region of the TM_{10} mode is predominantly capacitive field energy. Thus, for a certain critical thickness of the ENG shell, a condition of resonance can occur whenever the stored capacitive and inductive energies are equal.

The transmission line solution also shows that the inner and outer boundaries between the ENG shell and free space are highly reflective. The power transmission across these boundaries is predominantly reactive power with relatively little real power. The shell acts like a high-Q resonator.

Our analysis also explains why the frequency response of this system shows a much broader impedance bandwidth than would be expected based on the high energy-based Q of the system.

We construct an equivalent circuit with lumped elements consisting of a large capacitive reactance in series with a small inductive reactance and a small series radiation resistance to represent the field conditions outside the shell.

The circuit elements representing the field behavior within the shell are a large negative capacitance arising from the negative permittivity, along with a small inductive reactance arising from the relatively small amount of normal stored magnetic energy within the shell.

The negative capacitance can also be viewed as a non-Foster inductive reactance having an inverse dependence on frequency. Such a reactance results in an equivalent impedance-related Q that is as small as a few percent of the intrinsic energy-based Q.

There is no simple relationship between the energy-based Q and the impedance half-power bandwidth when non-Foster reactance elements are present. The much greater impedance half-power bandwidth does not represent a violation of the Chu lower bound on the energy-based Q. The equivalent parallel resonant circuit is similar to the classical parallel *RLC* circuit.

The small radiation resistance r_a can be represented as a large parallel resistance having the classical value of $Q^2 r_a$, where Q is the energy-based Q. It is found that the equivalent parallel resistance can be as large as 600,000 Ω or more because the energy-based Q in examples considered is around 1000. Likewise, the equivalent reactance can be very large and passes through zero at resonance.

Our analysis does not reveal any useful set of conditions that would yield a reasonable value for the radiation resistance and a small radiation reactance. Thus, the thin ENG shell does not appear to be useful for the purpose of providing an acceptable impedance match for the small dipole antenna. See also Tretyakov et al. (2005).

C.2 IMPEDANCE TRANSFORMATION ALONG A SPHERICAL TRANSMISSION LINE

We can view space as a spherical transmission line and will consider the TM_{10} mode for which the transverse magnetic field is given by

$$H_\phi = C_1 h_1^2(kr) + C_2 h_1^1(kr) \tag{C.1}$$

where h_1^1 and h_1^2 are, respectively, the spherical Hankel functions of the first and second kind and of order 1. These are defined by

$$h_1^1(\rho) = -\left(\frac{j}{\rho^2} + \frac{1}{\rho}\right)e^{j\rho} \quad \text{and} \quad h_1^2(\rho) = \left(\frac{j}{\rho^2} - \frac{1}{\rho}\right)e^{-j\rho} \tag{C.2}$$

where $k = \omega\sqrt{\mu\varepsilon}$, $Z = \omega\sqrt{\mu/\varepsilon}$, and $\rho = kr$.

We have omitted throughout a factor $\sin\theta$ that describes the variation of the field with the polar angle. The transverse electric field is given by

$$E_\theta = [C_1 Z_w^+(\rho) h_1^2(\rho) + C_2 Z_w^-(\rho) h_1^1(\rho)] \tag{C.3}$$

where the wave impedances $Z_w^\pm$ are now defined by

$$Z_w^+(\rho) = \frac{1-\rho^2+j\rho}{j\rho-\rho^2} Z \tag{C.4a}$$

$$Z_w^-(\rho) = \frac{1-\rho^2-j\rho}{\rho^2+j\rho} Z = -(Z_w^+)^* \tag{C.4b}$$

Moreover, we can express C_2 in Equations C.1 and C.3 as $\Gamma_2 C_1$, where Γ_2 can be considered to be a reflection coefficient.

Equations C.1 and C.3 are similar to the transmission line equations for voltage and current but now the wave impedances depend on the variable ρ as well.

Consider now a problem for which the medium beyond $r = r_2$ is free space and the medium for r less than r_2 is a dielectric. Let k and k_0 be the propagation constants in the dielectric medium and free space, respectively, and Z_0 the intrinsic impedance of free space. Equations C.1 and C.3 represent the incident and reflected waves in the dielectric medium. In the free space region beyond $r = r_2$, there will be only outgoing waves, so we will have

$$H_\phi(k_0 r) = C_3 h_1^2(k_0 r) \tag{C.5a}$$

$$E_\theta = C_3 Z_w^+(k_0 r) h_1^2(k_0 r) \tag{C.5b}$$

The boundary conditions require equality of all tangential fields on either side of the common boundary at $r = r_2$, which implies that their respective ratios, defined as the *total wave impedance*, must also match:

$$\frac{Z_w^+(kr_2)h_1^2(kr_2) + \Gamma_2 Z_w^-(kr_2)h_1^1(kr_2)}{h_1^2(kr_2) + \Gamma_2 h_1^1(kr_2)} = Z_w^+(k_0 r_2) \tag{C.6}$$

which can be solved for Γ_2 to give

$$\Gamma_2 = \frac{Z_w^+(kr_2) - Z_w^+(k_0 r_2)}{Z_w^+(k_0 r_2) - Z_w^-(kr_2)} \frac{h_1^2(kr_2)}{h_1^1(kr_2)} \tag{C.7}$$

Once Γ_2 is evaluated, the amplitude C_3 can be determined in terms of C_1 from Equations C.1 and C.5a:

$$C_3 = \frac{h_1^2(kr_2) + \Gamma_2 h_1^1(kr_2)}{h_1^2(k_0 r_2)} C_1$$

On a second interface at $r = r_1$, we can express the transverse fields at r_1 by simply setting $r = r_1$ in Equations C.1 and C.3. The ratio of the transverse electric field to the transverse magnetic field at r_1 defines the total wave impedance at r_1.

If we have an incident and a reflected field in the region $r < r_1$, we can repeat the matching of the total wave impedances at this interface. We can then solve for the reflection coefficient at this interface as well as all the field amplitudes, except that of the incident wave in the region $r < r_1$, which is determined by a boundary condition at the source. There will be some straightforward algebra involved but the overall procedure based on transmission line analysis is quicker and easier than solving a system of simultaneous equations, which for the problem described above involves four unknown amplitudes that must be found in terms of that of the incident wave in the region $r < r_1$.

When there are several interfaces present, the total transverse field wave impedances are equated at each interface and the corresponding reflection coefficients solved for. The reflected and transmitted wave amplitudes at each interface are then easily solved for. This process is carried through until the input region is arrived at. In the input region, the only remaining unknown amplitude is that of the incident wave and this is determined by a boundary condition at the source. The use of transmission line theory avoids the need to solve a large system of simultaneous equations and also gives a much clearer interpretation of the solution in terms of incident, reflected, and transmitted waves.

C.3 WAVE SOLUTION WITHIN A SPHERICAL SHELL OF ENG MATERIAL

We now consider the wave solution within a spherical shell of material with a negative dielectric permittivity as shown in Figure C.1. The inner radius of the shell is r_1 and the outer radius is r_2. The shell material is referred to as epsilon negative material or ENG material (**E**psilon **N**e**G**ative). The propagation constant for this medium is given by

$$\omega\sqrt{\varepsilon\mu_0} = \omega\sqrt{-|\varepsilon|\mu_0} = -jk \tag{C.8}$$

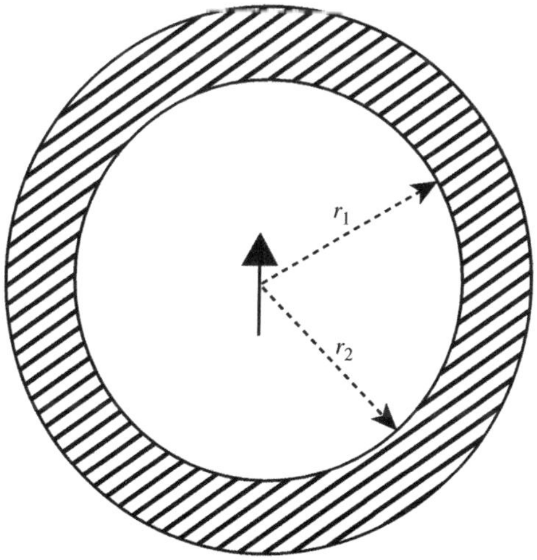

FIGURE C.1 A thin spherical shell of ENG material surrounding a small dipole antenna.

The equations for the field are the same as those given by Equations C.1 and C.2 provided we replace k by $-jk$. When this is done, we find that

$$h_1^2(-jk) = -j\left(\frac{1}{\rho^2} + \frac{1}{\rho}\right)e^{-kr} = f^+(\rho)$$

and

$$h_1^1(-jk) = j\left(\frac{1}{\rho^2} - \frac{1}{\rho}\right)e^{kr} = f^-(\rho)$$

where $\rho = kr$ and the new wave functions are represented by the symbols $f^+(\rho)$ and $f^-(\rho)$. The new expressions for the wave functions and wave impedances become

$$f^+(kr) = h_1^2(-jkr) \tag{C.9a}$$

$$f^-(kr) = h_1^1(-jkr) \tag{C.9b}$$

$$Z^+(\rho) = jZ\frac{1+\rho^2+\rho}{\rho^2+\rho} \tag{C.9c}$$

$$Z^-(\rho) = -jZ\frac{1+\rho^2-\rho}{\rho^2-\rho} \tag{C.9d}$$

where

$$\sqrt{\frac{\mu_0}{\varepsilon}} = j\sqrt{\frac{\mu_0}{|\varepsilon|}} = jZ$$

Note that in the ENG material the wave impedances are pure imaginary and the waves are evanescent.

We can solve for Γ_2 as outlined earlier or by making appropriate substitutions in Equation C.7. We replace the wave impedances $Z_W^+(kr_2)$ by $Z^+(kr_2)$ and $Z_W^-(kr_2)$ by $Z^-(kr_2)$.

In addition, the ratio of the two Hankel functions in Equation C.7 is replaced by the ratio $f^+(kr_2)/f^-(kr_2)$. For this problem, the solution for Γ_2 is

$$\Gamma_2 = \frac{Z^+(kr_2) - Z_W^+(k_0r_2)f^+(kr_2)}{Z_W^+(k_0r_2) - Z^-(kr_2)f^-(kr_2)} \tag{C.10}$$

At the inner surface of the ENG shell, $r = r_1$. The total wave impedance at this surface is the ratio of the transverse electric field to the transverse magnetic field at $r = r_1$. We have

$$H_\phi(kr_1) = C_1[f^+(kr_1) + \Gamma_2 f^-(kr_1)] \tag{C.11a}$$

$$E_\theta(kr_1) = C_1[Z^+(kr_1)f^+(kr_1) + \Gamma_2 Z^-(kr_1)f^-(kr_1)] \tag{C.11b}$$

We will call the total wave impedance at this inner surface $Z_{in}(kr_1)$. Using Equation C.11, we obtain

$$Z_{in}(kr_1) = \frac{Z^+(kr_1)f^+(kr_1) + \Gamma_2 Z^-(kr_1)f^-(kr_1)}{f^+(kr_1) + \Gamma_2 f^-(kr_1)} \tag{C.12}$$

If we have a small dipole antenna located at the origin and within the region $r \le r_1$, then the impedance presented to this dipole antenna at the surface at $r = r_1$ is that given by Equation C.12. The purpose of using a thin spherical shell of ENG material is to try and present a more favorable impedance to the dipole antenna so as to enhance the radiation from this antenna.

C.3.1 Power Transmission in ENG Media

Consider a field consisting of a forward and backward evanescent wave in ENG media. We assume that it consists of a TM_{10} mode only and we will omit the $\sin\theta$ factor. The transverse magnetic and electric fields are expressed as follows:

$$H_\phi = C[f^+ + \Gamma_2 f^-]$$
$$E_\theta = C[jX^+ f^+ + \Gamma_2 jX^- f^-]$$

where C is an amplitude factor and the wave impedances are

$$Z^+(kr) = jZ\frac{1 + k^2r^2 + kr}{k^2r^2 + kr} = jX^+$$

$$Z^-(kr) = -jZ\frac{1 + k^2r^2 - kr}{k^2r^2 - kr} = jX^-$$

Note that both X^+ and X^- are positive for kr less than 1. The reflection coefficient will be expressed as $\Gamma_2 = \Gamma_r + j\Gamma_i$. The complex Poynting vector is proportional to

$$|C|^2[jX^+f^+ + j\Gamma_2 X^- f^-][f^+ + \Gamma_2 f^-]^*$$

When we multiply by 1/2, restore the $\sin\theta$ factors, and integrate over the surface of a sphere with radius r, we obtain

$$P = |C|^2[jX^+f^+ + j\Gamma_2 X^- f^-][f^+ + \Gamma_2 f^-]^* \frac{4\pi r^2}{3} \tag{C.13}$$

From the definitions for f^+ and f^-, we see that $jX^+f^+ = |X^+f^+|$ and $jX^-f^- = -|X^-f^-|$. The real and imaginary parts of the Poynting vector are thus given by

$$\begin{aligned} P_\mathrm{r} &= \frac{4\pi r^2}{3}|C|^2\Gamma_\mathrm{i}|f^+f^-|(X^- - X^+) \\ &= |C|^2\Gamma_\mathrm{i}|f^+f^-|\frac{8\pi k^2 r^4}{3(1-k^2r^2)} = |C^2|\Gamma_\mathrm{i}\frac{8\pi}{3k^2}Z_0 \end{aligned} \tag{C.14a}$$

$$\begin{aligned} P_\mathrm{i} &= \frac{4\pi r^2}{3}|C^2|\left[X^+|f^+|^2 - \Gamma_\mathrm{r}(X^+ + X^-)|f^+f^-| + (\Gamma_\mathrm{r}^2 + \Gamma_\mathrm{i}^2)X^-|f^-|^2\right] \\ &\approx \frac{16\pi}{3}|C|^2\frac{1}{k^3 r}Z_0 \end{aligned} \tag{C.14b}$$

Note that the real power across any spherical surface of radius r is a constant as it must be in order to conserve power. For a single outward or inward propagating evanescent wave, there is no real power flow because the wave impedances are pure imaginary. Real power transfer between the input interface and the output interface occurs only if there is a reflected wave. Furthermore, the real power depends directly on the imaginary part of the reflection coefficient Γ_2. Since Γ_r is very close to unity, and Γ_i is very small as calculations show, we set it equal to 1 and also set $\Gamma_\mathrm{i} = 0$ to obtain the approximate expression for the reactive power. The ratio of real power transmitted to the reactive power is given approximately by $\Gamma_\mathrm{i}(kr)/2$, which is very small. When $r = r_2$, there will be very little real power transmitted across the interface compared with the amount of reactive power. This is an unfavorable situation as far as trying to improve the amount of radiated power from a small dipole antenna. The only parameter in the above analysis that is controlled by the source and input interface between free space and the DNG shell is the amplitude C of the field transmitted into the DNG shell.

C.4 DIPOLE ANTENNA SOURCE

In order to make a realistic evaluation of the effect of a thin spherical shell of ENG material on the radiation properties of a small dipole antenna, we must specify a realistic model of a dipole antenna. We could consider a small thin wire antenna with a small gap at the center. The boundary value problem of solving for the current and radiation from this type of antenna must be done numerically. Furthermore, this type of dipole would require that we include many spherical mode functions in the solution. The analysis would become very complex and would require a numerical solution, which has the disadvantage of obscuring what is happening physically. If we could use a suitable current distribution on the surface of a small sphere of radius a, we would have a problem with the required spherical symmetry that would allow an analytical solution involving only spherical TM_{10} modes to be constructed. We can obtain an equivalent source with the desired properties by use of a field equivalence principle first stated by Schelkunoff (1936). We assume that the surface of the sphere with radius a is a perfect magnetic conductor. Just outside this surface, we place an electric current sheet with a density equal to the total tangential magnetic field, both the primary field and the scattered field. Since the tangential magnetic field must vanish on the perfect magnetic conducting surface, this current sheet does support the total magnetic field in the space exterior to the sphere of radius a. In order to relate the current sheet density to a real source, we will assume that the real source is an infinitesimal Hertzian dipole at the origin that radiates a tangential magnetic field given by

$$H_\phi = A_p \sin\theta \, h_1^2(k_0 r)$$

where A_p is the known primary field amplitude, which is determined by the current in the Hertzian dipole, and $h_1^2(k_0 r)$ is the spherical Hankel function of the second kind and order 1. We choose a small sphere of radius a, which is a perfect magnetic conductor, on which we specify an electric current sheet of density given by

$$\mathbf{J}(\theta) = -J_0 \sin\theta \, \mathbf{a}_\theta = -A_p h_1^2(k_0 a) \sin\theta \, \mathbf{a}_\theta \qquad \text{(C.15)}$$

With this known specified current distribution, we can match the total tangential magnetic field—the sum of the primary and scattered fields—to the current density on the sphere using only the TM_{10} mode. The required boundary condition is

$$\mathbf{a}_r \times \mathbf{H} = \mathbf{J} \qquad \text{(C.16a)}$$

which is equivalent to

$$H_\phi = J_0 \qquad \text{(C.16b)}$$

where we have omitted the common factor $\sin\theta$.

An alternative source that also leads to an easy analytical solution is the use of a magnetic current sheet on a perfectly conducting metal sphere of radius a. This corresponds to using a specified voltage to drive the dipole antenna. The solution to this problem is presented later on in this paper.

In the region $a \le r \le r_1$, we must include both an outward and an inward propagating spherical wave since we must match the total field wave impedances at the surface $r = r_1$ and also satisfy the boundary condition given in Equation C.16b above.

Thus, we assume that

$$H_\phi = A\left[h_1^2(k_0 r) + \Gamma_1 h_1^1(k_0 r)\right] \qquad \text{(C.17a)}$$

$$E_\theta = A\left[Z_w^+(k_0 r) h_1^2(k_0 r) + \Gamma_1 Z_w^-(k_0 r) h_1^1(k_0 r)\right] \qquad \text{(C.17b)}$$

where A is an unknown amplitude constant and Γ_1 is an unknown reflection coefficient.

At $a = r_1$, these tangential fields must match those given by Equation C.11a. Note that the amplitude A is different from that of the primary field because the presence of a spherical shell outside the source region changes the amplitude of the outward radiated field from the source.

The reflection coefficient can be determined by using the impedance matching condition at the surface $r = r_1$, which requires that

$$\frac{E_\theta}{H_\phi} = \frac{Z_w^+(k_0 r_1) h_1^2(k_0 r_1) + \Gamma_1 Z_w^-(k_0 r_1) h_1^1(k_0 r_1)}{h_1^2(k_0 r_1) + \Gamma_1 h_1^1(k_0 r_1)} = Z_{in}(k r_1)$$

From this equation, we obtain

$$\Gamma_1 = \frac{Z_w^+(k_0 r_1) - Z_{in}(k r_1)}{Z_{in}(k r_1) - Z_w^-(k_0 r_1)} \frac{h_1^2(k_0 r_1)}{h_1^1(k_0 r_1)} \tag{C.18}$$

The remaining boundary condition at the source can be stated as

$$H_\phi = A\left[h_1^2(k_0 a) + \Gamma_1 h_1^1(k_0 a)\right] = J_0 \tag{C.19}$$

from which the constant A is determined in terms of the specified source strength. If the reflection coefficient Γ_1 is zero, there is no scattered field and the amplitude A becomes equal to that of the primary field.

C.5 ANTENNA RADIATION RESISTANCE AND REACTANCE

At the surface $r = a$, the complex Poynting vector is

$$\begin{aligned}\frac{1}{2} E_\theta H_\phi^* &= \frac{1}{2} E_\theta J_0^* \\ &= \frac{1}{2} \frac{Z_w^+(k_0 a) h_1^2(k_0 a) + \Gamma_1 Z_w^-(k_0 a) h_1^1(k_0 a)}{h_1^2(k_0 a) + \Gamma_1 h_1^1(k_0 a)} J_0 J_0^* \sin^2\theta\end{aligned}$$

where we have restored the $\sin\theta$ factors. We now replace J_0 by $I_0/2\pi a$, which is the total current measured at the equator of the source sphere. Next we multiply by $a^2 \sin\theta$ and integrate over the surface of the sphere. This produces a factor $8\pi a^2/3$. The total complex radiated power is then found to be

$$P = \frac{1}{3\pi} \frac{Z_w^+(k_0 a) h_1^2(k_0 a) + \Gamma_1 Z_w^-(k_0 a) h_1^1(k_0 a)}{h_1^2(k_0 a) + \Gamma_1 h_1^1(k_0 a)} I_0 I_0^* \tag{C.20}$$

Note that, when $a = r_1$, the total wave impedance at the boundary $r = r_1$ is the same on either side of the boundary. Thus, apart from the factor $1/3\pi$ and the current squared, we can substitute for the numerator and denominator the expression Z_{in} given by Equation C.12. This can be verified by substituting for Γ_1 from Equation C.18. For this case, the computation of the radiation resistance and radiation reactance can be found without any further elaborate calculation from the expression

$$P = \frac{1}{3\pi}\frac{Z^+(kr_1)f^+(kr_1)+\Gamma_2 Z^-(kr_1)f^-(kr_1)}{f^+(kr_1)+\Gamma_2 f^-(kr_1)}I_0 I_0^* \tag{C.21}$$

The radiation resistance and reactance calculated from the complex Poynting vector are smaller by a factor $2/3\pi$ than those given by Z_{in} in Equation C.12. We only need the expression given by Equation C.20 when a is less than r_1.

Consider the special case when the small spherical dipole radiates into a free space environment.

In this case $\Gamma_1 = 0$, so the complex radiated power is given by

$$\begin{aligned} P &= \frac{1}{3\pi}Z_w^+(k_0 a)I_0 I_0^* \\ &= \frac{Z_0}{3\pi}\left[\frac{(k_0 a)^2}{1+(k_0 a)^2} - \frac{j}{k_0 a(1+k_0^2 a^2)}\right]I_0 I_0^* \end{aligned} \tag{C.22}$$

The radiation resistance R_a is obtained by equating $\frac{1}{2}I_0 I_0^* R_a$ to the real part of the complex radiated power. For small values of $k_0 a$, we get

$$R_a = 80\pi^2\left(\frac{2a}{\lambda_0}\right)^2 \tag{C.23}$$

where λ_0 is the wavelength. This agrees with the classical expression for the radiation resistance of a small dipole antenna of length $2a$. The capacitive reactance of the small dipole antenna is large since it is proportional to $1/k_0 a$.

C.6 VOLTAGE-DRIVEN DIPOLE ANTENNA

For later reference, we point out that calculations show the real part of Γ_2 to be slightly greater than unity with a small imaginary part, while the real part of Γ_1 is slightly less than unity with a very small imaginary part. In Equation C.21, both f^+ and f^- are pure real quantities for ENG media. Thus, the real part of the denominator in Equation C.21 can vanish. In Equation C.20, the magnitude of the denominator can also become very small when the outer radius r_2 has the same value that makes the real part of the denominator in Equation C.20 vanish. This leads to a very large increase in the radiation resistance. The radiation reactance goes through a very rapid change from large positive values to very large negative values in the vicinity of where the radiation resistance peaks. This is not a normal resonance phenomenon. It is primarily an interface effect that causes a large impedance mismatch between the outer boundary of the ENG shell and the free space region when that outer boundary is located in the near-zone field of the dipole antenna. After transformation back to the surface at $r=r_1$, we can find a critical value of r_2 that makes the real part of the denominator in Equation C.21 for the complex power to vanish. At the point $r=r_1$, the tangential magnetic field, apart from the amplitude factor J_0, is almost zero and the input impedance is almost an open circuit. This makes it very difficult to force a significant current into the dipole antenna. Note that this phenomenon is not dependent on the interface to the ENG shell at $r=r_1$. The inner boundary of the ENG shell could be located at a radius $r_0 < r_1$ and the total wave impedance in the outward direction at $r = r_1$ is still given by Equation C.12. The vanishing of the real part of the denominator does not depend on wave interaction between the inner and outer boundaries of the ENG shell.

If the dipole antenna is driven by a fixed current, then the amount of radiated power will be large. But the voltage that would have to be applied to drive the current into the dipole antenna would be unrealistically large. The local electric field would be large and any losses present in the media near the dipole antenna would result in significant losses. In view of the large increase in the input impedance, it would be more realistic to consider driving the dipole antenna from a specified voltage source.

An equivalent voltage source can be postulated in the form of a magnetic current source outside of but adjacent to the surface of a metal sphere of radius a. The field equivalence principle can be used to relate the required strength of the magnetic current sheet to the strength of the Hertzian dipole at the origin. For this model of a small dipole radiator, we assume that adjacent to the sphere of radius a we have a magnetic current sheet given by

$$\mathbf{J}_{\mathrm{m}} = -\mathbf{a}_\phi J_{\mathrm{m}0} \sin\theta = -\mathbf{a}_\phi A_{\mathrm{p}} Z_{\mathrm{w}}^{+}(k_0 a) h_1^2(k_0 a) \sin\theta \tag{C.24}$$

where A_{p} is the amplitude of the primary electric field radiated by the Hertzian dipole.

This model allows us to also account for the losses associated with the electric current on the metal sphere, which we now assume to have a finite conductivity. The electric field along the surface of the metal sphere is given by

$$E_\theta = J_{\mathrm{s}} Z_{\mathrm{s}} = J_{\mathrm{s}} \frac{1+j}{\sigma \delta_{\mathrm{s}}} \tag{C.25}$$

where J_{s} is the current density equal to $-H_\phi$, σ is the conductivity of the sphere, and δ_{s} is the skin depth. The above equation defines the surface impedance Z_{s}.

For this model of the dipole antenna, the boundary condition at the surface of the small sphere is

$$E_\theta + J_{\mathrm{s}} Z_{\mathrm{s}} = E_\theta + H_\phi Z_{\mathrm{s}} = J_{\mathrm{m}0} \tag{C.26}$$

Just outside the magnetic current source, we have

$$E_\theta = A\left[Z_{\rm w}^+(k_0a)h_1^2(k_0a) + \Gamma_1 Z_{\rm w}^-(k_0a)h_1^1(k_0a)\right] \tag{C.27a}$$

$$H_\phi = A\left[h_1^2(k_0a) + \Gamma_1 h_1^1(k_0a)\right] \tag{C.27b}$$

where we have dropped the $\sin\theta$ factor and A is an unknown amplitude constant. When we use these expressions in Equation C.26, we obtain

$$A = \frac{J_{\rm m0}}{[Z_{\rm w}^+(k_0a) + Z_{\rm s}]h_1^2(k_0a) + \Gamma_1[Z_{\rm w}^-(k_0a) + Z_{\rm s}]h_1^1(k_0a)} \tag{C.28}$$

We will express the equivalent voltage driving the small dipole antenna as being equal to the total magnetic current across the equator of the small sphere, thus

$$V_{\rm m} = 2\pi a J_{\rm m0}$$

The complex radiated power at $r = a$ is obtained from the integral of the Poynting vector over the surface of the sphere. It is given by

$$P = \frac{1}{3\pi}\frac{\left[Z_{\rm w}^+ h_1^2 + \Gamma_1 Z_{\rm w}^- h_1^1\right]\left[h_1^2 + \Gamma_1 h_1^1\right]^*}{\left|[Z_{\rm w}^+ + Z_{\rm s}]h_1^2 + \Gamma_1[Z_{\rm w}^- + Z_s]h_1^1\right|^2} V_{\rm m}V_{\rm m}^* \tag{C.29}$$

All expressions for $Z_{\rm w}^{\pm}$ and $h_1^{1,2}$ are evaluated at k_0a.

Another useful formula for the complex power can be formed by using Equation C.26 to obtain

$$E_\theta H_\phi^* = H_\phi^* J_{\rm m0} - H_\phi H_\phi^* Z_{\rm s} \tag{C.30}$$

The term on the left-hand side is proportional to the outward radiated electromagnetic complex power. The first term on the right-hand side is proportional to the total power supplied by the source. The last term on the right-hand side has a negative sign and gives the complex power delivered to the complex surface impedance of the conducting metal sphere.

We will substitute into the above equation the expressions for the field components as given by Equation C.27a and use Equation C.28 to obtain

$$\frac{1}{2}E_\theta H_\phi^* = \frac{1}{2}\frac{J_{\rm m0}J_{\rm m0}^*\left[h_1^2 + \Gamma_1 h_1^1\right]^*}{\left\{[Z_{\rm w}^+ + Z_s]h_1^2 + \Gamma_1[Z_{\rm w}^- + Z_s]h_1^1\right\}^*} - \frac{1}{2}\frac{J_{\rm m0}J_{\rm m0}^*\left|h_1^2 + \Gamma_1 h_1^1\right|^2 Z_s}{\left|[Z_{\rm w}^+ + Z_s]h_1^2 + \Gamma_1[Z_{\rm w}^- + Z_s]h_1^1\right|^2}$$

Again, all expressions for $Z_{\rm w}^{\pm}$ and $h_1^{1,2}$ are evaluated at k_0a.

We now use $V_{\mathrm{m}} = 2\pi a J_{\mathrm{m0}}$, restore the $\sin\theta$ factors, and integrate over the surface of the sphere of radius $r = a$ to obtain the following alternative expression for the complex power balance equation:

$$P = \frac{1}{3\pi}\frac{\left[h_1^2 + \Gamma_1 h_1^1\right]^* V_m V_{\mathrm{m}}^*}{\left\{[\mathrm{Z}_{\mathrm{w}}^+ + \mathrm{Z}_{\mathrm{s}}]h_1^2 + \Gamma_1[\mathrm{Z}_{\mathrm{w}}^- + \mathrm{Z}_{\mathrm{s}}]h_1^1\right\}^*} - \frac{1}{3\pi}\frac{\left|h_1^2 + \Gamma_1 h_1^1\right|^2 \mathrm{Z}_{\mathrm{s}} V_m V_{\mathrm{m}}^*}{\left|[\mathrm{Z}_{\mathrm{w}}^+ + \mathrm{Z}_{\mathrm{s}}]h_1^2 + \Gamma_1[\mathrm{Z}_{\mathrm{w}}^- + \mathrm{Z}_{\mathrm{s}}]h_1^1\right|^2} \tag{C.31}$$

The left-hand side represents the outward radiated complex power. The first term on the right-hand side represents the total complex power supplied by the voltage source while the last term represents the power dissipated in losses in the conducting sphere and reduces the amount of radiated power.

If we neglect the losses from the current on the metal sphere, that is, set Z_{s} to zero, the above equation simplifies to

$$P = \frac{1}{3\pi}\frac{\left[h_1^2 + \Gamma_1 h_1^1\right]^*}{\left[\mathrm{Z}_{\mathrm{w}}^+ h_1^2 + \Gamma_1 \mathrm{Z}_{\mathrm{w}}^- h_1^1\right]^*} V_m V_{\mathrm{m}}^* \tag{C.32}$$

This expression is the reciprocal of the complex conjugate of that given earlier by Equation C.20 for the current-driven dipole. Hence, the radiation admittance of the dipole antenna is the reciprocal of the complex conjugate of the radiation impedance. This expression shows that, for the voltage-driven dipole, the radiated power becomes very small whenever the numerator becomes small. This means that the admittance of the dipole becomes very small. It would not be useful as an antenna under these conditions.

In the absence of the ENG shell, $\Gamma_1 = 0$. From Equation C.24,

$$V_m = 2\pi a J_{\mathrm{m0}} = A_p 2\pi a \mathrm{Z}_{\mathrm{w}}^+(k_0 a) h_1^2(k_0 a) = \mathrm{Z}_{\mathrm{w}}^+(k_0 a) I_0 \tag{C.33}$$

where the last step was the substitution for A_{p} from Equation C.13 and using $I_0 = 2\pi a J_0$. When we use this result in Equation C.29 with $\Gamma_1 = 0$, the expression for the complex power becomes equal to that given by Equation C.22. Thus, the two different sources lead to consistent results. From Equation C.29, it becomes clear that, whenever the outer radius of the ENG shell is chosen to coincide with the value needed for resonance, the required voltage needed to produce any significant radiated power becomes very large.

Our computations show that for an ENG shell with $\varepsilon = -3\varepsilon_0$, $r_1 = 10$ mm, and $r_2 = 16.0112$ mm at a frequency of 300 MHz, the radiation impedance was $(519{,}685 - j705)\,\Omega$. A minute change in r_2 will make the radiation reactance vanish. Thus, in order to radiate 1 W of power, a voltage equal to 1019.5 V would be required. A small change in the value of r_2 will dramatically reduce the value of the radiation resistance and substantially increase the radiation reactance. The impedance properties of the antenna surrounded by a thin shell of ENG material are so sensitive to small variations in the thickness of the shell that they would pose difficult manufacturing problems. If the thickness and frequency are chosen away from the resonance point, then the impedance properties of the shell enclosed dipole are not significantly better than those for a dipole in free space. The ratio of radiation reactance to radiation resistance is still large and thus poses a difficult impedance matching problem. This ratio is of order 100. If we increased the size of the dipole to that of the outer diameter of the shell, say to 34 mm, then the reactance to resistance ratio for the dipole in free space is also of order 100.

Since the constant voltage source leads to results that are the dual of those for the constant current source, we will not consider the constant voltage source any further.

C.7 NUMERICAL RESULTS

In this section, we present some numerical results that will illustrate the general features of the spherical shell of ENG material. We assume for simplicity that the radius a of the source sphere is equal to the inner radius of the spherical shell. We have chosen this to be 10 mm, which corresponds to that considered in Ziolkowski and Erentok (2006). In Figure C.2, the radiation resistance R_a and reactance X_a are plotted as a function of the outer radius of the spherical shell having $\varepsilon = -3\varepsilon_0$.

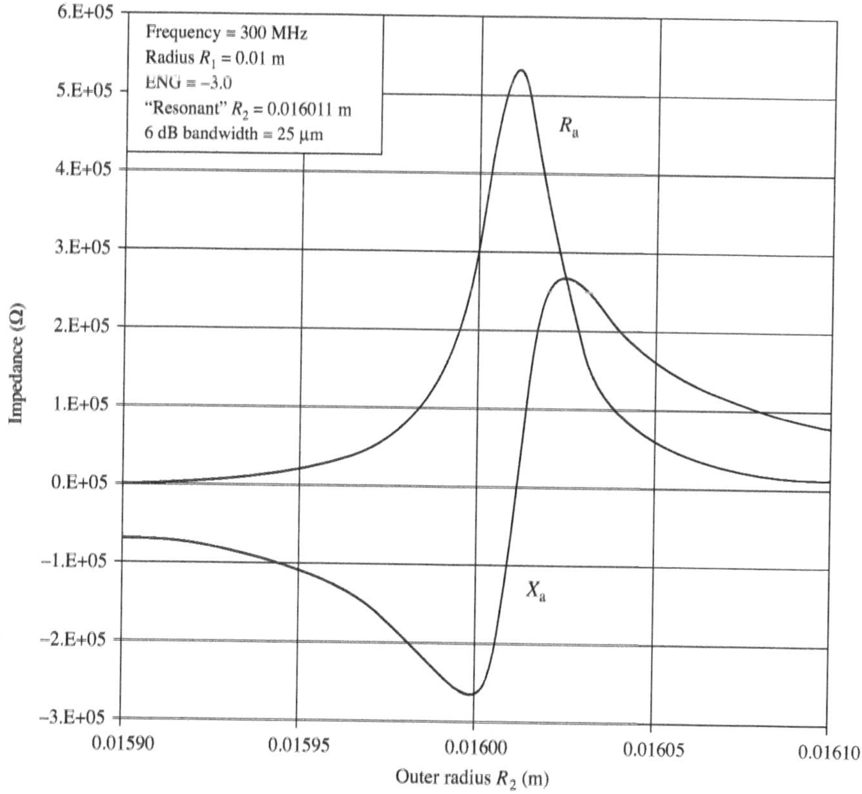

FIGURE C.2 Plot of radiation resistance R_a and reactance X_a as a function of the outer radius of the spherical shell. Inner radius is 10 mm; $\varepsilon = -3\varepsilon_0$.

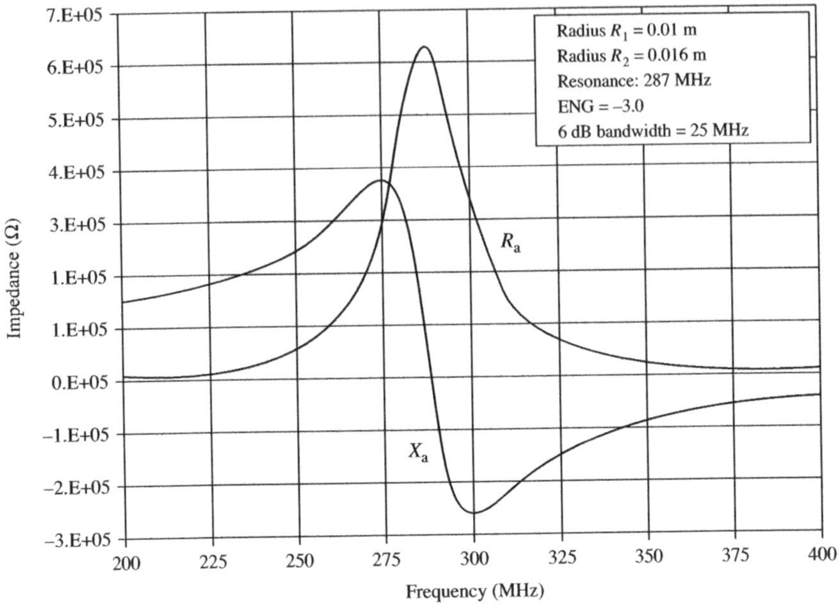

FIGURE C.3 Plot of R_a and X_a as a function of frequency for a shell with inner radius of 10 mm and outer radius of 16 mm; $\varepsilon = -3\varepsilon_0$.

The frequency is chosen as 300 MHz. This plot shows how extremely sensitive the condition for resonance is to the thickness of the shell. The 6 dB width of the resonance-like curve of R_a versus the outer radius of the shell is .025 mm. In Figure C.3, R_a and X_a are shown as a function of frequency for a shell with inner and outer radii equal to 10 and 16 mm, respectively, at a frequency of 300 MHz. If the antenna is driven from a constant current source, then the half-power point occurs when the resistance and reactance are equal. For a standard resonance curve, this point occurs where the resistance is one-half of the peak value. The 6 dB width of the curve for the resistance is approximately 25 MHz and the maximum radiation resistance is about $5.197 \times 10^5\,\Omega$ at the resonant frequency. The energy-based Q is 982 and the effective Q_e is 300/25 = 12. The peak positive value of the radiation reactance is approximately $3.84 \times 10^5\,\Omega$. The maximum negative value of the reactance is approximately $-2.56 \times 10^5\,\Omega$. The large impedance bandwidth is caused by the presence of a non-Foster inductive reactance that is present in the equivalent circuit for the ENG shell, as we show later on.

C.8 EQUIVALENT CIRCUIT FOR THE ENG SPHERICAL SHELL RESONATOR

In this section, we will show that the thin spherical shell with ENG material behaves like a resonator. The closest classical circuit that exhibits the essential features of the

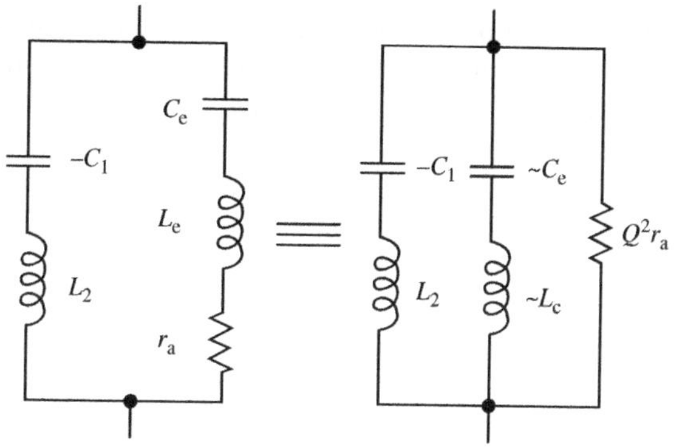

FIGURE C.4 A classical parallel *RLC* resonant circuit.

ENG shell resonator is the parallel combination of an inductor L and a capacitor C in series with a small resistance r_a as shown in Figure C.4. It is well known from classical network theory of simple resonant circuits that the ratio of the capacitive reactance to the series resistance is the Q of the circuit and that when the Q is large we can replace the small series resistance with a large parallel resistance R given by $R = Q^2 r_a$. The input impedance of this circuit is given by the expression

$$Z_{\text{in}} = \frac{j\omega LR}{R(1-\omega^2/\omega_0^2)+j\omega L} = \frac{R}{1+j(R/\omega L)\left[(\omega-\omega_0)(\omega+\omega_0)/\omega_0^2\right]} \tag{C.34a}$$

By introducing the Q given by $R/j\omega L$, this equation can be expressed in the form

$$Z_{\text{in}} = \frac{R[1-jQ(2\Delta\omega/\omega_0)]}{\left[1+4Q^2(\Delta\omega/\omega_0)^2\right]} \tag{C.34b}$$

This impedance function has a frequency response similar to that of the thin ENG spherical shell. The essential concepts underlining this analogy are developed below.

Consider the complex Poynting vector theorem given by

$$\begin{aligned} P &= \frac{1}{2}\oiint_a E_\theta H_\phi^* \, dS - \frac{1}{2}\oiint_\infty E_\theta H_\phi^* \, dS \\ &= 2j\left[\frac{\omega\mu_0}{4}\oiiint_V H_\phi H_\phi^* \, dV - \frac{\omega\varepsilon}{4}\oiiint_V E_\theta E_\theta^* \, dV\right] \end{aligned} \tag{C.35}$$

Note that we are evaluating the complex power directed into the volume enclosed. The first term is the complex power flowing through the surface at $r = a$. The second term is the negative of the real power radiated through a spherical surface at infinity.

The volume integrals are classically interpreted as the difference between the time average of the stored magnetic and electric energy in the enclosed volume. In general, such an interpretation would be correct only for conventional lossless and nondispersive media.

However, even if these terms are not expressions for physical stored energy, they are still valid for the interpretation of the inductive and capacitive reactances of the system. Since in ENG media epsilon is negative, we will not make any energy interpretation of these two terms but we will refer to the integral of the squared electric field in the ENG shell, where epsilon is negative, as inductive energy and when combined with the integral involving the square of the magnetic field we see that within the ENG shell we have only inductive energy.

For TM_{10} modes in free space, the stored energy in the near-zone field is predominantly electric, so the net energy stored is capacitive. We can now anticipate that for a suitable thickness of the ENG shell a condition of resonance can occur where the inductive and capacitive energies are equal. We have noted earlier that there is very little real power transmitted across the interface between the ENG shell and free space. Thus, the boundaries of the ENG shell are highly reflective and one can anticipate that this could make the structure behave like a high-Q resonator.

A further consideration of the wave impedance of the spherical TM_{10} mode in the near zone shows that the resistive part is proportional to $(k_0 r)^2$ and the reactive part is given by $-j/k_0 r$ to a good accuracy in the near-zone field region. This corresponds to a capacitive reactance in series with a small resistance. The inductive loading from the ENG shell then provides an equivalent parallel inductive reactance and thus we obtain a structure that should behave similarly to the simple circuit introduced at the beginning of this section.

In Equation C.21 for the complex power within the ENG shell, the square of the magnitude of the denominator is given by $|f^-|^2[(f^+/f^- + \Gamma_r)^2 + \Gamma_i^2]$, where Γ_r and Γ_i, respectively, are the real and imaginary parts of the reflection coefficient Γ_2. Since f^+/f^- is negative, the real part of the denominator can vanish.

The general behavior of the denominator is similar to that of the LC circuit. There is, however, one significant difference and that is due to the part of what we called inductive energy within the ENG shell that arises from the integral of the square of the electric field multiplied by the negative epsilon.

A positive epsilon leads to a capacitive reactance having an inverse dependence on frequency. A negative epsilon corresponds to a negative capacitance or an inductive reactance having an inverse dependence on frequency.

This type of reactance, as a function of frequency, has a negative slope and is called a non-Foster element. Thus, instead of a pure inductive reactance $j\omega L$, we should use $j\omega L_2 + j/\omega C_1$ in the lumped circuit model described by Equation C.34a, where the first term arises from the conventional stored magnetic energy and the second term represents the inductive energy associated with the negative capacitance. The primary effect of this is to make the apparent Q of the circuit much smaller. For this reason, the plot of R_{in} and X_{in} as a function of frequency shows a much broader half-power bandwidth than that predicted by the Q based on the energy definition of stored energy divided by the radiated power.

This shows that the definition of Q based on the half-power bandwidth associated with the input impedance of the circuit disagrees with the definition based on energy when a non-Foster reactance element is present in the circuit. An extreme example is a negative capacitance equal to the capacitance of the dipole antenna. These two, in series or parallel, would make the reactance equal to zero for all frequencies and the resultant bandwidth of the dipole antenna would be large. This phenomenon has been known for many years.

There are several papers and patents devoted to the use of active circuits to produce equivalent negative inductive and capacitive reactances and the use of these to obtain broadband matching of short dipole antennas (Skahill et al. 2000; Quirin, 1971; Sussman-Fort, 2006).

It is generally accepted that the Q of a spherical mode based on the energy stored outside the smallest circumscribing sphere that encloses the antenna provides a lower bound on the Q of the antenna. The physical argument for this hypothesis is that the total stored energy will be greater than or equal to that in the region outside the circumscribing sphere, since a physical system cannot store negative energy within the circumscribing sphere.

The lower bound is given by the approximate result derived by Chu 1948 or the more accurate result derived by Collin and Rothschild (1964) as well as many others. For a high-Q situation and the TM_{10} mode, this Q is given by $1/(k_0 r_2)^3$. For our example, this Q is around 1000.

On the other hand, the half-power bandwidth was found to be around 0.1, so clearly the classical relation that the half-power bandwidth is closely approximated by the reciprocal of the Q is not valid. The reason for this unusual behavior of our example is the presence of a non-Foster reactance in the equivalent circuit. In the next section, we will determine an equivalent circuit for the dipole antenna encased by a thin shell of ENG material.

Since the parameters $k_0 r$ and kr are small, we can expand all functions that occur in Equation C.12 for Z_{in} in power series of these parameters and retain only the low-order terms. This will lead to a simpler expression for Z_{in} that can be interpreted as an equivalent circuit. The branch on the left consists of the non-Foster inductive reactance, which can also be considered to be a negative capacitance. This reactance will be denoted by $j/\omega C_1$. A second inductive reactance, $j\omega L_2$, arising from the conventional stored magnetic energy within the ENG shell, is connected in series with the previous one. In parallel with this branch is the series connection of the capacitance, inductance, and resistance, which accounts for the stored electric field energy, stored magnetic field energy, and radiated power in the free space region outside the ENG shell. The impedance of this branch is represented by $j\omega L_e + r_a - j/\omega C_e$. The power series expansions that are required to establish the equivalent circuit are given in Section C.10.

The power series expansion of the numerator of Z_{in} is given by

$$\frac{2}{3\pi}[f^+Z^+ + \Gamma_2 f^- Z^-] = \frac{-2Zf^-}{3\pi}(ka)^2\left[1+(ka)^2\right]$$
$$\times\left\{\left[\frac{4}{3}+\frac{1}{6}\left(\frac{r_2}{a}\right)^3-\left(\frac{2}{5}+\frac{1}{10}\frac{r_2^5}{a^5}\right)(ka)^2\right]+j\frac{3}{8}\frac{r_2^3}{a^3}(k_0 r_2)^3\right\} \quad \text{(C.36a)}$$

where we have multiplied by the factor $2/3\pi$ to make Z_{in} correspond to Z_a obtained from the complex power given by Equation C.21. The power series expansion of the denominator of Z_{in} is given by

$$
\begin{aligned}
f^+ + \Gamma_2 f^- &= f^- \left[\frac{f^+}{f^-} + \Gamma_2\right] \\
&= -(ka)^3 f^- \left[\left(\frac{2}{3} - \frac{1}{6}\frac{r_2^3}{a^3}\right) + \left(\frac{2}{5} + \frac{1}{10}\frac{r_2^5}{a^5}\right)(ka)^2 - j\frac{3}{8}\frac{r_2^3}{a^3}(k_0 r_2)^3\right]
\end{aligned}
\tag{C.36b}
$$

With the above approximations and also approximating $1+(ka)^2$ by unity, the antenna input impedance can be expressed as the ratio of Equation C.36a to Equation C.36b and is

$$
R_a + jX_a = -j\frac{2Z}{3\pi}\frac{\left[\frac{4}{3} + \frac{1}{6}\left(\frac{r_2^3}{a^3}\right) - \left(\frac{2}{5} + \frac{1}{10}\frac{r_2^5}{a^5}\right)(ka)^2 + j\frac{3}{8}\frac{r_2^3}{a^3}(k_0 r_2)^3\right]}{\left[\left(\frac{2}{5} + \frac{1}{10}\frac{r_2^5}{a^5}\right)(ka)^3 - \left(\frac{1}{6}\frac{r_2^3}{a^3} - \frac{2}{3}\right)ka - j\frac{3}{8}\frac{r_2^3}{a^3}(k_0 r_2)^3 (ka)\right]}
\tag{C.37}
$$

For the equivalent circuit shown in Figure C.7, the input impedance is given by

$$
Z_{eq} = -j\frac{(1+\omega^2 L_2 C_1)(1-\omega^2 L_e C_e) + j\omega C_e r_a (1+\omega^2 L_2 C_1)}{\omega(C_e - C_1) + \omega^3 C_1 C_e (L_2 + L_e) - j\omega^2 C_e C_1 r_a}
$$

We can rewrite this expression in the form

$$
Z_{eq} = \frac{[(1+\omega^2 L_2 C_1)/\omega C_1] + [(1+\omega^2 L_2 C_1)(1-\omega^2 L_e C_e)/\omega^2 C_e C_1 r_a] + [(1+\omega^2 L_2 C_1)/\omega C_1]}{1 - j[(C_1 - C_e)/\omega C_e r_a C_1][1 - \omega^2 [C_e C_1/(C_1 - C_e)](L_e + L_2)]}
$$

We now note that the quality factor of the circuit is given by $Q = \omega C_e r_a$ and can be assumed to remain constant over the narrow operating bandwidth of the circuit. We also introduce the resonant frequency given by $\omega_0^2 = (C_1 - C_e)/C_1 C_e (L_2 + L_e)$ and the approximation $(\omega^2 - \omega_0^2)/\omega_0^2 = 2(\Delta\omega/\omega_0)$.

We are now able to express the equivalent circuit input impedance in the form

$$
\begin{aligned}
Z_{eq} &= \frac{(1/\omega C_1)(1+\omega^2 L_2 C_1)(1+j) + (C_e/C_1)R(1-\omega^2 L_e C_e)}{1 + jQ_e(2\Delta\omega/\omega_0)} \\
&\approx \frac{C_e/C_1}{1 + jQ_e(2\Delta\omega/\omega_0)} R
\end{aligned}
\tag{C.38}
$$

where we have introduced the equivalent quality factor $Q_e = [(C_1 - C_e)/C_1]Q$.

The equivalent parallel resistance R is given by $R = r_a/\omega^2 C_e^2 r_a^2 = Q^2 r_a$. The equivalent circuit quality factor arises from the non-Foster reactive element in the equivalent circuit. It makes the half-power bandwidth of the circuit larger than the one that would have occurred if the bandwidth were governed by the energy-based quality factor. For the equivalent circuit parameters derived from Equation C.37, the effective quality factor can be as small as 5% or less of the energy-based value.

In order to identify the values of the equivalent circuit parameters, we will express the antenna impedance given by Equation C.37 in a form like that in Equation C.38 for the equivalent circuit. Thus,

$$R_a + jX_a = \frac{2Z}{3\pi}\frac{\left[\frac{4}{3}+\frac{1}{6}\left(\frac{r_2^3}{a^3}\right)-\left(\frac{2}{5}+\frac{1}{10}\frac{r_2^5}{a^5}\right)(ka)^2+j\frac{3}{8}\left(\frac{r_2^3}{a^3}\right)(k_0r_2)^3\right]}{\frac{3}{8}\frac{r_2^3}{a^3}(k_0r_2)^3(ka)\left[1+j\frac{\left(\frac{2}{5}+\frac{1}{10}\frac{r_2^5}{a^5}\right)(ka)^2-\left(\frac{1}{6}\frac{r_2^3}{a^3}-\frac{2}{3}\right)}{\frac{3}{8}\frac{r_2^3}{a^3}(k_0r_2)3}\right]}$$

$$= \frac{2Z}{3\pi}\frac{\left[\frac{4}{3}+\frac{1}{6}\left(\frac{r_2^3}{a^3}\right)-\left(\frac{2}{5}+\frac{1}{10}\frac{r_2^5}{a^5}\right)(ka)^2+j\frac{3}{8}\left(\frac{r_2^3}{a^3}\right)(k_0r_2)^3\right]}{\frac{3}{8}\frac{r_2^3}{a^3}(k_0r_2)^3(ka)\left\{1-j\frac{\frac{1}{6}\frac{r_2^3}{a^3}-\frac{2}{3}}{\frac{3}{8}\left(\frac{r_2^3}{a^3}\right)(k_0r_2)^3}\left[1-\frac{\left(\frac{2}{5}+\frac{1}{10}\frac{r_2^5}{a^5}\right)(ka)^2}{\left(\frac{1}{6}\frac{r_2^3}{a^3}-\frac{2}{3}\right)}\right]\right\}}$$

We now identify the following parameters:

$$Q_e = \frac{\frac{1}{6}\frac{r_2^3}{a^3}-\frac{2}{3}}{\frac{3}{8}\frac{r_2^3}{a^3}(k_0r_2)^3} = \frac{\frac{1}{6}\frac{r_2^3}{a^3}-\frac{2}{3}}{\frac{3}{8}\frac{r_2^3}{a^3}}Q = \frac{C_1-C_e}{C_1}Q \tag{C.39a}$$

$$\frac{\omega^2}{\omega_0^2} = \frac{\left(\frac{2}{5}+\frac{1}{10}\frac{r_2^5}{a^5}\right)(ka)^2}{\frac{1}{6}\frac{r_2^3}{a^3}-\frac{2}{3}} \tag{C.39b}$$

When we equate Equation C.39b to unity, it becomes an equation for the resonant frequency as determined by ka.

By introducing the above definitions into the equation given above for the antenna impedance, it becomes

$$Z_a = R_a + jX_a$$
$$= \frac{2Z_0}{3\sqrt{3}\pi} \frac{\left[\frac{4}{3}+\frac{1}{6}\frac{r_2^3}{a^3}-\left(\frac{2}{5}+\frac{1}{10}\frac{r_2^5}{a^5}\right)(ka)^2\right]\left[\frac{8}{3}\frac{a^3}{r_2^3}\frac{Q}{ka}\right]+j\frac{1}{ka}}{1+jQ_e\frac{2\Delta\omega}{\omega_0}} \tag{C.40}$$

When we retain only the dominant term in the numerator, we obtain

$$Z_a = \frac{2Z_0}{3\sqrt{3}\pi} \frac{\left[\frac{4}{3}+\frac{1}{6}\frac{r_2^3}{a^3}\right]\left[\frac{8}{3}\frac{a^3}{r_2^3}\frac{Q}{ka}\right]}{1+jQ_e\frac{2\Delta\omega}{\omega_0}}$$
$$= \frac{16Z_0}{27\pi} \frac{\left[\frac{4}{3}\frac{a^2}{r_2^3}+\frac{1}{6}\frac{r_2}{a}\right]Q^2(k_0r_2)^2}{1+jQ_e\frac{2\Delta\omega}{\omega_0}} \tag{C.41}$$

This equation gives a good approximation to the behavior of the ENG shell–dipole antenna system. Note that the radiation resistance given by the numerator without the Q^2 factor is almost the same as that for a dipole antenna of total length $2r_2$ and radiating into free space. For the parameters $a = 10$ mm, $r_2 = 16.0$ mm, frequency $= 287$ MHz, and $\varepsilon = -3\varepsilon_0$, the exact analytical solution shows that the ENG shell is resonant, the bandwidth is approximately 25 MHz., and the maximum radiation resistance is about $6.79 \times 10^5\ \Omega$. The energy-based Q is 1124 and the effective Q_e is $287/25 = 11.5$.

From the equivalent circuit, we found that, for $r_2 = 16$ mm, the resonant frequency was 288 MHz, the Q was 1112, and the effective Q_e was 11.3. The approximate equivalent circuit gives results that are in good general agreement with those provided by the analytical solution. Figure C.5 shows a plot of R_a and X_a as a function of frequency for $r_2 = 16$ mm, based on the equivalent circuit. Since the bandwidth is quite broad, the variation of Q and Q_e was taken into account.

The effective Q is critically dependent on the dimensions of the ENG shell. An examination of Equation C.39a shows that the effective Q becomes zero when $r_2/a = 4^{1/3}$, that is, $r_2 \approx 1.587a$. This condition means that the negative capacitance C_1 and the external capacitance C_e are equal. Under these conditions, the frequency response of the circuit is much flatter and the half-power bandwidth becomes large.

In Figure C.6, a plot of R_a and X_a, obtained from the exact analytical solution, is given for $r_2 = 1.59$ mm. This plot shows the much greater bandwidth of the circuit, as predicted from the equivalent circuit.

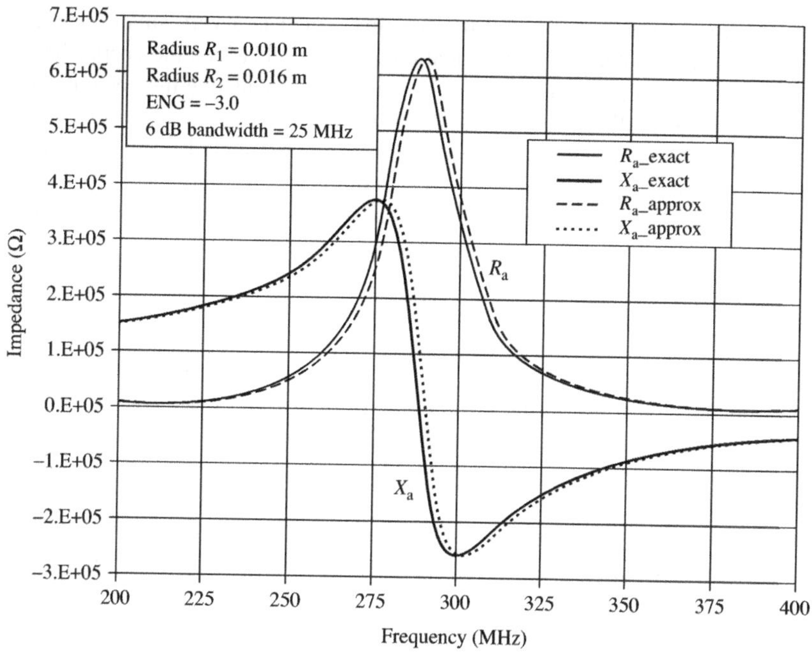

FIGURE C.5 Plot of R_a and X_a obtained from Equation C.41 for the equivalent circuit.

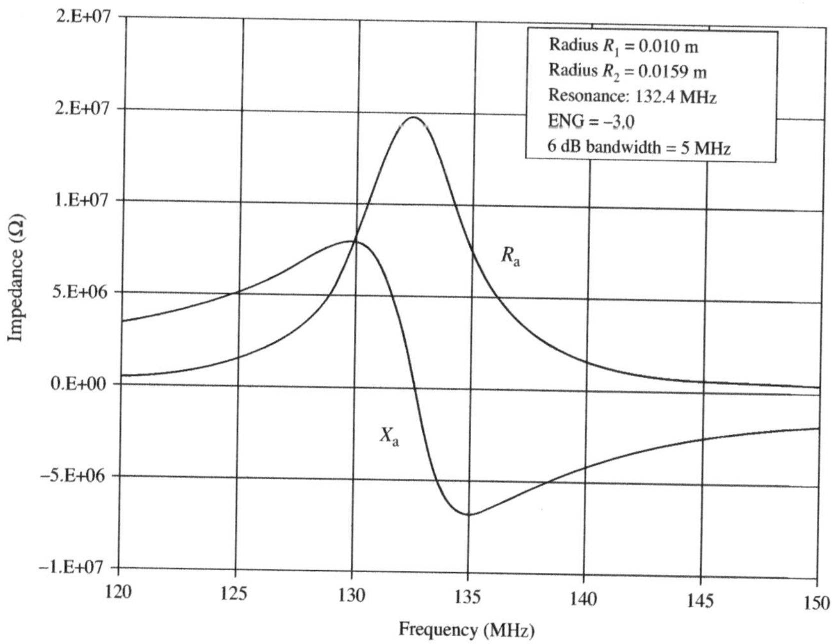

FIGURE C.6 Plot of R_a and X_a showing the broadband behavior when the outer radius of the shell is 15.9 mm. Exact analytical solution.

When we change r_2 from 16.0 to 16.1 mm, which is a very small change, the resonant frequency obtained from Equation C.39b is 383 MHz, the energy-based Q is 464, and the effective Q is 8.7. This illustrates how sensitive the behavior of the ENG shell is to small changes in its dimensions.

The equivalent circuit is not needed for the numerical determination of the behavior of the dipole antenna enclosed within a thin spherical shell of ENG material. The analytical solution is best suited for that purpose. However, the equivalent circuit does provide a clearer understanding of why the thin shell of ENG material behaves in the way that it does, in particular why the impedance bandwidth is much larger than one would expect from a consideration of the energy-based Q.

The above results are all based on the assumption that the ENG material is dispersionless. This is not a valid assumption. If we use the same dispersion model as in Ziolkowski and Erentok (2006), then the negative permittivity is given by $\varepsilon = \varepsilon_0(1 - \omega_p^2/\omega^2)$, which has a frequency derivative given by $\partial(\omega\varepsilon)/\partial\omega = \varepsilon_0(1 + \omega_p^2/\omega^2)$. Thus, the susceptance is now no longer a non-Foster element.

The frequency behavior of this susceptance is somewhat different from that of an ideal capacitor. Nevertheless, one can expect that the effective Q for the dipole–ENG shell system will be comparable to the intrinsic energy-based value of Q.

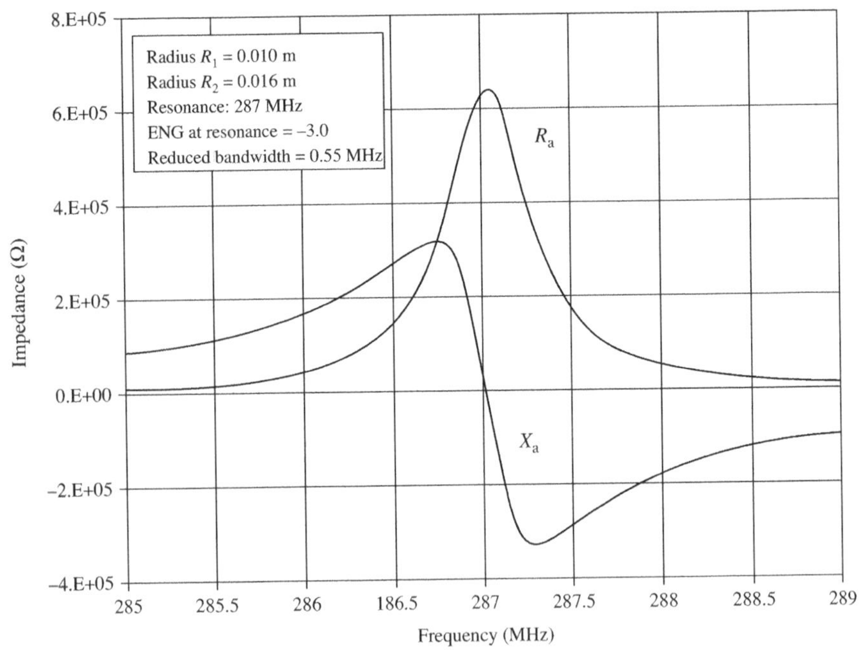

FIGURE C.7 Plot of R_a and X_a for a shell with inner radius of 10 mm and outer radius of 16 mm, but with a dispersive ENG material. At the resonant frequency of 287 MHz, $\varepsilon = -3\varepsilon_0$. Note the much narrower bandwidth. For the dispersive material, there is no non-Foster reactance present in the equivalent circuit.

In Figure C.7, a plot of the radiation resistance and reactance as a function of frequency is given for the case where $a = r_1 = 10$ mm, $r_2 = 16$ mm, and $f_p = 574$ MHz making $\varepsilon = -3\varepsilon_0$ for $f_p = 287$ MHz.

The plot shows that the width of the resistance curve at the 6 dB point is approximately 0.55 MHz. This corresponds to an effective Q of 522, which is approximately one-half of the intrinsic energy-based Q. From this one example, it seems reasonable to expect that for any physically realizable ENG material the dispersion will not allow a broad impedance bandwidth to be obtained. From this fact alone, one would not expect the use of a thin shell of DNG material to be useful for the purpose of matching a small dipole antenna to free space. If we choose a dipole antenna with a total length equal to 32 mm, the outer diameter of the ENG shell, the intrinsic energy-based Q is only 123. This is less than that of a dipole enclosed within a thin shell of ENG material that exhibits dispersion. It would be easier to match the dipole alone using a conventional matching network.

C.9 CONCLUSIONS

Using transmission line theory, we developed an exact analytical solution for the radiation impedance of a dipole antenna modeled as a current sheet on a small spherical surface, surrounded by a thin spherical shell of ENG material. This approach avoids the need to solve a system of simultaneous equations for the wave amplitudes in each region.

We found that the radiation resistance and reactance are critically dependent on the thickness of the shell. We also developed an equivalent circuit for the dipole–sphere system.

The negative permittivity of the material of the shell implies a non-Foster inductive reactance (negative capacitive susceptance). The presence of a non-Foster reactance greatly reduces the Q of the circuit. Even though the intrinsic energy-based Q is around 1000, the effective Q is only around 8–12. Consequently, the impedance bandwidth of the system is much broader than would be expected from the energy-based value of the quality factor.

Realistic materials with negative permittivity must be frequency dispersive. When a simple dispersion model is used to characterize the ENG material, the non-Foster reactance vanishes and the system displays a bandwidth corresponding to approximately one-half of the energy-based quality factor Q.

The analysis for this problem, in terms of reflected and transmitted waves along with the equivalent circuit, provides a clearer understanding of the properties exhibited by a dipole antenna surrounded by a thin shell of ENG material than the one obtained by the numerical solution presented earlier by Ziolkowski and Erentok (2006). Many of the same features found by these authors were also found from our analysis but in a more quantitative way.

We did not find any particular dimension for the spherical shell of ENG material that would make it useful as a device to match a small dipole antenna to free space.

It became quite clear that neglecting the frequency dispersion of realistic metamaterials leads to erroneous conclusions on how useful some devices utilizing metamaterials might be.

C.10 ADDENDUM

In this section, we list a number of formulas for various quantities used in the analysis, as well as the power series expansions needed for the determination of the equivalent circuit.

$$Z_{\mathrm{w}}^{+}(k_0 r) = \frac{1 - k_0^2 r^2 + jk_0 r}{jk_0 r - k_0^2 r^2} Z_0$$

$$= Z_0 \frac{k_0^2 r^2}{1 + k_0^2 r^2} - Z_0 \frac{j}{k_0 r(1 + k_0^2 r^2)}$$

$$Z_{\mathrm{w}}^{-}(k_0 r) = -[Z_{\mathrm{w}}^{+}(k_0 r)]^*$$

$$h_1^1(k_0 r) = -\left(\frac{j}{k_0^2 r^2} + \frac{1}{k_0 r}\right) e^{jk_0 r}$$

$$= \left[\frac{\sin k_0 r}{k_0^2 r^2} - \frac{\cos k_0 r}{k_0 r}\right] - j\left[\frac{\cos k_0 r}{k_0^2 r^2} + \frac{\sin k_0 r}{k_0 r}\right]$$

$$h_1^2(k_0 r) = [h_1^1(k_0 r)]^*$$

$$\frac{h_1^2(k_0 r)}{h_1^1(k_0 r)} = \frac{k_0 r - j}{k_0 r + j} e^{-2jk_0 r} = -e^{-2j(k_0 r - \arctan k_0 r)}$$

$$\frac{f^+(kr)}{f^-(kr)} = -\frac{1 + kr}{1 - kr} e^{-2kr}$$

$$Z = \sqrt{\frac{\mu_0}{|\varepsilon|}}$$

$$Z^+(kr) = +jZ \frac{1 + k^2 r^2 + kr}{k^2 r^2 + kr}$$

$$Z^-(kr) = -jZ \frac{1 + k^2 r^2 - kr}{k^2 r^2 - kr}$$

$$f^+(kr) = -j\left(\frac{1}{k^2r^2} + \frac{1}{kr}\right)e^{-kr}$$

$$f^-(kr) = +j\left(\frac{1}{k^2r^2} - \frac{1}{kr}\right)e^{kr}$$

The following power series expansions have been derived:

$$f^+(ka)Z^+(ka) = \frac{Z}{(ka)^3}\left(1 + \frac{(ka)^2}{2} - \frac{2(ka)^3}{3} + \frac{3(ka)^4}{8} - \frac{2(ka)^5}{15}\right)$$

$$f^-(ka)Z^-(ka) = -\frac{Z}{(ka)^3}\left(1 + \frac{(ka)^2}{2} + \frac{2(ka)^3}{3} + \frac{3(ka)^4}{8} + \frac{2(ka)^5}{15}\right)$$

$$\Gamma_2 = 1 + \frac{1}{6}(kr_2)^3 - \frac{1}{10}(kr_2)^5 + j\frac{3}{8}(k_0r_2)^3(kr_2)^3$$

$$\Gamma_2 Z^- = -Z\left[\frac{2}{3} - \frac{1}{6}\frac{r_2^3}{a^3} + \left(\frac{2}{5} + \frac{1}{10}\frac{r_2^5}{a^5}\right)(ka)^2 - j\frac{3}{8}\frac{r_2^3}{a^3}(k_0r_2)^3\right]$$

$$\frac{2}{3\pi}[f^+Z^+ + \Gamma_2 f^-Z^-]$$

$$= -\frac{2}{3\pi}(ka)^2\left[1 + (ka)^2\right]Zf^-\left\{\left[\frac{4}{3} + \frac{1}{6}\frac{r_2^3}{a^3} - \frac{2}{5} + \frac{1}{10}\frac{r_2^5}{a^5}(ka)^2\right] + j\frac{3}{8}\frac{r_2^3}{a^3}(k_0r_2)^3\right\}$$

The power series expansion of the denominator of Z_{in} is

$$f^-\left[\frac{f^+}{f^-} + \Gamma_2\right]$$

$$= -(ka)^3 f^-\left[\left(\frac{2}{3} - \frac{1}{6}\frac{r_2^3}{a^3}\right) + \left(\frac{2}{5} + \frac{1}{10}\frac{r_2^5}{a^5}\right)(ka)^2 - j\frac{3}{8}\frac{r_2^3}{a^3}(k_0r_2)^3\right]$$

REFERENCES

Chu, L. J. Physical Limitations of Omni-Directional Antennas. *J. Appl. Phys.*, Vol. 19, 1948, pp. 1163–1175.

Collin, R. E. and Rothschild, S. Evaluation of Antenna *Q*. *IEEE Trans.*, Vol. AP-12, January 1964, pp. 23–27.

Quirin, J. D. *A Study of High Frequency Solid State Negative-Impedance Converters for Impedance Loading of Dipole Antennas,* Master's Thesis, University of Illinois, Urbana-Champaign, 1971.

Schelkunoff, S. A. Some Equivalence Theorems of Electromagnetics and Their Applications to Radiation Problems. *Bell Syst. Tech. J.*, Vol. 15, 1936, pp. 92–112.

Skahill, G., Rudish, R. M., and Pierro, J. A. *Apparatus and Method for Broadband Matching of Electrically Small Antennas*, U.S. Patent No. 6,121,940, September 19, 2000.

Smith, D. R., Padilla, W. J., Vier, D. C., Nemat-Nasser, S. C., and Schultz, S. Composite Medium with Simultaneous Negative Permeability and Permittivity. *Phys. Rev. Lett.*, Vol. 84, No. 18, 2000, pp. 4184–4187.

Special Issue of the IEEE Transactions on Microwave Theory and Applications on the topic Metamaterial Structures, Phenomena, and Applications, Vol. 53, No. 4, April 2005.

Sussman-Fort, S. E. Matching Network Design Using Non-Foster Impedances. *Int. J. RF Microwave Comput.-Aided Eng.*, Vol. 16, No. 2, February 2006, pp. 135–142.

Tretyakov, S. A., Maslovski, S. I., Sochava, A. A., and Simovski, C. R. The Influence of Complex Material Coverings on the Quality Factor of Simple Radiating Systems. *IEEE Trans.*, Vol. AP-53, No. 3, March 2005, pp. 965–970.

Ziolkowski, R. W. and Erentok, A. Metamaterials Based Efficient Electrically Small Antennas. *IEEE Trans.*, Vol. AP-54, No. 7, July 2006, pp. 2113–2130.

Ziolkowski, R. W. and Kipple, A. D. Application of Double Negative Materials to Increase the Power Radiated by Electrically Small Antennas. *IEEE Trans.*, Vol. AP-51, No. 10, November 2003, pp. 2626–2640.

APPENDIX D

FREQUENCY DISPERSION LIMITS RESOLUTION IN VESELAGO LENS

ROBERT E. COLLIN
Life Fellow Member, IEEE, Highland Heights, OH, USA

D.1 INTRODUCTION

In 1968, Veselago introduced the concept of a material having simultaneous negative values of epsilon and mu. He discussed a number of properties of such media, such as negative refraction, left-handed wave solutions, and a flat lens configuration, among other features (Veselago, 1968). Veselago pointed out that any medium with negative mu or epsilon would have to be frequency dispersive in order for the field energy to be positive. Many years later, Pendry considered a flat slab lens made from material with the relative values of both epsilon and mu equal to -1 (Pendry, 2000). He showed that a propagating plane wave incident upon this lens would be perfectly matched at the interface; that is, the reflection coefficient would be zero and the transmission coefficient through the slab would equal 1 whenever the relative values of epsilon and mu were both equal to -1. In the case of an incident evanescent wave, the reflection and transmission coefficients become infinite when the index of refraction becomes equal to -1, so the standard method of solving for the reflected and transmitted waves cannot be used. In order to overcome this difficulty, Pendry began with relative values of epsilon and mu different from -1 and expressed the solution as a series of multiple reflected waves within the slab. This series is a geometric series that was summed and then the limit was taken as the relative values of epsilon and mu approached -1. The result showed that the overall transmission through the slab for the evanescent waves was in the form of a single exponentially growing wave within the slab. This was a somewhat surprising result but was generally considered to be a correct result by many people carrying out research on negative index of refraction media. It is interesting to note that in Pendry's original work he assumed the presence of a decaying wave in the slab but after summing the multiple reflected wave series

Small Antenna Handbook, Robert C. Hansen and Robert E. Collin.

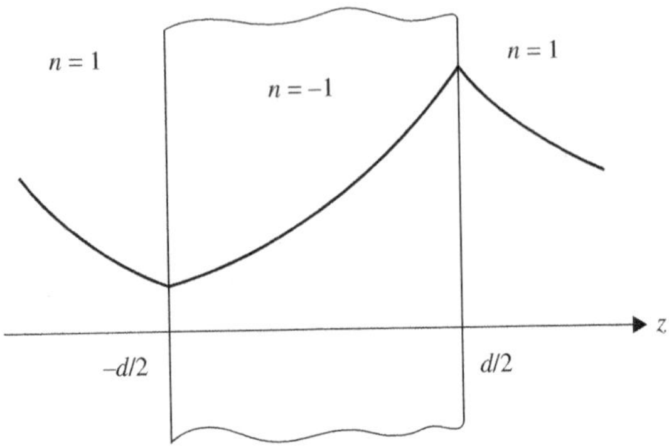

FIGURE D.1 Electric field distribution for Pendry's solution for an evanescent incident wave on a slab with an index of refraction equal to -1.

and letting the relative values of epsilon and mu become equal to -1 this part of the solution was cancelled out. At the frequency for which the index of refraction is -1 in the slab and 1 outside the slab, the evanescent wave attenuation constants are the same in both media. Consequently, the exponential growth of the waves within the slab compensates for the exponential decay outside the slab and results in the amplitude of all evanescent waves to be restored to their original values at the image plane as illustrated in Figure D.1. From this result, Pendry concludes that such a lens would reproduce the original scene on the object plane with infinite resolution at the image plane.

There is a flaw in Pendry's method that was overlooked, which is that as the relative values of epsilon and mu become quite close to -1 in value the magnitude of the ratio of successive terms in Pendry's multiple reflected wave series becomes greater than 1 and the geometric series does not converge. However, there is a mathematical solution that satisfies the boundary conditions and that is consistent with Pendry's solution for a lossless lens. The final result he obtained may be easily demonstrated in the following way. Assume that the relative values of epsilon and mu are equal to -1 and let an exponentially decaying evanescent wave be incident on the first interface between free space and the negative index of refraction medium. It is then easy to show that the boundary conditions on the tangential electric and magnetic fields at the first interface can be satisfied by assuming that in the negative index of refraction slab the field consists of a single exponentially growing wave. Likewise it can be shown that at the output interface the boundary conditions can be satisfied by assuming an exponentially decaying wave on the output side, which confirms Pendry's result. These three waves together will satisfy the boundary conditions that require the tangential electric and magnetic fields to be continuous across the two interfaces. This solution is a solution of the source-free Maxwell's equations and must be a resonant mode even though the evanescent wave on the input side was the incident evanescent wave in Pendry's solution.

Pendry's solution requires the field to be a steady-state time harmonic oscillation of infinite duration, at the frequency for which the index of refraction is exactly equal to -1. If we view this solution as a resonant mode, then it is not a proper physical solution since the field grows exponentially away from both sides of the first interface. For a single interface, a surface wave that decays away from both sides of the interface can also be supported by the surface. This is a proper physical mode solution. A proper physical solution must vary in a continuous manner when the physical parameters that characterize the problem change. Pendry's solution fails this test as can be seen by noting that if the relative values of epsilon, mu, or the index of refraction change from the exact value of -1 the boundary conditions are no longer satisfied because the tangential magnetic field will no longer be continuous across each interface, the latter requiring the attenuation constants to be the same in the slab as in the surrounding medium and the relative value of mu to equal -1. Thus, any change in epsilon and mu, even the addition of small loss, will cause a change in the solution such that the tangential magnetic field is no longer continuous across the interfaces. If Pendry's solution is viewed as a resonant mode, then it also fails the uniqueness test for field solutions since the field becomes infinite at infinity.

When the relative values of epsilon and mu do not equal -1, then the transfer function obtained by Pendry before the limit of setting the relative values of epsilon and mu equal to -1 must be used. This transfer function, which describes the electric field at the output interface of the lens in terms of the field at the input interface, has the following form for an incident evanescent wave (Pendry, 2000):

$$\tau_r(k_x, \omega) = \frac{tt' e^{-\alpha d}}{1 - r'^2 e^{-2\alpha d}} = \frac{4\mu\alpha_0 \alpha e^{-\alpha d}}{(\mu\alpha_0 + \alpha)^2 - (\mu\alpha_0 - \alpha)^2 e^{-2\alpha d}}$$

where μ and ε, respectively, are the relative values of the permeability and permittivity of the lens slab material. The attenuation constants in the free space region and within the lens are given by $\alpha_0 = \sqrt{k_x^2 - k_0^2}$ and $\alpha = \sqrt{k_x^2 - \mu\varepsilon k_0^2}$, where k_x is the transverse wave number of the evanescent wave and $k_0 = \omega/c$ is the free space wave number. When the relative values of epsilon and mu become equal to -1, then $\tau_r = e^{\alpha d}$, which cancels the corresponding decay $e^{-\alpha_0 d}$ in the field between the object plane and the first interface of the lens and that from the output interface of the lens to the image plane. This results in the amplitudes of all evanescent waves to be restored at the image plane and results in a lens with perfect resolution. However, the field within the slab becomes divergent for large values of the transverse wave number. This divergent behavior raises some questions as to whether or not Pendry's solution is a valid physical solution. In our analysis, we arrive at the conclusion that it is not a complete solution by itself.

Several authors have derived modifications to Pendry's transfer function given above by introducing small losses in epsilon and mu or by considering a small perturbation in epsilon or mu away from the value of -1 for large values of the transverse wave number (Smith et al., 2003; Merlin, 2004; Milton et al., 2005; Marqués et al., 2008; Ruppin, 2001; Haldane, 2002; Gómez-Santos, 2003;

Grbic, 2007; Yaghjian and Hansen, 2006; de Wolf, 2005, 2006; Chew, 1995). This has the effect of eliminating the field divergence problem but reduces the resolving power of the lens, thus limiting the resolution that can be obtained.

Various authors have recognized that the negative index slab can support resonant surface wave modes and that these modes play an important role in the behavior of the lens (Smith et al., 2003; Merlin, 2004; Milton et al., 2005; Marqués et al., 2008; Ruppin, 2001; Haldane, 2002; Gómez-Santos, 2003; Grbic, 2007; Yaghjian and Hansen, 2006; de Wolf, 2005, 2006). The excitation of these resonant surface wave modes does not appear in the strict steady-state solutions. Gómez-Santos considered an input sinusoidal signal that was turned on at $t = 0$ and turned off at $t = \tau$ (Gómez-Santos, 2003). He proposed modeling the lens as two coupled mechanical resonators, with resonant frequencies corresponding to those of the even and odd surface wave modes that can exist on the negative index slab. The resonant frequencies are well separated for small values of the wave attenuation constants, which occur for small values of k_x, but merge together as k_x approaches infinity. From the solution to the coupled oscillator problem, Gómez-Santos showed that the length of time for the oscillations to build up to the steady-state value was proportional to the reciprocal of the resonant frequency separation $\Delta\omega$ of the two modes. In the limit, the amplitude of the response of the second oscillator, corresponding to the output interface of the lens, was found to be proportional to $(\Delta\omega t)^2 e^{-2\alpha d}$ in the initial phase of the buildup of the oscillations. Thus, in the limit of infinitely large transverse wave numbers, the oscillations would never build up to infinite values in any finite time interval. This mechanism was proposed by Gómez-Santos to eliminate the singularity in Pendry's solution. On this basis, he concluded that Pendry's solution was acceptable.

Grbic also considered a time-domain solution and included frequency dispersion in epsilon and mu (Grbic, 2007). He chose a cosinusoidal input signal of semi-infinite duration that had a frequency spectrum proportional to $1/(\omega^2 - \omega_0^2)$, where ω_0 is the frequency at which the relative values of epsilon and mu equal -1. He obtained a result similar to that obtained by Gómez-Santos. Neither Grbic nor Gómez-Santos include the branch cut integrals that occur in the inverse Fourier or Laplace transform evaluations. A similar input signal, but with a finite duration, was considered by Yaghjian and Hansen (2006). They also analyzed the effect of losses on the resolving power of the lens. For the case of a sinusoidal signal turned on at $t = -t_0$ and turned off at $t = t_0$, their spectral function given by Equation 36 in their paper should have been expressed as a spectral function that is applicable for $(L+2d)/c - t_0 < t < t_0 + (L+2d)/c$ and a spectral function applicable for $t > (L+2d)/c + t_0$; that is, causality requires the output to be delayed by the propagation time delay. In the first time interval, the spectral function exhibits poles at $\omega = \pm\omega_0$ and would produce a dominant wave at the frequency ω_0, corresponding to Pendry's solution, plus the excitation of the even and odd surface waves that would interfere with the dominant wave. After the signal is turned off, their spectral function as given by their Equation 36 is applicable. We will also analyze the problem using a sinusoidal signal of finite duration but evaluate the inverse Fourier transform using the residues at the poles plus branch cut integrals, and thus obtain a more complete solution.

An approximate solution to the lens problem when the dispersion obeys the Smith–Kroll model was developed by de Wolf (2005, 2006). He found approximate expressions for the resonant surface wave modes but did not identify these as resonant surface wave modes. He considered the input signal spectrum to be a narrow band of frequencies with finite density and obtained an expression for the slab transmission coefficient by integrating the transmission coefficient over a narrow band of frequencies, but did not evaluate the solution as an inverse Fourier transform that would have included a $e^{j\omega t}$ time factor. Thus, his solution is not in the form of resonant surface wave modes. However, he did find from his approximate solution that for a narrowband signal spectrum the field in a slab lens did not grow exponentially for slabs with a thickness greater than some minimum value, which typically was very small. Chew has also examined this problem but he uses a strict steady-state solution and thus encounters wave solutions that diverge (Chew, 2005). However, by introducing loss along the lines other earlier investigators followed, the field divergences can be eliminated. There have been a number of studies of other lens configurations that depart from the specific configuration considered by Pendry but these will not be reviewed here.

Several researchers have built periodic structures to simulate a negative index of refraction medium and constructed flat slab lenses and attempted to demonstrate super-resolution, but with only limited success (Eleftheriades, 2007; Grbic and Eleftheriades, 2003; Eleftheriades et al., 2002; Sanada et al., 2004).

Even though a number of authors have expressed the view that the resonant surface wave modes will play an important role in the behavior of Veselago's lens, it does not appear that anyone has expressed the complete field solution for Veselago's lens, and explicitly including the excitation of the even and odd resonant surface wave modes within the slab as well as exterior to the slab, probably because they do not appear directly in a steady-state solution when epsilon and mu are not frequency dependent. The fact that for Pendry's solution the boundary conditions at the first interface can be satisfied by a single exponentially growing wave on each of the two sides of the interface is a troubling result because we are at liberty to assume that for the slab of negative index of refraction media the thickness of the slab can be made arbitrarily large and we then have the capability to create an electric field with enormous intensity. This can hardly be accepted as being a valid physical result. The original solution for the transmission factor or transfer function from the object plane through the negative index of refraction slab and to the image plane was dependent on the relative values of epsilon and mu being exactly equal to -1. Furthermore, the incident field was assumed to be a steady-state single-frequency oscillation. The operation of the lens depended critically on the excitation of a resonant mode that was intimately tied to the two interfaces of the lens. The assumption of a strict steady-state solution is the primary cause of the divergence associated with Pendry's solution when the losses are set equal to zero. Clearly, an input signal of semi-infinite or infinite duration is nonphysical.

In view of the above considerations, we were led to consider the effect of frequency dispersion, which as Veselago stated, is necessary. We therefore examined a simple model where both epsilon and mu were considered to have a simple dependence on frequency, as was used for an example in Veselago's paper (1968). In the analysis given below, we first consider the problem of a single interface that is found to support a resonant surface wave mode for which the field decays in an exponential manner away from both sides of the interface. The results of this analysis showed that at the frequency ω_e, where the relative values of epsilon and mu become equal to -1, a pole in the frequency response of a single interface occurs. This leads to a discrete mode with frequency ω_e, which undergoes exponential decay, not growth, in the negative index of refraction medium. If the incident field consists of a sinusoidal signal at the frequency ω_e for which the relative values of epsilon and mu are equal to -1, and of finite duration, then a double pole occurs in the response function for which the time response is proportional to $te^{j\omega_e t}$. Thus, this example suggests that if frequency dispersion is taken into account, the dilemma of a single exponentially growing wave on the output side of the first interface in Veselago's lens will be eliminated.

We next consider the two-interface problem on which a pair of resonant coupled surface wave modes are excited. When the incident field has a continuous frequency spectrum with finite density, our solution results in a set of proper surface waves or resonant modes on the slab and does not require the introduction of loss or small perturbations in epsilon or mu in order to avoid the short transverse wavelength divergence. This leads to a proper solution for the Veselago lens, but this solution does not support the exponential growth of the evanescent waves within the slab and thus for this type of incident field no super-resolution is possible. We also consider an incident field consisting of a single discrete frequency sinusoidal oscillation of finite duration. For this case, the field at the image plane is found to consist of a driven mode, corresponding to Pendry's solution, at the frequency of the incident field along with the resonant even and odd surface wave modes, plus fields with continuous frequency spectra that arise from branch cut integrals. For the lossless slab, the unavoidable excitation of the even and odd surface wave modes at their resonant frequencies produces interference at the image plane that makes it difficult, if at all even possible, to coherently reconstruct the amplitudes of the evanescent waves at the image plane. We also found that a continuous spectrum of interfering propagating waves would also be produced at the image plane, which will cause blurring of the image. This solution is based on the evaluation of the excitation of the resonant surface wave modes in terms of the residues at the poles and sheds new insight into the operation of the flat lens. When the excited surface waves are included as part of the solution, the divergence of the field for large transverse wave numbers is eliminated.

The existence of surface wave poles can be accounted for by using standard Fourier or Laplace transform techniques and thus obtaining finite responses in terms of the residues at the poles. There is a very large body of literature dealing with radiation from various kinds of sources over layered media that dates back to the 1909 classical work by Sommerfeld for dipole radiation over a lossy earth (Ishimaru, 1991; Chew, 1995; Felson and Marcuvitz, 1973). The many techniques developed in that research can be applied equally well to the Veselago's lens problem. This is a very realistic approach since many scenes that one might want to image are best represented by a stochastic process that has a broad frequency spectrum. A quantity of interest is the optical coherence function and this is best dealt with using Fourier transforms (Papoulis, 1968).

D.2 ANALYSIS: SINGLE INTERFACE

Consider the interface at $z = 0$ between free space and a negative index of refraction medium as shown in Figure D.1. The medium parameters of the output half-space will be assumed to be given by $\varepsilon\varepsilon_0$ and $\mu\mu_0$, where

$$\varepsilon = 1 - \frac{2\omega_e^2}{\omega^2} \tag{D.1a}$$

$$\mu = 1 - \frac{2\omega_e^2}{\omega^2} \tag{D.1b}$$

For this medium, the relative values of both epsilon and mu become equal to -1 at the frequency ω_e and remain negative for $\omega < \sqrt{2}\omega_e$. In the analysis given below, we will make use of Gabor's concept of an analytic signal (Born and Wolf, 1964). Consider a real time signal $f(t)$ and its Fourier transform

$$F(\omega) = \int_{-\infty}^{\infty} f(t)e^{-j\omega t}\, dt$$

The original signal can be recovered using the inverse Fourier transform, thus

$$f(t) = \frac{1}{2\pi}\int_{-\infty}^{\infty} F(\omega)e^{j\omega t}\, d\omega = \frac{1}{\pi}\mathrm{Re}\int_{0}^{\infty} F(\omega)e^{j\omega t}\, d\omega$$

because $F(-\omega) = F^*(\omega)$, which is the complex conjugate function. The integral without the real part designation is called the analytic signal. The advantage gained by using the analytic signal is that we only need to include positive frequencies. The real part of the analytic signal gives the physical solution.

We will assume that the source illumination comes from a finite-width aperture that is located at $z = -a$ and extends from minus to plus infinity along the y-direction. The electric field will be assumed to be in the y-direction and its intensity is a function of x only. The Fourier transform of the spatial intensity will be represented by $A(k_x, \omega)$ and its frequency spectral density by $S(\omega, k_x)$. In general, the aperture field will be a function of the radian frequency ω and the frequency spectrum of each spatial component may be different. Thus, we show the spatial spectral density and the frequency spectral density as a function of both ω and k_x. The incident field will be assumed to be that of a S polarized wave (TE wave) with a y-directed electric field and a magnetic field with x and z components. Let the incident electric field of an evanescent wave be given by the Fourier transform

$$E_{\rm i} = \frac{1}{2\pi}\int_{-\infty}^{\infty} A(k_x,\omega)S(\omega,k_x)e^{-jk_x x-\alpha_0(\omega)(z+a)}e^{j\omega t}\,d\omega \tag{D.2a}$$

The corresponding Fourier transform representation of the output signal will be

$$E_{\rm o} = \frac{1}{2\pi}\int_{-\infty}^{\infty} A(k_x,\omega)S(\omega,k_x)T(\omega)e^{-jk_x x-\alpha_0 a-\alpha(\omega)z}e^{j\omega t}\,d\omega \tag{D.2b}$$

where the transmission coefficient across the interface for an S polarized wave is $T(\omega)$. On the input side, there will be a reflected evanescent decaying wave with a reflection coefficient $R(\omega)$. The reflection and transmission coefficients are given by (Pendry, 2000)

$$R(\omega) = \frac{\mu\alpha_0-\alpha}{\mu\alpha_0+\alpha} \tag{D.3a}$$

$$T(\omega) = \frac{2\mu\alpha_0}{\mu\alpha_0+\alpha} = 1+R(\omega) \tag{D.3b}$$

where $\alpha_0 = \sqrt{k_x^2-k_0^2}$, $\alpha = \sqrt{k_x^2-\mu\varepsilon k_0^2}$, and $k_0 = \omega/c$. (Note that Pendry uses the $e^{-i\omega t}$ time dependence.) The transmission coefficient can be expressed in the following form:

$$T(\omega) = \frac{2\mu\alpha_0(\mu\alpha_0-\alpha)}{\mu^2\alpha_0^2-\alpha^2} = \frac{2\mu\alpha_0(\mu\alpha_0-\alpha)}{[1-(2\omega_{\rm e}^2/\omega^2)]^2(k_x^2-k_0^2)-k_x^2+[1-(2\omega_{\rm e}^2/\omega^2)]^2k_0^2}$$

which can be simplified to the form

$$T(\omega) = -\frac{2\mu\alpha_0(\mu\alpha_0-\alpha)\omega^4}{4\omega_{\rm e}^2k_x^2(\omega^2-\omega_{\rm e}^2)} \tag{D.4}$$

We see that $T(\omega)$ has poles at $\omega = \pm\omega_{\rm e}$. When ω becomes very large, both α_0 and α become equal to $j\omega/c$. Hence, for $t < (a+z)/c$, the contour of integration can be closed in the lower half of the complex ω plane and there will be no contribution to the output field before $t = (a+z)/c$, which is required by causality. We will assume that the spectral density function $S(\omega,k_x)$ vanishes sufficiently fast for large values of ω so that the integral converges in the upper half of the ω plane and we choose a contour of integration that runs below the poles at $\pm\omega_e$. The pole contribution to the integral may be evaluated by residue theory and gives

$$\begin{aligned} E_o &= -2\pi jA(k_x,\omega_e)\frac{\mu\alpha_0(\mu\alpha_0-\alpha)\omega_e}{8\pi k_x^2}S(\omega_e,k_x)e^{j\omega_e t}e^{-jk_x x-\alpha_0 a-\alpha z}+\text{c.c.} \\ &= -jA(k_x,\omega_e)\frac{k_x^2-k_0^2}{2k_x^2}\omega_e S(\omega_e,k_x)e^{j\omega_e t}e^{-jk_x x-\alpha_0 a-\alpha z}+\text{c.c.},\quad t>(a+z)/c \end{aligned} \tag{D.5a}$$

where all terms are evaluated for $\omega=\omega_e$ and c.c. represents the complex conjugate term. The first term by itself represents the analytic signal that we designate as $\widetilde{E}_o$, thus

$$\widetilde{E}_o = -jA(k_x,\omega_e)\frac{k_x^2-k_0^2}{2k_x^2}\omega_e S(\omega_e,k_x)e^{j\omega_e t}e^{-jk_x x-\alpha_0 a-\alpha z},\quad t>(a+z)/c \tag{D.5b}$$

This solution is a discrete exponentially decaying wave at the frequency ω_e. The discrete wave represents some type of resonance, which could be said to be due to surface plasmons (Pendry, 2000).

From Equation D.3b it is clear that the tangential electric field will be continuous across the interface. The tangential magnetic field on the input side is proportional to

$$[1-R(\omega)]\frac{j\alpha_0}{\omega\mu_0}A(k_x,\omega) = \frac{j2\alpha_0\alpha(\mu\alpha_0-\alpha)}{\omega\mu_0[(\mu\alpha_0)^2-\alpha^2]}A(k_x,\omega) \tag{D.6a}$$

while on the output side it is proportional to

$$[1+R(\omega)]\frac{j\alpha}{\omega\mu\mu_0}A(k_x,\omega) = \frac{j2\mu\alpha_0\alpha(\mu\alpha_0-\alpha)}{\omega\mu\mu_0[(\mu\alpha)^2-\alpha^2]}A(k_x,\omega) \tag{D.6b}$$

These two expressions are equal and have the same residues at the poles. However, it should be noted that there is no pole associated with the incident field. The reflection coefficient and the transmission coefficient have the same residues. The tangential electric and magnetic fields of this resonant mode are continuous across the interface and the field has exponential decay away from the interface on both sides. Thus, this discrete frequency mode that is excited is clearly a resonance effect. Once this mode is excited by the incident field, it will continue to oscillate, eventually decaying to zero because of losses that usually will be present. The mode may also be viewed as a surface wave that is bound to the interface and its electric field decays in an exponential manner away from both sides of the interface.

An interesting variation of the above results is obtained if we consider a sinusoidal signal of finite duration and with a frequency equal to the resonant frequency of the surface wave. Thus, consider the input signal consisting of a sinusoidal oscillation

$\sin \omega_e t$ at the frequency ω_e, which is turned on at $t = 0$ and turned off at $t = \tau$. For this signal, the spectral function is given by

$$\begin{aligned} S(\omega) &= \frac{e^{j(\omega_e-\omega)\tau}-1}{2(\omega-\omega_e)} - \frac{e^{-j(\omega+\omega_e)\tau}-1}{2(\omega+\omega_e)} \\ &= je^{j(\omega_e-\omega)\tau/2}\frac{\tau}{2}\frac{\sin(\omega-\omega_e)\tau/2}{(\omega-\omega_e)\tau/2} - je^{-j(\omega+\omega_e)\tau/2}\frac{\tau}{2}\frac{\sin(\omega+\omega_e)\tau/2}{(\omega+\omega_e)\tau/2} \end{aligned} \quad \text{(D.7)}$$

The part of $S(\omega)$ that involves the terms depending on τ does not contribute to the output for $t < \tau + (a+z)/c$. Prior to that time, the spectral function has a pole at the resonant frequency of the surface wave. Thus, the system has a double pole at $\omega = \omega_e$. The time response for a double pole is $-te^{j\omega_e t}$ and thus the response for the single interface is given by

$$E_o = jA(k_x, \omega_e)\frac{(k_x^2-k_0^2)\omega_e}{4k_x^2}e^{-jk_x x-\alpha_0 a-\alpha z}(te^{j\omega_e t}) + \text{c.c.}, \quad (a+z)/c < t < \tau + (a+z)/c \quad \text{(D.8)}$$

This type of response is well known for a lossless resonant circuit when excited by a sinusoidal signal of finite duration and at the resonant frequency of the circuit.

Later on we will show that for a lossless slab there will be two surface wave modes whose frequencies coalesce at $\omega = \omega_e$, when the transverse wave number k_x approaches infinity, to produce a triple pole with a time response proportional to $t^2 e^{j\omega_e t}$. This result was obtained by Gómez-Santos (2003) from a model of the slab as a pair of coupled mechanical resonators. The signal output does not become infinite since the signal is turned off at $t = \tau$. When the signal input is turned off, the frequency spectrum no longer has a pole and is of the form that makes the output become small because of a rapidly oscillating spectral function.

The integrand also has branch points at $\omega = \pm k_x c$ and $\omega = \pm\sqrt{k_x^2 c^2 + 2\omega_e^2}$, where c is the velocity of light in free space. Thus, in addition to the pole contributions, there are additional contributions to the field from branch cut integrals. The branch cut integrals give rise to a continuous spectrum of waves and these fields will satisfy the boundary conditions at the interface. The branch cut integrals will not give rise to exponentially growing fields because the branches for which the real part of α_0 and α are greater than zero must be chosen in order to ensure that the fields remain finite as z approaches infinity. The analysis to include the branch cut integrals would be similar to that described by Stratton for propagation in a dispersive medium but will not be pursued in this paper for the single-interface problem (Stratton, 1941).

D.3 SOLUTION FOR THE TWO-INTERFACE PROBLEM

The Veselago's flat lens involves two interfaces, one at $z = -d/2$ and the other at $z = d/2$, as shown in Figure D.2. Each interface by itself can support a resonant

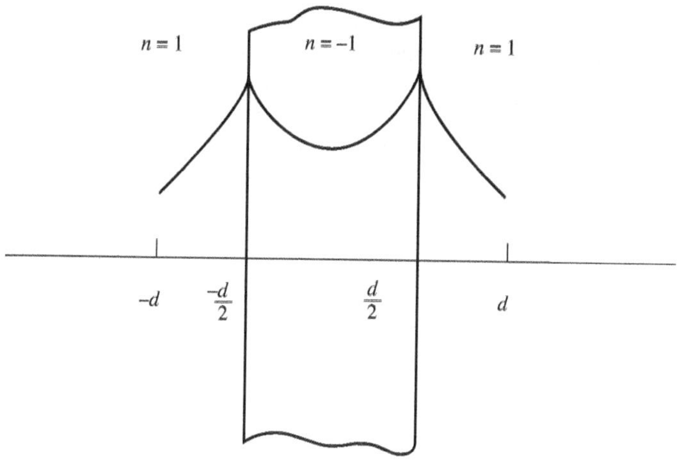

(a)

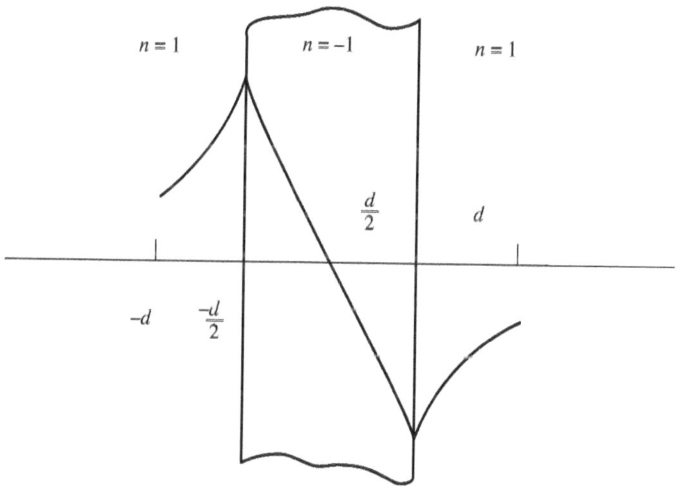

(b)

FIGURE D.2 (a) The even surface wave mode field distribution in a flat slab lens. (b) The odd surface wave mode field distribution in a flat slab lens.

surface plasmon mode, that is, surface wave. When the spacing between the two interfaces is finite, these two modes will interact and the result will be two new perturbed modes with resonant frequencies that lie above and below that of the mode supported by a single interface. An analogy with a similar waveguide problem will help to clarify the phenomenon involved. Consider a rectangular waveguide that is operated below cutoff. Let a short section of this waveguide be filled with a high dielectric constant material so that it forms a resonant cavity. Since a dielectric slab can support even and odd surface wave modes at discrete frequencies, these represent the resonant modes of the cavity. The response function of this cavity will have poles at the resonant frequencies of the cavity. If a spectrum of evanescent waves with a continuous frequency spectrum is incident on this cavity, this incident field will excite a finite response in the resonant cavity at its resonant frequencies. If we assumed that a steady-state sinusoidal field at the resonant frequency of one of the cavity modes was incident upon it, the response would be infinite. But with a field having a finite frequency spectral density, the response of the cavity is also finite and is determined by the residues at the poles that lie in the frequency range of the incident field spectrum. This same phenomenon occurs with a slab of negative index material sandwiched between two regions with a positive index of refraction. When we include frequency dispersion, the overall transmission factor through the slab of negative index material exhibits poles corresponding to the new frequencies for the two coupled resonators. Thus, an incident field with a continuous but finite frequency spectrum will excite these resonances in addition to a continuous spectrum of transmitted waves. These resonant responses do not exhibit growing exponential waves. If the Veselago lens was to be used to image the aperture field distribution at $z = -d$ at an image plane located at $d/2$ beyond the second interface, then the excitation of the surface plasmons would represent an artifact that should not be present in the image since the aperture field does not contain discrete frequency components (by assumption) (Smith et al., 2003). When frequency dispersion is taken into account and the analysis is carried out in a more complete manner, rather than assuming a steady-state sinusoidal incident field of infinite duration, the results are different from what Pendry obtained.

Our solution has a number of features similar to what Gómez-Santos included but differs in some important aspects. If we assume an incident field with a continuous frequency spectrum that overlaps both resonant frequencies, and with a finite spectral density, the response of the lens can be evaluated in terms of the residues at the two poles corresponding to the surface wave mode resonances. This results in a field distribution within the lens that does not grow exponentially and is quite different from what Pendry had found. For a lossless slab and a sinusoidal input signal of finite duration, and at the frequency for which the relative values of epsilon and mu equal -1, the response does not become exponentially large when the transverse wave number approaches infinity because of the excitation of the surface wave modes. The interference produced by the excited surface wave modes destroys the super-resolution properties of the lens. We also show that an interfering signal that blurs the output at the image plane for the propagating waves also occurs. This interfering signal arises from branch cut integrals that are part of the complete solution.

The analysis given below will provide the details that support the above description and conclusions. We will take advantage of the symmetry inherent in the problem and construct the even and odd mode solutions in separate steps. We can superimpose the two solutions so as to obtain the solution for a field incident from one side of the slab only. This has the advantage that it provides simpler expressions to evaluate for the residues at the poles. We will express the even solution, for one component of the spatial spectrum $A(k_x, \omega)$ of the electric field, in the form

$$E_e = \frac{1}{2\pi} \int_{-\infty}^{\infty} \Psi_e(\omega) e^{-jk_x x} e^{j\omega t}\, d\omega \tag{D.9}$$

where

$$\begin{aligned}\Psi_e(\omega) &= C_1 e^{-\alpha_0(z+d)} + C_2 e^{\alpha_0(z+d/2)}, \quad -d < z < -d/2\\ &= C_3 \cosh \alpha z, \quad -d/2 < z < d/2\\ &= C_1 e^{\alpha_0(z-d)} + C_2 e^{-\alpha_0(z-d/2)}, \quad d/2 < z\end{aligned}$$

The corresponding spectral function for the magnetic field is (we omit a factor $1/j\omega\mu_0$)

$$\begin{aligned}\Psi_{he}(\omega) &= -\alpha_0 C_1 e^{-\alpha_0(z+d)} + \alpha_0 C_2 e^{\alpha_0(z+d/2)}, \quad -d < z < -d/2\\ &= \frac{\alpha}{\mu} C_3 \sinh \alpha z, \quad -d/2 < z < d/2\\ &= C_1 \alpha_0 e^{\alpha_0(z-d)} - \alpha_0 C_2 e^{-\alpha_0(z-d/2)}, \quad d/2 < z\end{aligned}$$

We now match the fields at the interfaces and solve for the amplitude constants to obtain

$$C_2 = C_1 \frac{\mu\alpha_0 \cosh(\alpha d/2) - \alpha \sinh(\alpha d/2)}{\mu\alpha_0 \cosh(\alpha d/2) + \alpha \sinh(\alpha d/2)} e^{-\alpha_0 d/2} \tag{D.10a}$$

$$C_3 = C_1 \frac{2\mu\alpha_0 e^{-\alpha_0 d/2}}{\mu\alpha_0 \cosh(\alpha d/2) + \alpha \sinh(\alpha d/2)} \tag{D.10b}$$

where $\alpha_0 = \sqrt{k_x^2 - k_0^2}$ and $\alpha = \sqrt{k_x^2 - \mu^2 k_0^2}$. For the odd mode solution, the electric field spectral functions are chosen as

$$\begin{aligned}\Psi_o(\omega) &= C_1 e^{-\alpha_0(z+d)} + D_2 e^{\alpha_0(z+d/2)}, \quad -d < z < -d/2\\ &= -D_3 \sinh \alpha z, \quad -d/2 < z < d/2\\ &= -C_1 e^{\alpha_0(z-d)} - D_2 e^{-\alpha_0(z-d/2)}, \quad d/2 < z\end{aligned}$$

The magnetic field spectral functions are

$$\begin{aligned}\Psi_{ho}(\omega) &= -\alpha_0 C_1 e^{-\alpha_0(z+d)} + \alpha_0 D_2 e^{\alpha_0(z+d/2)}, \quad -d < z < -d/2 \\ &= -\frac{\alpha}{\mu} D_3 \cosh \alpha z, \quad -d/2 < z < d/2 \\ &= -C_1 \alpha_0 e^{\alpha_0(z-d)} + \alpha_0 D_2 e^{-\alpha_0(z-d/2)}, \quad d/2 < z\end{aligned}$$

The solutions for the amplitude constants are

$$D_2 = C_1 \frac{\mu\alpha_0 \sinh(\alpha d/2) - \alpha \cosh(\alpha d/2)}{\mu\alpha_0 \sinh(\alpha d/2) + \alpha \cosh(\alpha d/2)} e^{-\alpha_0 d/2} \tag{D.11a}$$

$$D_3 = C_1 \frac{2\mu\alpha_0 e^{-\alpha_0 d/2}}{\mu\alpha_0 \sinh(\alpha d/2) + \alpha \cosh(\alpha d/2)} \tag{D.11b}$$

where all amplitude constants are functions of k_x and ω. The superposition of the two solutions gives the expressions for the electric field spectral functions in the three regions. However, we will leave the solutions in the form of the even and odd modes since the expressions we need to evaluate to obtain the residues at the poles are simpler. In the slab, the superposition of the even and odd modes reduces to the results obtained by Pendry when the frequency is set equal to ω_e and $D_3 = -C_3$; that is, a strict steady-state solution is assumed to exist. Since we still have to carry out the inversion of the Fourier transform, the final solution will be determined by the residues at the poles along with contributions from integration along the branch cuts.

The expressions for the pole locations are obtained by equating the denominators in Equations D.10a and D.11a to zero. These equations are, for the even and odd modes, respectively (Smith et al., 2003; Haldane, 2002; Gómez-Santos, 2003),

$$\coth \frac{\alpha d}{2} = -\frac{\alpha}{\mu\alpha_0} \tag{D.12a}$$

$$\tanh \frac{\alpha d}{2} = -\frac{\alpha}{\mu\alpha_0} \tag{D.12b}$$

An alternative form of these equations that are convenient to use for numerical calculations is

$$e^{-\alpha d} = \frac{\alpha + \mu\alpha_0}{\alpha - \mu\alpha_0} \tag{D.13a}$$

$$e^{-\alpha d} = \frac{\mu\alpha_0 + \alpha}{\mu\alpha_0 - \alpha} \tag{D.13b}$$

The first equation has a zero when $\omega = \omega_e^+ > \omega_e$, while the second equation has a zero when $\omega = \omega_e^- < \omega_e$. The negatives of these two frequencies are also zeroes. It is easy to interpret the form of the equations given above. The two resonant modes correspond to a mode with an even electric field distribution about the midplane of the slab and a second mode with an electric field distribution that is odd about the midplane of the slab. Equation D.12a is a statement of the equality of the field impedance seen when looking into the slab with an open circuit at the midplane to that seen looking outward from the interface. Similarly, Equation D.12b corresponds to setting the impedance seen looking into the slab with a short circuit at the midplane and equating this to the impedance looking out from the interface. This is an application of the well-known transverse resonance method that is used to solve for the surface waves on many microwave structures. For each value of the transverse wave number k_x, there is a pair of resonant modes with resonant frequencies that depend on k_x. For large values of k_x, the two interfaces are electrically far apart, so the interaction between the two modes is small and the two resonant frequencies will be close to ω_e. The electric field in the slab for the even mode is described by the function $\cosh \alpha z$ and by $-\sinh \alpha z$ for the odd mode, as illustrated in Figure D.2. Both of these modes will be excited when the input signal has a spectral width that extends from at least ω_e^- to ω_e^+. These modal solutions can be evaluated in terms of the residues at the poles.

We now superimpose the even and odd solutions to obtain the final solution with a field incident only from the object plane at $z = -d/2$. On the input side, the electric field is given by

$$E = \frac{1}{2\pi}\int_{-\infty}^{\infty} [2C_1 e^{-\alpha_0(z+d)} + (C_2 + D_2)e^{\alpha_0(z+d/2)}]e^{-jk_x x + j\omega t}\, d\omega$$

$$= \frac{1}{2\pi}\int_{-\infty}^{\infty} A(k_x,\omega)S(\omega)\left[2e^{-\alpha_0(z+d)} + \frac{\mu\alpha_0 \cosh(\alpha d/2) - \alpha \sinh(\alpha d/2)}{\mu\alpha_0 \cosh(\alpha d/2) + \alpha \sinh(\alpha d/2)} e^{\alpha_0 z}\right.$$

$$\left. + \frac{\mu\alpha_0 \sinh(\alpha d/2) - \alpha \cosh(\alpha d/2)}{\mu\alpha_0 \sinh(\alpha d/2) + \alpha \cosh(\alpha d/2)} e^{\alpha_0 z}\right] e^{-jk_x x + j\omega t}\, d\omega,$$

$$-d < z < -d/2 \qquad \text{(D.14a)}$$

which is an equation that shows that the constant $C_1 = A(k_x, \omega)S(\omega)$. Within the slab, the total electric field is

$$E = \frac{1}{2\pi}\int_{-\infty}^{\infty} A(k_x,\omega)S(\omega)\left[\frac{2\mu\alpha_0 e^{-\alpha_0 d/2}\cosh(\alpha z)}{\mu\alpha_0 \cosh(\alpha d/2) + \alpha \sinh(\alpha d/2)}\right.$$

$$\left. - \frac{2\mu\alpha_0 e^{-\alpha_0 d/2}\sinh \alpha z}{\mu\alpha_0 \sinh(\alpha d/2) + \alpha \cosh(\alpha d/2)}\right] e^{-jk_x x + j\omega t}\, d\omega,$$

$$-d/2 < z < d/2 \qquad \text{(D.14b)}$$

while at the image plane at $z = d$ the solution is given by

$$E_o = \frac{1}{2\pi}\int_{-\infty}^{\infty} A(k_x,\omega)S(\omega,k_x)\left[\frac{\mu\alpha_0\cosh(\alpha d/2)-\alpha\sinh(\alpha d/2)}{\mu\alpha_0\cosh(\alpha d/2)+\alpha\sinh(\alpha d/2)}e^{-\alpha_0 d}\right.$$

$$\left.-\frac{\mu\alpha_0\sinh(\alpha d/2)-\alpha\cosh(\alpha d/2)}{\mu\alpha_0\sinh(\alpha d/2)+\alpha\cosh(\alpha d/2)}e^{-\alpha_0 d}\right]e^{-jk_x x+j\omega t}\,d\omega, \quad z=d \tag{D.14c}$$

The residues can be found by the usual procedure of evaluating the frequency derivative of each denominator at the corresponding zero. For the first factor, the denominator vanishes when $\omega = \omega_e^+$, where $\mu\alpha_0\cosh(\alpha d/2) = -\alpha\sinh(\alpha d/2)$. The residue associated with this factor is

$$\begin{aligned}\mathrm{Res}(\omega_e^+) &= \left.\frac{-2\alpha\sinh(\alpha d/2)e^{-\alpha_0 d+j\omega t}}{\partial[\mu\alpha_0\cosh(\alpha d/2)+\alpha\sinh(\alpha d/2)]/\partial\omega}\right|_{\omega_e^+}\\ &= \left.\frac{-2\omega_e\alpha(1-e^{-\alpha d})e^{-\alpha_0 d}e^{j\omega_e^+ t}}{(A+B+C+D)+e^{-\alpha d}(E+F-GH)}\right|_{\omega_e^+}\\ &= \widehat{\mathrm{Res}}(\omega_e^+)e^{-\alpha_0(\omega_e^+)d+j\omega_e^+ t}\end{aligned} \tag{D.15a}$$

Similarly, the residue at the pole $\omega = \omega_e^-$ is given by

$$\begin{aligned}\mathrm{Res}(\omega_e^-) &= \left.\frac{2\alpha\cosh(\alpha d/2)e^{-\alpha_0 d+j\omega t}}{\partial[\mu\alpha_0\sinh(\alpha d/2)+\alpha\cosh(\alpha d/2)]/\partial\omega}\right|_{\omega_e^-}\\ &= \left.\frac{2\omega_e\alpha(1-e^{-\alpha d})e^{-\alpha_0 d}e^{j\omega_e^- t}}{(A+B+C+D)-e^{-\alpha d}(E+F-GH)}\right|_{\omega_e^-}\\ &= \widehat{\mathrm{Res}}(\omega_e^-)e^{-\alpha_0(\omega_e^-)d+j\omega_e^- t}\end{aligned} \tag{D.15b}$$

where $A = 4k_e^3\alpha_0/k_0^3$, $B = -\mu k_0 k_e/\alpha_0$, $C = -4\mu k_e^3/\alpha k_0$, $D = -\mu^2 k_0 k_e/\alpha$, $E = -\mu k_0 k_e/\alpha_0$, $F = 4\alpha_0 k_e^3/k_0^3$, $G = (\alpha-\mu\alpha_0)d-1$, $H = (4\mu k_e^3/\alpha k_0)+(\mu^2 k_0 k_e/\alpha)$, and $k_e = \omega_e/c$.

We note that A, B, and C are positive constants and that the sum $A+B+C+D$ is never zero. Thus, the residues at the image plane will be proportional to $e^{-\alpha_0 d}$. The form of the solution obtained from Equation D.14c is thus (we only need to evaluate the expressions for the poles at $\omega_e^\pm$ to obtain the analytic signal representation)

$$\widetilde{E}_o = \frac{2\pi j}{2\pi}\Big[A(k_x,\omega_e^+)S(\omega_e^+)e^{-jk_x x}\widehat{\text{Res}}(\omega_e^+)e^{-\alpha_0(\omega_e^+)d}e^{j\omega_e^+ t}$$

$$+ A(k_x,\omega_e^-)S(\omega_e^-)e^{-jk_x x}\widehat{\text{Res}}(\omega_e^-)e^{-\alpha_0(\omega_e^-)d}e^{j\omega_e^- t}\Big],$$

$$z = d, \quad t > 2d/c \tag{D.16}$$

Inside the slab, the solution is given by Equation D.14b and the contribution from the surface waves when evaluated in terms of the residues is given by

$$\widetilde{E} = j\Bigg[A(k_x,\omega_e^+)S(\omega_e^+)e^{-jk_x x}\widehat{\text{Res}}(\omega_e^+)e^{-\alpha_0(\omega_e^+)d/2+j\omega_e^+ t}\frac{\cosh(\alpha z)}{\cosh(\alpha d/2)}\Bigg|_{\omega_e^+}$$

$$+A(k_x,\omega_e^-)S(\omega_e^-)e^{-jk_x x}\widehat{\text{Res}}(\omega_e^-)e^{-\alpha_0(\omega_e^-)d/2+j\omega_e^- t}\frac{\sinh\alpha z}{\sinh(\alpha d/2)}\Bigg|_{\omega_e^-}\Bigg],$$

$$-d/2 < z < d/2, \quad t > (z+d)/c \tag{D.17}$$

The residues are given by Equation D.15a and the extra factors compensate for the difference in the numerators in Equations D.14b and D.14c. These resonant surface wave modes that are excited are proportional to $e^{-\alpha_0 d/2}$ at $z = d/2$ and do not exhibit exponential growth. In Figure D.3, we show the resonant frequencies for the even and odd surface wave modes and their residues as a function of k_x/k_e. Note that the residues remain bounded as k_x becomes large. The above solutions are also valid for $k_x < k_0$ with α_0 and α replaced by $j\beta_0 = \sqrt{k_0^2 - k_x^2}$ and $j\beta = \sqrt{\mu\varepsilon k_0^2 - k_x^2}$.

Let us, at this point, assume that the field in the aperture plane consists of a sinusoidal oscillation $\sin\omega_e t$ at the frequency ω_e, which is turned on at $t = 0$ and turned off at $t = \tau$. For this signal, the spectral function is given by Equation D.7 and is repeated below:

$$S(\omega) = \frac{e^{j(\omega_e-\omega)\tau}-1}{2(\omega-\omega_e)} - \frac{e^{-j(\omega+\omega_e)\tau}-1}{2(\omega+\omega_e)}$$

$$= je^{j(\omega_e-\omega)\tau/2}\frac{\tau}{2}\frac{\sin(\omega-\omega_e)\tau/2}{(\omega-\omega_e)\tau/2} - je^{-j(\omega+\omega_e)\tau/2}\frac{\tau}{2}\frac{\sin(\omega+\omega_e)\tau/2}{(\omega+\omega_e)\tau/2} \tag{D.18}$$

For notational convenience, we will let the function in the integrand in Equation D.14c be denoted by $F(\alpha_0,\omega)$, where

$$F(\alpha_0,\omega) = \left[\frac{\mu\alpha_0\cosh(\alpha d/2) - \alpha\sinh(\alpha d/2)}{\mu\alpha_0\cosh(\alpha d/2) + \alpha\sinh(\alpha d/2)} - \frac{\mu\alpha_0\sinh(\alpha d/2) - \alpha\cosh(\alpha d/2)}{\mu\alpha_0\sinh(\alpha d/2) + \alpha\cosh(\alpha d/2)}\right]e^{-\alpha_0 d}$$

$$= \frac{8\mu\alpha_0\alpha e^{-\alpha_0 d - \alpha d}}{(\mu\alpha_0+\alpha)^2 - (\mu\alpha_0-\alpha)^2 e^{-2\alpha d}} \tag{D.19}$$

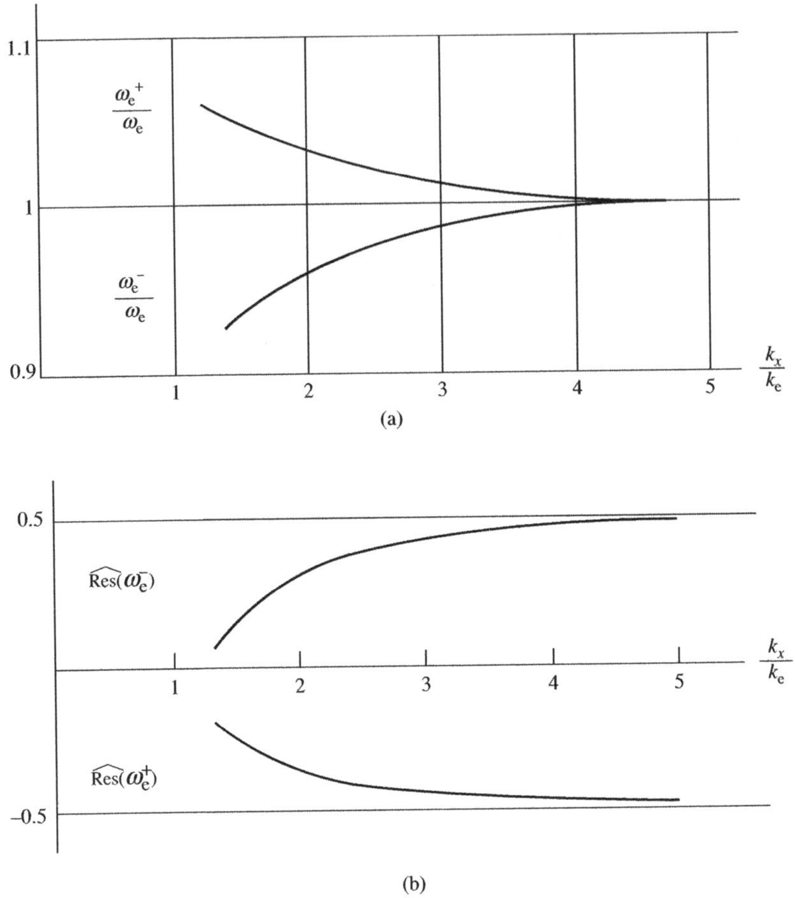

FIGURE D.3 A plot of the resonant frequencies for the even and odd surface wave modes, and their residues, as a function of $k_x/k_e = k_x c/\omega_e$.

which was obtained by combining the even and odd mode solutions in Equation D.14c. At the frequencies $\omega = \pm\omega_e$, where the relative values of epsilon and mu are equal to -1, this function equals 2. There are no poles at $\pm\omega_e$. However, the function $F(\alpha_0, \omega)$ has poles at $\pm\omega_e^{\pm}$, which are the resonant frequencies of the even and odd surface wave modes.

The inversion contour for the Fourier transform runs parallel to the real ω axis and just below the poles at $\pm\omega_e^+$, $\pm\omega_e^-$, and $\pm\omega_F$. From a consideration of the wave function $e^{j\omega t - j\omega\tau - j2\sqrt{k_0^2-k_x^2}d}$, we see for large values of ω that for $t < \tau + 2d/c$ the wave function will become small on the semicircle at infinity in the lower half of the complex ω plane. Thus, we can close the inversion contour in the lower half of the complex plane and since no singularities are enclosed the value of the integral will be zero. For the part of the spectral density function in Equation D.18 that does not depend on τ, the inversion contour can also be closed in the lower half of the complex ω plane when $t < 2d/c$ and will not give any contribution to the field. These conditions are simply the requirements of causality. Note that as ω becomes very large the propagation factor $\sqrt{\mu\varepsilon k_0^2 - k_x^2}$ in the slab becomes the same as that in free space because epsilon and mu approach the free space values, and hence there is no problem with time running backward as far as imposing the causality condition is concerned. The spectral density function of the input signal that depends on τ will give the output field at the image plane after the sinusoidal signal has been turned off. This signal would be of less interest since it is unlikely that measurements of the field at the image plane would be made after the illumination of the object has been turned off.

The evaluation of the fields in terms of the residues requires that the Fourier inversion integral be taken over a closed contour enclosing the poles and that the integrand be single valued within the contour. The expression in Equation D.14a is an even function of α and hence has branch points associated only with α_0. Suitable branch cuts are the lines joining the two branch points, corresponding to the zeroes $\omega = \pm k_x c = \pm\omega_x$ of α_0, to plus and minus infinity as shown in Figure D.4. The original inversion contour runs parallel to the real axis, from minus infinity to plus infinity, but below the poles at $\pm\omega_e^{\pm}$ and $\pm\omega_e$. This contour is closed by the contour shown in Figure D.4, which includes a contour running around the branch cut from $\omega_x = -k_x c$ to $-\infty$ and from $k_x c$ to ∞, and closed by a semicircle contour at infinity in the upper half of the complex $\omega + j\sigma$ plane. There is no contribution to the integrals from the semicircle contour. The poles are enclosed within the contour. The value of the integral in Equation D.14a thus consists of the terms corresponding to the residues at the poles plus integrals around the branch cuts but traversed in the opposite direction. In the absence of loss, the poles lie on the real ω axis. On the

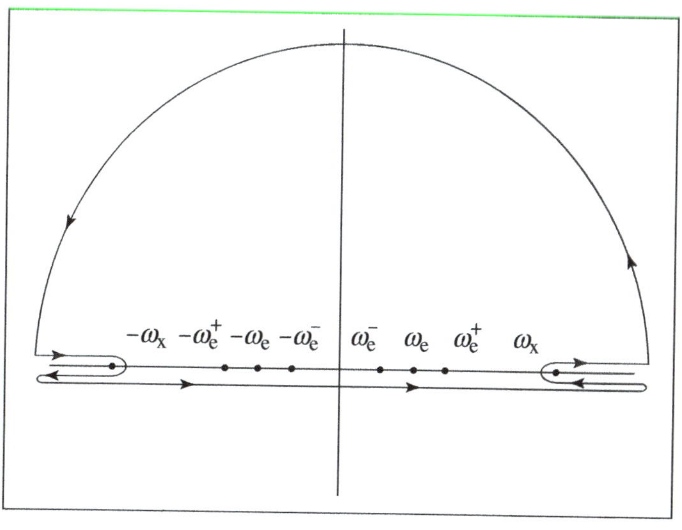

FIGURE D.4 The modified integration contour showing the contours around the branch cuts.

bottom side of the branch cut in the left half plane $\alpha_0 = j\sqrt{k_x^2 - k_0^2}$, but on the top side of this branch cut α_0 changes sign. On the branch cut in the right half plane $\alpha_0 = j\sqrt{k_x^2 - k_0^2}$ on the top side of the cut and equals the negative of this on the lower side of the cut. The integration along the top and bottom sides of the branch cuts can be combined and is given by

$$I_1 = \frac{1}{2\pi}\int_{-\infty}^{-\omega_x} A(k_x)\left[F\left(j\sqrt{k_0^2-k_x^2},\omega\right)-F\left(-j\sqrt{k_0^2-k_x^2},\omega\right)\right]$$
$$\left[\frac{1}{2(\omega+\omega_e)}-\frac{1}{2(\omega-\omega_e)}\right]e^{-jk_x x+j\omega t}\,d\omega$$
$$+\frac{1}{2\pi}\int_{\omega_x}^{\infty}\left|A(k_x)\right|\left[F\left(-j\sqrt{k_0^2-k_x^2},\omega\right)-F\left(j\sqrt{k_0^2-k_x^2},\omega\right)\right]$$
$$\left[\frac{1}{2(\omega+\omega_e)}-\frac{1}{2(\omega-\omega_e)}\right]e^{-jk_x x+j\omega t}\,d\omega \tag{D.20}$$

For the propagating waves, the solution at the image plane consists of the residue wave from the poles at $\pm\omega_e$ with a frequency of ω_e, plus the field from the branch cut 'integrals, which is given by Equation D.20. The desired wave is the pole wave at the frequencies $\pm\omega_e$, which is given by the residues at $\pm\omega_e$ (note that $F(j\sqrt{k_0^2-k_x^2},\omega_e) = 2$),

$$E_{op} = jA(k_x)e^{-jk_x x+j\omega_e t} + \text{c.c.}, \quad 2d/c < t < \tau + 2d/c \tag{D.21a}$$

where c.c. is the complex conjugate term. The contribution from the branch cut integrals is

$$-\frac{1}{2\pi}\int_{\omega_x}^{\infty} A(k_x)\left[\frac{4\mu\beta_0\beta e^{-j(\beta_0+\beta)d}}{(\mu\beta_0+\beta)^2-(\mu\beta_0-\beta)^2e^{-j2\beta d}}+\frac{4\mu\beta_0\beta e^{-j(\beta-\beta_0)d}}{(\mu\beta_0-\beta)^2-(\mu\beta_0+\beta)^2e^{-j2\beta d}}\right]$$
$$\frac{\omega_e e^{-jk_x x+j\omega t}}{\omega^2-\omega_e^2}\,d\omega + \text{c.c.}, \quad 2d/c < t < \tau + 2d/c \tag{D.21b}$$

where $\beta_0 = \sqrt{k_0^2-k_x^2}$ and $\beta = \sqrt{\mu\varepsilon k_0^2-k_x^2}$. For large values of k_0 and for $\varepsilon = \mu$, the integrand in the above expression becomes

$$A(k_x)\left\{\frac{-8j\sqrt{\varepsilon\mu}\,\sin[(\sqrt{\varepsilon\mu}+1)k_0 d]}{(\sqrt{\mu}+\sqrt{\varepsilon})^2}\right\}\frac{\omega_e}{\omega^2}e^{-jk_x x+j\omega t}$$
$$= -2jA(k_x)\sin\left[2(1-\omega_e^2/\omega^2)k_0 d\right]\frac{\omega_e}{\omega^2}e^{-jk_x x+j\omega t}$$

(When the complex conjugate of this expression is added) This result shows that the contribution from the branch cut integrals vanishes for large values of k_0.

If the frequency ω_e is greater than ω_x, then the point $\omega = \omega_e$ lies on the branch cut but the rest of the integrand vanishes for $\omega = \omega_e$, so there is no pole on the branch cut. There are no surface wave poles associated with the propagating waves.

For the evanescent waves, the solution consists solely of the pole waves. Thus, for $2d/c < t < \tau + 2d/c$, the sum of the dominant wave given by Equation D.21a and the excited surface waves is given by

$$E_o = \frac{j}{2}A(k_x)e^{-jk_x x}\left[2e^{j\omega_e t} + \frac{\omega_e e^{-\alpha_0(\omega_e^+)d}}{\omega_e^+ - \omega_e}\widehat{\mathrm{Res}}(\omega_e^+)e^{j\omega_e^+ t} - \frac{\omega_e e^{-\alpha_0(\omega_e^-)d}}{\omega_e - \omega_e^-}\widehat{\mathrm{Res}}(\omega_e^-)e^{j\omega_e^- t}\right] + \text{c.c.} \tag{D.22}$$

where c.c. is the complex conjugate term. The values of the residues are given by Equations D.15a and D.15b. The first wave at $\pm\omega_e$ corresponds to Pendry's solution for the loss-free lens. The other two terms are the resonant surface wave modes at $\pm\omega_e^{\pm}$ whose excitation cannot be avoided when frequency dispersion is included for epsilon and mu. These resonant modes will cause interference with the desired mode with frequency ω_e and thus will make it very difficult to obtain a coherent reconstruction of the evanescent wave amplitudes since the resonant frequencies of the surface wave modes vary with the transverse wave number. One can anticipate that some loss will be present in the slab and this will limit the lifetime of the resonant surface wave modes (surface plasmons), so the interference will die out. But as noted by a number of investigators, the presence of loss will limit the subwavelength resolution of the lens. With either scenario, the performance of the lens is reduced. In either case, the frequency dispersion or loss, or a combination of both, will avoid any field divergence for large values of the transverse wave number. It is also important to keep in mind that for large values of k_x the surface wave resonant frequencies are very close to ω_e, so the frequency of the incident field must be carefully controlled. If the frequency of the incident field is shifted to either ω_e^+ or ω_e^-, this will create a double pole with a time response proportional to $te^{j\omega_0 t}e^{-\alpha_0(\omega_0)d}$, where ω_0 equals ω_e^+ or ω_e^- instead of the desired dominant wave at ω_e.

It can be shown that for k_x greater than 5 the resonant surface wave frequencies are essentially equal, and furthermore to a high degree of accuracy (Gómez-Santos, 2003),

$$\frac{\omega_e}{\omega_e^+ - \omega_e}e^{-\alpha_0 d} \approx \frac{\omega_e}{\omega_e - \omega_e^-}e^{-\alpha_0 d} \approx 2 \tag{D.23}$$

Also as reference to Equations D.15a and D.15b shows, the residues are approximately equal to $-0.5\omega_e e^{-\alpha_0 d}$ and $0.5\omega_e e^{-\alpha_0 d}$ because A is the dominant coefficient in the denominator and very nearly equal to $4k_x$. Thus, we can express the solution in the form

$$E_o = jA(k_x)e^{-jk_x x}\left[e^{j\omega_e t} - e^{j(\omega_e^+ + \omega_e^-)t/2}\cos\Delta\omega t\right] + \text{c.c.}, \quad 2d/c < t < \tau + 2d/c \tag{D.24}$$

where $\Delta\omega = (\omega_e^+ - \omega_e^-)/2$. We can also make the approximation $(\omega_e^+ + \omega_e^-)/2 \approx \omega_e$. This is essentially the result obtained by Gómez-Santos by using a model of two coupled mechanical resonators (Gómez-Santos, 2003). For $\Delta\omega t$ small, the expansion of the cosine function gives

$$E_o \approx \frac{j}{2} A(k_x) e^{-jk_x x} (\Delta\omega t)^2 e^{j\omega_e t} = \frac{j}{2} A(k_x) e^{-jk_x x + j\omega_e t} (\omega_e t)^2 e^{-2\alpha_0 d} \tag{D.25}$$

which shows that it takes a considerable length of time for the field at the image plane to build up to its steady-state value when the decaying exponential factor is very small. For example, if $d = \lambda_0/4$ and $k_x = 20k_0$, then

$$t = \frac{1}{\Delta\omega} = \frac{1}{\omega_e} e^{20k_0\lambda_0/4} = 4.4 \times 10^{13} \frac{1}{\omega_e}$$

which for a frequency of 10 GHz gives t equal to 11.7 min. The result shown in Equation D.25 can also be obtained from a different approach. When k_x approaches infinity, $\Delta\omega$ approaches zero and ω_e^+ and ω_F^- coalesce to produce a triple pole given by

$$\frac{1}{4} e^{-\alpha_0 d} \left[\frac{\omega_e}{\omega - \omega_e^+} - \frac{\omega_e}{\omega - \omega_e^-} \right] \frac{1}{\omega - \omega_e} \to \frac{\omega_e}{4} e^{-\alpha_0 d} \frac{\omega_e^+ - \omega_e^-}{(\omega - \omega_e)^3} \to \frac{\omega_e^2}{4} e^{-2\alpha_0 d} \frac{1}{(\omega - \omega_e)^3}$$

which has the time response $(j/8)(\omega_e t)^2 e^{-2\alpha_0 d} e^{j\omega_e t}$. Since the signal is turned off at a finite time τ, the field at the image plane vanishes as k_x approaches infinity. For finite values of k_x and $\Delta\omega t$ that is large, the field described by Pendry's solution is slowly modulated by the cosine factor. Since the resonant frequency of the surface wave modes depends on k_x, the field at the image plane is not coherent in frequency and thus it would be virtually impossible to achieve a coherent reconstruction of the evanescent wave amplitudes at the image plane. If features as small as one-tenth of a wavelength were to be observed, values of k_x up to about $10k_e \approx 10k_0$ would have to be retained. The interference from the excited resonant surface waves would make it unlikely that any useful coherent reconstruction of the evanescent waves with these values of k_x could be achieved.

After the signal has been turned off, the spectral function $S(\omega)$ that must be used is

$$S(\omega) = je^{j(\omega_e - \omega)\tau/2} \frac{\tau}{2} \frac{\sin(\omega - \omega_e)\tau/2}{(\omega - \omega_e)\tau/2} - je^{-j(\omega + \omega_e)\tau/2} \frac{\tau}{2} \frac{\sin(\omega + \omega_e)\tau/2}{(\omega + \omega_e)\tau/2}$$

For this spectrum, there are no poles at $\pm\omega_e$. The rapid oscillations of the spectral function for large values of $|\omega - \omega_e|$ and $|\omega + \omega_e|$ will ensure that the branch cut integrals are small.

Various authors have shown that losses in the negative index slab would also reduce the resolution capability of the lens even if there was no frequency dispersion in epsilon and mu. The excited resonant surface wave modes will decay to zero because of losses that will be present, even though we did not include losses in the above analysis that focused on the limitations of the loss-free Veselago lens because of frequency dispersion in epsilon and mu.

When the losses in the lens material are small, the new surface wave eigenvalues can be found using a perturbation method based on the Newton–Raphson method. Consider the eigenvalue equation (Equation D.13a) and let

$$f(\omega) = \mu\alpha_0 - \alpha + (\mu\alpha_0 + \alpha)e^{\alpha d}$$

We now assume that the loss in the material is the same for epsilon and mu and thus let

$$\varepsilon = 1 - \frac{2\omega_e^2}{\omega(\omega + j\gamma)} \approx 1 - \frac{2\omega_e^2}{\omega^2} + \frac{j2\gamma\omega_e^2}{\omega^3} = \mu$$

where γ is the loss parameter and is considered to be very small relative to ω_e. Since the root for this equation is very close to ω_e, the first approximation to the root when loss is included is given by

$$f(\omega) = f(\omega_e) + \frac{\partial f}{\partial \omega}\bigg|_{\omega_e}(\omega - \omega_e) = 0 \tag{D.26a}$$

which gives

$$\omega = \omega_e - \frac{f(\omega_e)}{\partial f(\omega)/\partial \omega|_{\omega_e}} \tag{D.26b}$$

This expression can be evaluated and when only the first-order terms in γ are retained and the transverse wave number k_x is assumed to be large it is found that $\omega = \omega_e^+ + j(\gamma/2)$, and similarly $\omega = \omega_e^- + j(\gamma/2)$, for the eigenvalues of the even and odd surface wave modes when a small loss is included. This result is the same as that found by Grbic (2007). When these values for the eigenvalues are used in Equation D.19, it is found that for large k_x

$$F(\alpha_0, \omega_e) \approx \frac{2}{1 + (\gamma/\omega_e)^2 e^{2k_x d}} \tag{D.27}$$

which is now the residue for the dominant wave. The solution for the excited surface waves will now be

$$E_o = \frac{j}{2}A(k_x)e^{-jk_x x}\left[\frac{\omega_e e^{-\alpha_0(\omega_e^+)d}}{\omega_e^+ - \omega_e + j\gamma/2}\widehat{\text{Res}}(\omega_e^+)e^{j\omega_e^+ t - \gamma t} - \frac{\omega_e e^{-\alpha_0(\omega_e^-)d}}{\omega_e - \omega_e^- - j\gamma/2}\widehat{\text{Res}}(\omega_e^-)e^{j\omega_e^- t - \gamma t}\right]$$

We now make use of Equation D.23 to obtain

$$\frac{\omega_e}{(\omega_e^+ - \omega_e)[1 + j\gamma/2(\omega_e^+ - \omega_e)]} = \frac{2e^{\alpha_0 d}}{1 + j\omega_e/[(\omega_e^+ - \omega_e)Q]} = \frac{2e^{\alpha_0 d}}{1 + je^{\alpha_0 d}/Q}$$

where Q is the quality factor ω_e/γ for the surface wave resonator. A similar expression will hold for the odd surface wave mode. In place of Equation D.22, the solution for the evanescent waves for large values of k_x is now given by

$$E_o \approx jA(k_x, \omega_e)e^{-jk_x x}\left[\frac{e^{j\omega_e t}}{1 + e^{2\alpha_0 d}/Q^2} - \frac{1}{2[1 + je^{\alpha_0 d}/Q]}e^{-\gamma t/2}(e^{j\omega_e^+ t} + e^{j\omega_e^- t})\right] + \text{c.c.} \quad \text{(D.28)}$$

for $2d/c < t < \tau + 2d/c$.

For the propagating waves where $k_x < k_0$, the result given in Equation D.21a should be replaced by

$$jA(k_x, \omega_e)e^{-jk_x x}\left[\frac{(1 - j/Q)e^{j\omega_e t}}{1 + e^{j2\beta d}k_x^2/(\beta_0^2 Q^2)}\right] + \text{c.c.}, \quad 2d/c < t < \tau + 2d/c \qquad \text{(D.29)}$$

which is valid for small losses and $|k_x| < k_0$. This term, together with the branch cut integrals, when integrated over $|k_x| < k_0$, gives the total image field arising from the propagating waves at the image plane. As noted earlier, the propagating wave spectrum given by the branch cut integrals produces some blurring of the image, an artifact that was not present in Pendry's ideal lens solution. Equation D.29 shows that losses will also produce some blurring of the image.

The loss reduces the amplitudes of the residues by a substantial amount. In addition, it can be seen that the excited surface wave modes will decay quite rapidly. For example, if the quality factor or Q of the surface wave mode resonances equals 10^4, then the surface wave modes become negligible in less than a microsecond if the frequency is equal to 10 GHz. Hence, interference from the excited surface wave modes is not likely to be a serious factor in reducing the performance of a lossy Veselago lens. However, the factor multiplying the dominant pole wave will become small whenever the factor $(\gamma/\omega_e)e^{\alpha_0 d} = Q^{-1}e^{\alpha_0 d} \approx Q^{-1}e^{k_x d} > 1$. For the above example, this occurs for $k_x > 9.2/d$. If $d = \lambda_0/4$, this corresponds to $\lambda_x > 0.17\lambda_0$. The amplitude of features smaller than this will be reduced by a factor of more than $1/2$ at the image plane. This reduction in the amplitudes of the evanescent waves increases exponentially with k_x. Merlin showed that a small perturbation σ in the relative value of ε reduced the resolution of the lens in accordance with a formula like that in Equation D.28 with $j2\gamma/\omega_e$ replacing σ (Merlin, 2004). Thus, if the frequency of the incident field drifts away from the value ω_e, this will be equivalent to a change in the relative values of epsilon and mu from -1 and can produce a significant reduction in the resolution of the lens. Hence, in practice, if a Veselago lens could be constructed, it will be the losses and the frequency stability of the source that illuminates the object that will limit the resolution, not the interference from the excited surface wave modes.

We can now understand what happens when a steady-state sinusoidal incident field at the frequency ω_e is assumed. For this case, the frequency spectral function can be represented by a delta function, that is, $S(\omega) = 2\pi\delta(\omega - \omega_e)$, and thus the inverse Fourier transform results in the field solutions being evaluated at the frequency ω_e, where $\varepsilon = \mu = -1$, which gives Pendry's solution. When frequency dispersion is neglected, then the only pole is a double pole that occurs when k_x becomes infinite. Although Pendry's solution has some of the characteristics of a resonant mode, its resonant frequency is not clearly defined except perhaps through the condition that the relative values of epsilon and mu must equal -1. This is the cause for the divergent behavior of Pendry's solution since it corresponds to a steady-state sinusoidal signal being applied to a resonant system at its resonant frequency. When frequency dispersion is included and an incident field with finite frequency spectral density is assumed, then the response is obtained in terms of the residues at the surface wave poles and this response is finite even when k_x becomes infinite. For the case of a sinusoidal signal turned on at $t = 0$ and later turned off, the frequency spectrum contains a pole term at the frequency of the sinusoidal signal instead of a delta function spectral term. The use of a sinusoidal signal of finite duration reveals much richer physical phenomena associated with Veselago's lens that is completely missed in a steady-state sinusoidal solution.

D.4 CONCLUSIONS

An analysis of transmission through a flat slab lens was carried out. For the first case considered, it was assumed that the input field had a continuous frequency spectrum with finite density, and that both epsilon and mu exhibited frequency dispersion. The field excited in the lens was expressed in terms of the even and odd resonant surface wave modes whose amplitudes were evaluated in terms of the residues at the poles. For this case, it was found that there were no exponentially growing evanescent waves in the slab. However, when the incident field was chosen as a sinusoidal signal with finite duration, a dominant wave at the frequency ω_e, at which the relative values of epsilon and mu were equal to -1, was also excited, but due to interference from the excited surface wave modes a coherent reconstruction of the evanescent wave amplitudes was not possible. As a consequence of this result, a lossless Veselago flat lens with super-resolution is not physically possible. When small loss is included in the material parameters, the excited surface wave modes decay away in a very short period of time and their interference effects become negligible. The resolution of the Veselago lens in now limited by the loss, a result previously established by a number of investigators, and/or a signal frequency that deviates from that for which the relative values of epsilon and mu are exactly equal to -1. A new result that others had not found was the existence of a continuous spectrum of propagating waves that arise from branch cut integrals and that will blur the image of these waves at the image plane. It was also concluded that Pendry's solution for a lossless lens was not a continuous function of the physical parameters and hence did not constitute a proper physical solution. The analysis presented in this appendix is a classical one and gives a solution that satisfies the required conditions for a proper physical solution.

REFERENCES

Born, M. and Wolf, E. *Principles of Optics*, 2nd ed., Macmillan, New York, 1964.

Chew, W. C. *Waves and Fields in Inhomogeneous Media*, IEEE Press Series on Electromagnetic Waves, IEEE Press, Piscataway, NJ, 1995.

Chew, W. C. Some Reflections on Double Negative Materials. *Prog. Electromagn. Res.*, PIER Vol. 51, 2005, pp. 1–26.

de Wolf, D. A. Transmission of Evanescent Wave Modes Through a Slab of DNG Material I. *IEEE Trans.*, Vol. AP-53, January 2005, pp. 270–274.

de Wolf, D. A. Transmission of Evanescent Wave Modes Through a Slab of DNG Material II. *IEEE Trans.*, Vol. AP-54, January 2006, pp. 263–264.

Eleftheriades, G. V. Enabling RF/Microwave Devices Using Negative-Refractive-Index Transmission-Line (NRI-TL) Metamaterials. *IEEE Antenn. Propag. Mag.*, Vol. 49, No. 2, April 2007.

Eleftheriades, G. V., Tyler, A. K., and Kremer, P. C. Planar Negative Refractive Index Media Using Periodically L-C Loaded Transmission Lines. *Trans. IEEE*, Vol. MTT-50, December 2002, pp. 2702–2712.

Felson, L. B. and Marcuvitz, N. *Radiation and Scattering of Waves*, Prentice Hall, Englewood Cliffs, NJ, 1973.

Gómez-Santos, G. Universal Features of the Time Evolution of Evanescent Modes in a Left-Handed Perfect Lens. *Phys. Rev. Lett.*, Vol. 90, No. 7, February 2003, 077401.

Grbic, A. Left-Handed Lens Metrics. *Int. Symp. on Electromagnetic Theory (EMTS),* Ottawa, Canada, July 26-28, 2007, 4 pp. Available online at http://www.ursi.org/B/EMT_2007/04-18/2-Grbic118.pdf.

Grbic, A. and Eleftheriades, G. Negative Refraction, Growing Evanescent Waves, and Sub-Diffraction Imaging in Loaded Transmission-Line Metamaterials. *Trans. IEEE*, Vol. MTT-51, No. 12, December 2003, pp. 2297–2305.

Haldane, F. D.Electromagnetic Surface Modes at Interfaces with Negative Refractive Index Make a "Not-Quite-Perfect" Lens, 2002. Available online at http://arxiv.org/abs/condmat/0206420.

Ishimaru, A. *Electromagnetic Wave Propagation, Radiation, and Scattering*, Prentice Hall, Englewood Cliffs, NJ, 1991.

Marqués, R., Martin, F., and Sorolla, M. *Metamaterials with Negative Parameters: Theory, Design, and Microwave Applications*, Wiley, New York, 2008.

Merlin, R. Analytical Solution of the Almost-Perfect-Lens Problem. *Appl. Phys. Lett.*, Vol. 84, February 2004, pp. 1290–1292. See also erratum in *Appl. Phys. Lett.,* Vol. 85, No. 11, September 2004.

Milton, G. W., Nicorovici, N.-A. P., Mcphedran, R. C., and Podolskiy, V. A. A Proof of Superlensing in the Quasistatic Regime, and Limitations of Superlenses in this Regime Due to Anomalous Resonances. *Proc. R. Soc.*, Vol. A2005, October 2005, pp. 3999–4034.

PAPOULIS, A. *Systems and Transforms with Applications to Optics*, McGraw-Hill, New York, 1968.

PENDRY, J. B. Negative Refraction Makes a Perfect Lens. *Phys. Rev. Lett.*, Vol. 85, No. 18, October 2000, pp. 3966–3969.

RUPPIN, R. Surface Polaritons of a Left-Handed Material Slab. *J. Phys.: Condens. Matter*, Vol. 13, 2001, pp. 1811–1818.

SANADA, A., CALOZ, C., AND ITOH, T. Planar Distributed Structures with Negative Refractive Index. *Trans. IEEE*, Vol. MTT-52 No. 4, April 2004, pp. 1252–1263.

SMITH, D. R. ET AL. Limitations on Subdiffraction Imaging with a Negative Refractive Index Slab. *Appl. Phys. Lett.*, Vol. 82, No. 10, March 2003, pp. 1506–1508.

STRATTON, J. A. *Electromagnetic Theory*, McGraw-Hill, New York, 1941. Chapter 5.

VESELAGO, V. G. The Electrodynamics of Substances with Simultaneously Negative Values of ε and μ. *Sov. Phys. Usp.*, Vol. 10, January–February 1968, pp. 509–514.

YAGHJIAN, A. D. AND HANSEN, T. B. Plane-Wave Solutions to Frequency-Domain and Time-Domain Scattering from Magnetodielectric Slabs, *Phys. Rev. E*, Vol. 73, 2006, 046608.

AUTHOR INDEX

Chapter 1

Small Antenna Handbook, Robert C. Hansen and Robert E. Collin.

Chapter 2

Chapter 3

Chapter 4

Chapter 5

Chapter 6

Chapter 7

SUBJECT INDEX

Small Antenna Handbook, Robert C. Hansen and Robert E. Collin.

Printed and bound by CPI Group (UK) Ltd, Croydon, CR0 4YY
16/09/2026
14963413-0004